Participation in Fisheries Governance

Reviews: Methods and Technologies in Fish Biology and Fisheries

VOLUME 4

Series editor:

Jennifer L. Nielsen

U.S. Geological Survey
Biological Resources Division
Anchorage, Alaska

This book is dedicated to all those men and women who risk their lives in lifeboat and air-sea rescue bids to save fishers in peril.

Acknowledgements

This book originated in a Workshop on 'Fisheries Governance' that I organised under the auspices of the Environmental Politics Specialist Group of the UK Political Studies Association, which was held at the University of Newcastle during 4-6 September 2003. I am grateful for the financial support for the Workshop which was provided by both the Political Studies Association and Newcastle University, and for the organisational help provided by my colleague, Dr Derek Bell. I would also like to acknowledge my debt to the European Fifth Framework Project on the European Fisheries Ecosystem Plan (EFEP) under the leadership of Professor Frid, which stimulated me to investigate the issues surrounding participation in fisheries governance. My special thanks go to Jenny Hatchard, who worked with me for three years as a Research Associate on EFEP, for her invaluable work in formatting this book for publication.

CONTENTS

FIGURES AND TABLES/BOXES

Figures

Tables/Boxes

ACRONYMS

4M	Multi-species, Multi-fleet, Multi-area Modelling-package
ACFA	Advisory Committee on Fisheries and Aquaculture, EU
ACFM	Advisory Committee on Fishery Management, ICES
AGAC	Atlantic Groundfish Advisory Committee, Canadian fishing industry
AMA	Area Management Agreement
ASCOBANS	Agreement on the Conservation of Small Cetaceans of the Baltic and North Seas
ASF	Associated Fisheries of Maine
ASFC	Association of Sea Fisheries Committees, England and Wales
BFA	British Fishermen's Association
BTF	Back to the Future project
CAFSAC	Canadian Atlantic Fisheries Scientific Advisory Committee
CBD	Convention on Biological Diversity, UN
CCW	Countryside Council for Wales
CFP	Common Fisheries Policy, EU
CEFAS	Centre for Environment, Fisheries and Aquaculture Science, UK
CPUE	Catch-per-Unit-Effort
CSAS	Canadian Science Advisory Secretariat
CSTA	Council of Science and Technology Advisors, Canada
CUS	Coasts Under Stress project
DAFS	Department of Agriculture and Fisheries, Scotland
DEFRA	Department for Environment, Food and Rural Affairs, UK
DflD	Department for International Development, UK
DFO	Department of Fisheries and Oceans, Canada
DG Fisheries/DG Fish	Directorate-General for Fisheries, European Commission
DSCC	Deep Sea Conservation Coalition
DTI	Department of Trade and Industry, UK
EA	Environment Agency, UK
EBA	Ecosystem-based Approach
EC	European Community
EEAC	European Environment Advisory Council
EEZ	Exclusive Economic Zone
EFA	Exclusive Fishing Area
EFEP	European Fisheries Ecosystem Plan project
EFH	Essential Fish Habitat
EIA	Environmental Impact Assessment
EMS	European Marine Site
EN	English Nature
ENCFAN	European Nature Conservation and Fisheries Advisory Network
ENGO	Environmental Non-governmental Organisation
ENRC	Environment and Natural Resources Committee, Australia

EP	European Parliament
EPAP	Ecosystem Principles Advisory Panel, US
EPLPC	Eastport Peninsula Lobster Protection Committee, Newfoundland
EU	European Union
F	Fishing Mortality
FAL	Fishermen's Association Limited, UK
FAO	Food and Agriculture Organisation, UN
FASG	Fisheries and Aquaculture Strategy Groups, Wales
FEAP	Federation of European Aquaculture Producers
FEK	Fishers' Ecological Knowledge
FEP	Fisheries Ecosystem Plan
FIFG	Financial Instrument for Fisheries Guidance, EU
FK	Fishers' Knowledge
FMP	Fishery Management Plan
FRCC	Fisheries Resource Conservation Council, Canada
FS	Fisheries Science
GBRMP	Great Barrier Reef Marine Park
IBEC	Indian Bay Ecosystem Corporation, Newfoundland
ICCAT	International Commission for the Conservation of Atlantic Tunas
ICES	International Council for the Exploration of the Sea
ICZM	Integrated Coastal Zone Management
IEEP	Institute for European Environmental Policy
IQ	Individual Non-Transferable Quota
ITQ	Individual Transferable Quota
ISE	Iterative Stakeholder Engagement
JNCC	Joint Nature Conservation Committee, UK
LEK	Local Ecological Knowledge
LFA	Lobster Fishing Area, Newfoundland
LFAACs	Lobster Fishing Area Advisory Committees, Newfoundland
LK	Local Knowledge
LME	Large Marine Ecosystem
M	Natural Mortality
MAFF	Ministry of Agriculture, Fisheries and Food, UK
MAGP	Multi-Annual Guidance Programme, EU
MCS	Marine Conservation Society
MLS	Minimum Landing Size
MNR	Marine Nature Reserve (UK)
MP	Member of Parliament, UK
MPA	Marine Protected Area
MSC	Marine Stewardship Council
MSVPA	Multi-species Virtual Population Analysis
MSY	Maximum Sustainable Yield
NAC	Norwegian Allowable Catch
NAFO	North Atlantic Fisheries Organisation
NAMA	Northwest Atlantic Marine Alliance, US

NCA	Nature Conservation Agency
NEFMC	New England Fisheries Management Council
NEPA	National Environmental Policy Act, US
NERC	Natural Environment Research Council, UK
NFFO	National Federation of Fishermen's Organisations, England and Wales
NGO	Non-governmental Organisation
NMFS	National Marine Fisheries Service, US
NPAFC	North Pacific Anadromous Fish Commission
NSCFP	North Sea Commission Fisheries Partnership
NSM	New Social Movement
NSRAC	North Sea Regional Advisory Council
NTZ	No-Take-Zone
NWNWSFC	North Western and North Wales Sea Fisheries Committee
OSPAR	Commission for the Protection of the Marine Environment of the North-East Atlantic
OY	Optimum Yield
PAR	Participatory Action Research
PMSU	Prime Minister's Strategy Unit, UK
PO	Producers' Organisation
PP	Precautionary Principle
PVIS	*Productschap Vis* – Netherlands Fish Product Board
RAC	Regional Advisory Council, EU
RAP/ZAP/NAP	Regional/Zonal/National Advisory Processes, CSAS, Canada
RCEP	Royal Commission on Environmental Pollution, UK
RFERACs	Regional Fisheries, Ecology and Recreation Advisory Committees, UK
RMC	Regional Management Council, US
RSE	Royal Society of Edinburgh
RSPB	Royal Society for the Protection of Birds
SAC	Special Area of Conservation, EU
SAP	Special Access Permit
SARC, STAR and SEDAR	Committees of the National Marine Fisheries Service Standing Committee on Research and Statistics, US
SEA	Strategic Environmental Assessment
SEERAD	Scottish Executive Environment and Rural Affairs Department
SFA	Sustainable Fisheries Act, US
SFCs	Sea Fisheries Committees of England and Wales
SFF	Scottish Fishermen's Federation
SFI	Sea Fisheries Inspectorate, UK
SIFAG	Scottish Inshore Fisheries Advisory Group
SNH	Scottish Natural Heritage
SPA	Special Protected Area, EU
SSB	Spawning Stock Biomass
SSSI	Site of Special Scientific Interest, UK
STECF	Scientific, Technical and Economic Committee for Fisheries, EU

SWFPO	South Western Fish Producers' Organisation
SWSFC	South Wales Sea Fisheries Committee
TAC	Total Allowable Catch
TEK	Traditional Ecological Knowledge
TK	Traditional Knowledge
TSF	Trawler Survival Fund, Maine
TWG	Tripartite Working Group, Scotland
UK	United Kingdom
UN	United Nations
US/USA	United States of America
USCOP	US Commission on Ocean Policy
VMS	Vessel Monitoring System
VPA	Virtual Population Analysis
WAG	Welsh Assembly Government
WGNSSK	Working Group on the North Sea and Skaggerak, ICES
WTO	World Trade Organisation
WWF	Worldwide Fund for Nature

NOTES ON THE CONTRIBUTORS

Dean Bavington is a doctoral candidate in the Geography and Environmental Studies Department at Wilfrid Laurier University in Waterloo, Canada. He studied Biology and Philosophy during his undergraduate degree programme, and completed a Masters in Environmental Studies at York University, Canada. He received an Organization of American States Fellowship to study environmental ethics and restoration ecology at the University of Northern Arizona in 1999 and was a Fulbright Scholar at the University of California, Berkeley in 2000. At his home in St. John's, Newfoundland he is currently completing a dissertation entitled: *Handling Fish and People: Managerial Ecology in Newfoundland and Labrador Cod Fisheries*.

Clare Coffey is a Senior Fellow at the Institute for European Environmental Policy in Brussels, where she undertakes policy analysis and advocacy work for a wide range of international, EU and national organisations. Her current focus is on EU environmental governance in general, although she maintains a strong interest in environmental integration within the Common Fisheries Policy.

Alyne Elizabeth Delaney (Ph.D. University of Pittsburgh, 2003) is an environmental anthropologist specializing in common property theory. As a researcher at the Institute for Fisheries Management and Coastal Community Development (IFM) in Hirtshals, Denmark, her current projects include investigating uses and applications of scientific knowledge in fisheries policy (EU-funded PKFM project), as well as analysis of fisheries institutions in the Lower Mekong River Basin (Cambodia, Laos, Vietnam). Her doctoral dissertation was an exploration of fishing rights in coastal Japan, particularly as they were affected by industrialisation, mechanisation, and environmental degradation.

Euan Dunn (PhD) is Head of Marine Policy at the Royal Society for the Protection of Birds, Europe's largest wildlife charity and UK partner of BirdLife International. He is responsible for wide-ranging policy development and advocacy on the interactions between fisheries and marine birds, nationally and internationally. At the European level, a key focus of his work is promoting environmental integration in the practice of the Common Fisheries Policy. He is the Environmental NGOs' representative on the Advisory Committee on Fisheries and Aquaculture (ACFA) and he represents BirdLife International on the North Sea Regional Advisory Council. He also manages the RSPB team that coordinates BirdLife International's Global Seabird Programme, a key focus of which is the Save the Albatross Campaign, entailing working with fishermen, governments and international organisations to adopt mitigation measures which reduce seabird mortality in fisheries.

N Clare Eno is the Senior Fisheries Policy Officer at the Countryside Council for Wales (CCW) where she has worked for 6 years building capacity within the organisation in relation to formulating marine policy and advising on marine activities. Since completion of her doctorate at Cambridge she worked first on marine environmental science (in the Red Sea and then at the MAFF fisheries laboratory) and for the last 16 years on marine nature conservation – research, advice and policy – with all of the British statutory nature conservation agencies. She has focused particularly on interactions between fisheries and nature conservation at a national and European level, and her recent work has included facilitating the integration of fishers and other stakeholders to help deliver an ecosystem-based approach to fisheries management, especially around Wales and the Irish Sea.

Chris Frid is Professor of Marine Biology at the University of Liverpool. A benthic ecologist by training he has worked extensively on characterising human impact on marine systems and developing effective strategies to reduce them. He is a member of, and has previously chaired, the ICES Working Group on the Ecosystem Effects of Fishing Activities. He has also recently co-ordinated the work of the European Fisheries Ecosystem Plan (EFEP) project, which has developed a Fisheries Ecosystem Plan for the North Sea.

Rob van Ginkel is an associate professor at the University of Amsterdam's Department of Sociology and Anthropology. In 1993, he obtained his Ph.D. from the same university with a dissertation on transformations in two fishing villages on the Dutch island of Texel. His research interests are mainly in the field of maritime cultures, fisheries and fishing communities. He is a co-founder of *Maritime (Anthropological) Studies* and the Centre for Maritime Research (MARE) in Amsterdam.

Mark Gray is the Sea Fisheries Liaison Officer at the Countryside Council for Wales (CCW). He provides advice to the fishing industry and fisheries regulators on fisheries and nature conservation related casework, develops partnership projects and commissions, and contributes to science and policy research. He has been employed by MAFF and WWF (UK) on fisheries research contracts and by the Environment Agency, Wales as a fisheries scientist, and he has served as a Senior EU Fisheries Observer on Portuguese vessels fishing off the Canadian Grand Banks. He has contributed to a number of fisheries governance papers and his publications include 'The Coastal Fisheries of England and Wales, Part III: A Review of the Status 1992-1994', *Fisheries Research Technical Report* No. 100 (Lowestoft: MAFF Directorate of Fisheries Research 1995).

Tim S Gray is Emeritus Professor of Political Thought in the School of Geography, Politics and Sociology at the University of Newcastle upon Tyne, UK. During the last 15 years, he has specialised in environmental politics, with a particular focus on fisheries. He has edited a book on *The Politics of Fishing*, and he has published papers on the EU's Common Fisheries Policy, the strategy of the environmental movement in relation to the fishing industry, and the relation between fishers' knowledge and fisheries science. During 2001-2004, he took part in the interdisciplinary EU-funded European Fisheries Ecosystem Plan (EFEP) project, led by Chris Frid, in which, together with Jenny Hatchard and Knut Mikalsen, he was responsible for the social science component.

Madeleine Hall-Arber, Ph.D. has focused her research on fishing communities since 1975 when she devoted her summer fieldwork as a Brandeis University graduate student to going out on the commercial fishing vessels of Provincetown, MA. The goal of her research on the impacts of regulatory change on fishing communities is to help managers and the communities identify ways to mitigate the impacts of management decisions. Her published work on New England fishing communities serves as the basis for describing the human environment for several fishery management plans. She also works closely with fishing industry members on collaborative research projects.

Jenny Hatchard is based in the School of Geography, Politics and Sociology at the University of Newcastle, UK. She has recently completed a three-year consultation with

North Sea fisheries stakeholders, contributing to the EU-funded European Fisheries Ecosystem Plan project (www.efep.org). Her publications include papers with Katherine Brookfield and Tim Gray on 'The concept of fisheries-dependent communities' (*Fisheries Research*), and with Tim Gray on elements of the reform of the EU's Common Fisheries Policy (*Marine Policy* and *MAST*). She is currently working to complete her doctorate, entitled *Fishing Lines: A Discourse Analysis of the Rhetoric of Commercial Marine Fishing*.

Tony Hawkins is chair of the North Sea Commission Fisheries Partnership, which brings fishers and scientists together, and is rapporteur for the newly formed North Sea Regional Advisory Council. He is based in the Department of Geography at the University of Aberdeen, Scotland. He is also managing director of Loughine Ltd, an environmental consultancy and a Director of the Dee Salmon Fishery Trust.

Hans-Kristian Hernes (PhD) is associate professor in the Department of Political Science, University of Tromsø where he teaches political theory and public policy. He has done research on relations between state and interest groups in Norwegian fisheries, and is currently involved in research on the politics of fisheries, and democracy and indigenous rights.

Svein Jentoft is professor of sociology and organisation studies at the Norwegian College of Fisheries Science, University of Tromsø. He has published books and articles within the area of industrial organisation, community development and resource management related to fisheries. Among his recent books are *Fish for Life: Interactive Governance for Fisheries* (Amsterdam University Press 2005 with Jan Kooiman *et al*) and *Fisheries Development: The Institutional Challenge* (Eburon 2004 with Bjørn Hersoug and Poul Degnbol).

Paul Knapman is presently based in St John's, Newfoundland working as a fisheries consultant with Ocean Resource Conservation Associates EU (ORCA-EU), specialising in fisheries and environmental integration on both sides of the Atlantic. He has previously worked for the Department for Environment, Food and Rural Affairs (DEFRA) as a Fisheries Officer; for English Nature (EN) as their Senior Fisheries Policy Advisor; and for the Eastern Sea Fisheries Joint Committee (ESFJC), as their Clerk and Chief Fishery Officer.

Knut H. Mikalsen is professor of politics at the University of Tromsø, Norway. He has recently contributed to two EU-funded projects on the politics of fisheries management – the development of a European Fisheries Ecosystem Plan (EFEP) with, among others, Chris Frid, Tim Gray and Jenny Hatchard; and, a project focusing on the decentralisation of fisheries management responsibilities with Luc van Hoof. He is currently engaged in research on stakeholder involvement in Norwegian and European fisheries.

Grant Murray is currently a post-doctoral fellow with the Coasts Under Stress Programme at Memorial University, St. John's, Newfoundland. His interdisciplinary research involves the combination of elements from both natural and social sciences, and includes the social aspects of marine protected areas, the environmental history of Quintana Roo, Mexico, and the development of an environmental history of the fisheries of the Labrador and northwest Newfoundland coasts through the combination of local ecological knowledge and available social and natural science information.

Barbara Neis is co-director of SafetyNet, A Research Alliance in Workplace Health and Safety in Marine and Coastal Work at Memorial University, St. John's, Newfoundland. She is also Professor in the Department of Sociology. She has researched many different aspects of the Newfoundland and Labrador fisheries. Her current areas of research focus include occupational asthma from snow crab, fishing vessel safety, and local ecological knowledge and science.

Tim Oliver is the editor of the UK and Ireland fishing industry weekly trade newspaper Fishing News, a post he has held for 18 years. Before that he was a reporter on the paper for three years. He is a former distant water fisherman, sailing mainly from Hull, his home-town, and he holds a mate's certificate. He left fishing in 1975 when the distant water trawling industry collapsed in the wake of the 'cod wars' with Iceland and the advent of 200-mile limits. He took a degree at Hull University and was a researcher/lecturer in fisheries at the Hull College of Higher Education before joining Fishing News in 1983.

Rosemary Ommer (PhD) is the Director of Special Projects at the University of Victoria, and she is Project Director of Coasts Under Stress (CUS), a 5-year, $6.2 million interdisciplinary research project investigating restructuring and its effects on the health of the environment, communities and individuals, on the east and west coasts of Canada. She holds her Ph.D. in geography from McGill University, and an MA in historical geography from Memorial University of Newfoundland (MUN), and she has researched and taught in Atlantic Canada since the early 1970s. She taught economic history at MUN from 1982 until 1999, and was Research Director of Memorial's Institute for Social and Economic Research from 1990-1996. She is the author and/or editor of several books including, most recently, *The Resilient Outport*, an interdisciplinary look at the East Coast fishery crises published by ISER books in 2002.

Jake Rice is Director of Assessment and Advice with the Canadian Science Advisory Secretariat (CSAS) of the Department of Fisheries and Oceans. CSAS is responsible for peer review and provision of science advice for the fisheries management, oceans management and policy sectors of DFO. Before taking this position he was a research scientist and science manager on both the East and West Coasts of Canada. He has served in several positions with ICES, including Chief Scientist, and on the Science Advisory Board for NOAA. His research has centred on ecosystem dynamics and the application of ecosystem approaches in management.

David Schneider is a professor in the Ocean Sciences Centre at Memorial University in St. John's, Newfoundland, Canada. Since 1978 he has authored over 70 journal articles, 20 review articles and book chapters, and 60 miscellaneous publications on oceanography, the biology of large marine organisms, and fisheries. He has taken an active role in interdisciplinary projects throughout his career, beginning with study of the dynamics of the shelf ecosystem in the Bering Sea, and most recently in Coasts Under Stress, a study of social and environmental restructuring in coastal communities in Canada.

Selina Stead is based in the School of Marine Science and Technology at the University of Newcastle, UK. She is a Board Director of the European Aquaculture Society and an editor of *Aquaculture International*. Her earlier research looked at how environmental changes affect the behaviour, feeding, growth and sexual development of freshwater and marine finfish. This led to work on the aquaculture and fisheries industries where she became interested in linking natural and social science theory to help develop practical approaches to coastal governance. Current research includes applications of fishermen's knowledge to traditional and numerical methods used in fisheries science and management in the UK and overseas.

David Symes is Reader Emeritus in Geography at the University of Hull, UK. He has published extensively in the fields of rural social exchange and fisheries management and jointly edited several volumes on aspects of fisheries development in Europe. These include *Fisheries Management in Crisis* (1996), *Property, Rights and Regulatory Systems in Fisheries* (1998), *Alternative Management Systems for Fisheries* (1999), *Fisheries Dependent Regions* (2000) – all published by Blackwell Science – and *Inshore Fisheries Management* (2001), published by Kluwer. Recently he has been acting as advisor to nature conservation agencies in the UK on matters relating to environmental integration and the regionalisation of fisheries management.

Kelly Vodden is a PhD candidate in the Department of Geography, Simon Fraser University (SFU), Instructor in SFU's Centre for Sustainable Community Development and consultant to government and non-governmental organisations. Her academic and professional endeavours have focused on sustainable community economic development and community-based resource management initiatives, particularly in coastal/fishing communities.

Douglas Clyde Wilson (PhD Michigan State University 1996) is a Senior Researcher at the Institute for Fisheries Management and Coastal Community Development in Hirtshals, Denmark. He is an environmental and natural resource sociologist with fifteen years experience in fisheries management research in Africa, Europe and North America. His research interests include the institutional aspects of the governance of fisheries and other commons, the social impacts of environmental policies, and the sociology of knowledge as applied to resource management. He serves as the editor of the *Common Property Resource Digest*; as an Associate Editor of *Society and Natural Resources*; and as a member of the ICES Working Group on Fisheries Systems.

James Wilson is a senior resource economist, recognised for his work on complex systems and common pool resources and he has three decades of experience working with the fisheries of Maine. He received his Ph.D. in Economics from the University of Wisconsin in 1971 and is coordinator of the dual MS degree program in Marine Sciences and Policy at the University of Maine. Currently, he co-chairs the Downeast Initiative, a community-based coalition of fishermen, scientists, and state government, and he was a driving force in developing the Portland Fish Exchange (a daily auction for groundfish in Maine which has raised prices and increased fish quality significantly), and the Maine Fishermen's Forum (an annual conference which brings together a diversity of fishermen for vital discussions). He has chaired the Scientific and Statistical Committee of the New England Fisheries Council, chaired the committee that implemented Maine's lobster zone system of co-management, was vice-president of the Maine Fishermen's Forum, and is a member of the Resilience Alliance, an international group of scientists that addresses problems in natural resource management.

CHAPTER 1: THEORISING ABOUT PARTICIPATORY FISHERIES GOVERNANCE

TIM S GRAY
School of Geography, Politics and Sociology, Newcastle University, Newcastle upon Tyne, NE1 7RU, UK

Abstract

This edited book is about participation in fisheries governance, which is an issue that has become fashionable during the last decade, partly because of dissatisfaction with the performance of fisheries management systems across the world; partly because of the increasing interest in the notion of 'governance' as a substitute for 'government' in a variety of policy sectors; and partly because of the growing popularity of the concept of stakeholder participation in all areas of decision-making. The purpose of this introductory chapter is to establish a theoretical framework within which the participatory mode of governance may be best understood. First, I explore the conceptual issues raised by the notion of governance. Second, I analyse and discuss the three main ways in which the notion of governance has been applied to fisheries management – the hierarchical mode; the market mode; and the participatory mode - focusing especially on the four sub-types of the participatory mode: industry self-governance; co-management; community partnership; and environmental stewardship. Third, I discuss the wider implications of the three different modes. Finally, I provide a synopsis of the chapters in the book, showing how they all focus in one way or another on the central imperative of contemporary fisheries governance – how to make greater use of participation in order to improve the quality of decision-making.

1.1 Introduction

It is a commonplace that many of the world's commercial fisheries are in a state of crisis. As Blyth *et al* (2003:409) point out, in 2000, the Food and Agriculture Organisation (FAO) reports that 72-75 per cent of the world's major fish stocks are either "over-exploited, fully exploited, rebuilding or depleted". A recent report by the highly respected UK Royal Commission on Environmental Pollution (RCEP 2004: paras 1.7-1.8) refers to a "Crisis in the marine environment", claiming that the seas "are being depleted of fish and other living creatures at an alarming and unsustainable rate." Much of the blame for this crisis is levelled at the way in which fisheries are managed (RCEP 2004:para 5.109; van Vliet and Dubbink 1999:13; Jentoft *et al* (1999:239). Symes and Phillipson (1999:59) are in no doubt where the blame lies – with the 'top-down' or hierarchical mode that characterises 'conventional' management systems, but Holden (1994) argues for a reinforcement of the hierarchical mode. Other critics claim that only a suitably managed market system can deliver a sustainable fishing industry. However, a strong body of opinion favours a much more participatory mode of governance, linked to environmental imperatives to curb chronic over-fishing.

In this chapter, I examine the theoretical foundations and practical implications of the three main modes of fisheries governance – the still dominant hierarchical mode, and its two main rival modes, the market mode and the participatory mode, dividing the latter into its four main types: industry self-governance; co-management; community partnership; and environmental stewardship. My argument is that, although in the real

T. S. Gray (ed.), Participation in Fisheries Governance, 1–25.

world we will find a mixture of all three modes of governance in which the hierarchical mode plays a leading part, there is increasing emphasis on the participatory mode of fisheries governance. In the last section of the chapter, I introduce the subsequent chapters in the book, showing where they fit into this schema or taxonomy of modes of fisheries governance, and pointing out their contribution to our understanding of the participatory mode. But first, I analyse the concept of 'governance'.

1.2 The meaning of the term 'governance'

The term 'governance' is ambiguous, spawning a variety of meanings (Pierre and Peters 2000:7). Political scientists such as Rhodes (1996:652) have associated it with the minimal state, the hollowing out of the state, public-private partnerships, corporatism, new public management, and policy networks. Often 'governance' is contrasted with 'government': during the 1990s, it became fashionable to denote a shift from the hard-nosed concept of **government**, with its connotations of a legally-based, centralised, sovereign state authority, formally elected, and possessing constitutional powers (including the right to exercise coercive force), ruling over a specific territory by means of an exclusive elite; to the soft-nosed concept of **governance**, with its connotations of a more informally-based, decentralised, shared, collective and inclusive decision-making structure, with multiple levels of engagement. According to Rhodes (1996:652-653), governance is less about making and enforcing authoritative decisions, than about extending decision-making outwards to embrace a wider public, thereby creating a culture of mutual respect between governors and governed. If government is founded on **consent**, governance is founded on **consensus**. Pierre and Peters argue that there has been a "gradual shift from 'government' towards governance" (2000:25), and that the "governing state has been replaced by an enabling state that governs to a large extent by co-ordinating and facilitating other powerful actors in society" (2000:12). However, in my view, 'governance' has not so much replaced government, as supplemented it, by adding more consensual processes for accomplishing its ends (Rosenau 1992:4).

Another governance issue is about the distinction between governance as a structure of decision-making, and governance as a set of principles. So far, I have been assuming that governance simply refers to *structures* (such as hierarchical, market or participant structures), but the literature also alludes to *principles* of governance, such as transparency, the rule of law, and equity. The European Commission, in its definition of governance, refers to principled elements, as we can see from the so-called "Roadmap" of the 2002 Common Fisheries Policy (CFP) Reform process (EC 2002:23 footnote 14): "Governance means rules, processes and behaviour that affect the way in which powers are exercised, particularly as regards openness, participation, accountability, effectiveness and coherence." Some of these governance principles are, of course, directly connected to structures – such as the principle of participation. But others, such as the principle of the rule of law, are largely independent of structures. In what follows, I will include both structures and principles in my analysis of modes of fisheries governance.

1.3 Three modes of fisheries governance

Different writers suggest different typologies for modes of governance (Pierre and Peters 2000:14; Kooiman 1999a and 2003). But the most persuasive typology is that of van Vliet and Dubbink (1999:14), who suggest the following three modes: hierarchical governance; market governance; and participatory governance, and it is this typology that I have adopted.

1.3.1 HIERARCHICAL GOVERNANCE

Hierarchical governance is the 'state-centric' or 'directive' mode of fisheries governance, featuring a principal role for the state (van Vliet and Dubbink 1999:22). This is currently the most common mode of fisheries governance, though its dominance is now being challenged by both the other modes. Features of the hierarchical style of governance include its top-down structure, and its emphasis on legality, political legitimacy, centralisation, bureaucracy, interventionism, command-and-control, scientific elitism and exclusivity, and sense of public responsibility. Part of the rationale of the hierarchical mode is that fisheries are a public resource – an important element of the national heritage – and therefore, like other public resources such as air space, are a prime responsibility of the state. Fisheries cannot be either privatised or communalised, because that would signify that fish can be exclusively owned by either individuals or groups, whereas they are the property of the whole nation.

The ideological underpinning of hierarchical governance is captured in John Dryzek's account of the environmental discourse which he calls "administrative rationalism", or "leave it to the experts", which places emphasis upon problem solving by a public-spirited elite of bureaucrats and scientists (Dryzek 1997: chapter 4; Frid, this volume). Decision-making is administration rather than politics, and the decision makers are the expert few, not the mass public. The psychological underpinning of hierarchical governance is Hobbesian – that human nature is self-centred and egoistical, and that the only way to avoid "the tragedy of the commons" (Hardin 1968) is to institute strict measures of control, backed up by force. Typically, this requires fish quotas, days-at-sea, decommissioning, satellite surveillance, and inspectors on boats and in ports to check that catches and landings do not break the rules. In other words, the stick rather than the carrot is necessary to discipline fishers' behaviour that puts fish stocks at risk.

An example of the hierarchical mode of fisheries governance is the UK system (Symes and Phillipson 1999:70-71), where the most important decisions are made by a central government department – the Department for Environment, Food and Rural Affairs (DEFRA). Further up the chain of command is another example of hierarchical governance – the European Union's (EU) CFP (Symes (1999a:5; Kooiman 1999b:160,166; Hawkins, this volume). It is true, Kooiman concedes, that national governments in the European Fisheries Council can, and often have, resisted the cuts in quotas (total allowable catches or TACs) proposed by the European Commission, but he argues that national governments do not have much influence over policy decisions. The fishing industry has even less influence. Even the much vaunted 2002 CFP reform process, with all its emphasis on consultation and transparency, was perceived by the industry to have been conducted in a very hierarchical fashion, as an editorial in *Fishing News* (27/9/02:2) makes clear:

> One of the most striking aspects of the CFP reform package that is currently being drawn up is just how little input the fishing industry has into the detailed proposals. These are being worked out almost entirely behind the scenes by member state and Commission officials.

Moreover, in Article 11 of the proposed new Constitution for the EU (EC 2004), fisheries policy, under the "conservation of marine biological resources", has been made one of four areas (the other three areas being "customs union, commercial policy, monetary policy") where the EU will have "exclusive competences" (*Fishing News* 14/2/03:2), thus ruling out the possibility of devolving management powers to the newly launched Regional Advisory Councils (RACs) (*Fishing News* 1/10/04:5).

By contrast to most commentators and the fishing industry, who are all highly critical of the hierarchical mode of governance, one of the most forceful advocates of hierarchical governance is the late Mike Holden, a senior official in the European Commission's Fisheries Directorate during the 1980s. Holden (1994:245ff) argues that the reason for the failure of the conservation objectives of the CFP is not because the CFP is *too* hierarchical, but because it is *insufficiently* hierarchical.

There are three main criticisms of hierarchical governance (van Vliet and Dubbink 1999:22). First, the state does not have a monopoly of knowledge about fisheries: other stakeholders have important contributions to make to our understanding of the marine ecosystem, the sheer complexity of which makes it impossible for a single body to grasp (Kooiman *et al* 1999:261). Second, the state does not have a monopoly of judgement about the right measures to introduce to deal with fisheries problems. Again, other stakeholders have much valuable advice to give on the utility of alternative measures. Third, the state does not have a monopoly of power to enforce its measures. It is almost impossible to prevent individuals and groups from undermining government policies, if these policies are unpopular. In a remarkably frank statement, a recent report from the British Prime Minister's Strategy Unit (PMSU 2004: para 3.5.7) sums up these criticisms of the top-down structure of the CFP as follows:

> Simple command-and-control policies will not work in complex, multi-jurisdictional, mixed fisheries. Currently, the quota control system implicitly assumes that stocks can be measured reasonably accurately and that the capacity exists to develop appropriate management measures and plans for all EU stocks centrally in the Commission. It assumes that the Fisheries Council can and will take the necessary detailed decisions to manage stocks. Furthermore, it is assumed that Member States can enforce the rules and that fishermen will obey them. This set of assumptions is for the large part flawed and does not reflect the reality of fisheries management in the EU.

Nevertheless, despite these weaknesses of the hierarchical mode, many writers insist that the state cannot be absent from fisheries governance (Kooiman 1999b:167; Pierre and Peters 2000:18,68). On this view, there will always be a need for at least some element of hierarchy, no matter what the prevailing mode of fisheries governance. According to Symes (1999b:32), the state supplies several vital functions which every fisheries management system requires, including "democratic accountability", "exclusive legal status in negotiations with third countries", and "legislative and revenue raising powers", and, we may add, coercive power to enforce the rules. This means, says Symes (1999a:32-33), at least as far as EU fisheries are concerned, that there is no prospect of a 'hollowing out' of the state. However, events have to some extent overtaken this prediction, and the 2002 CFP reform has addressed at least some of the above criticisms, as we shall see.

1.3.2 MARKET GOVERNANCE

Turning now to the second of the three modes of fisheries governance – market governance – notwithstanding Symes' assertion, we will find that part of the impetus towards it has come from the hollowing out trend in other policy areas. Markets empower ordinary people as consumers (Pierre and Peters 2000:19), and incentivise entrepreneurs as producers. Market governance is based on the natural forces of supply and demand, untrammelled by government interference, though supported by the legal security of private property rights. Dryzek (1997: chapter 6) characterises this mode as "economic rationalism", or "leave it to the market".

Market governance follows the classical economic theory of Adam Smith, in that it assumes that the pursuit of individual economic self-interest, within the legal framework of the protection of rights of life, liberty and property, will lead to the optimal benefit for everyone, by the so-called 'invisible hand mechanism'. On this neo-liberal theory, failure to achieve optimality is usually because of interference with the market mechanism by governments for ideological reasons. As Hayek (1944) argues, the workings of economic and social enterprises are so complicated that no-one can possible know how to run them, and so they should be left largely to run themselves. Further theoretical underpinnings of market governance include the methodological assumption that all social activity in the end boils down to individual decision-making (methodological individualism); the ethical assumption that each person knows best what is in his or her own interest (utilitarianism or philosophical radicalism); and the psychological assumption that people are rational in the choices that they make (rational choice theory).

Applying this theory to fisheries, instead of trying to replace the free market forces of supply and demand (as the hierarchical CFP does by adjusting fish price levels; imposing the principle of relative stability; designating special boxes, such as the Irish and Shetland Boxes; and creating the Hague Preferences), governments should adjust market carrots and sticks to reward self-interested behaviour that protects public resources, and punish self-interested behaviour that damages them, and then leave the forces of supply and demand to get on with it (van Vliet and Dubbink 1999:19-20).

Of the EU Member States, Spain is the most vociferous advocate of a shift towards the market mode of governance in the way in which the CFP is managed. For instance, Jose Fuertes (Director-General of the Vigo Fishing Vessel Owners' Cooperative) argued at the Public Hearing in Brussels in June 2001 on the CFP Reform Green Paper, that the fishing industry should be treated by the EU like other industries, in compliance with World Trade Organisation (WTO) principles, with guaranteed freedom of fisheries activity, non-discrimination between fishers of different nationalities, equal access to all markets, complete transferability of fishing rights, free competition, and anti-monopoly regulations (Wood and Ritchie 2001:2-3). The recent report from the British PMSU (2004: para 9.1) argues for a move away from the command-and-control model to a "central role for market-driven incentives and mechanisms whereby information can be used to influence decision-making by individual businesses."

In answer to critics such as Hardin (1968), who claim that the free market produces the tragedy of the commons, whereby the remorseless pursuit of self-interest leads to the destruction of common user resources, free marketeers say that the solution is not to abandon the market, but to structure it in such a way as to incentivise producers to take good care of scarce resources. "For the market to work, privatisation is essential" (Jentoft and McCay 2003:295). In the case of fisheries, this means introducing a system of individual transferable quotas (ITQs) (van Vliet and Dubbink (1999:15). There is an Aristotelian assumption here that people are much more likely to look after a resource that they themselves own, than a resource that is common to all (Sissenwine and Mace 2001:13). Cooperation between fishers is secured out of mutual self-interest, rather than because of either state coercion (hierarchical governance) or collective commitment to the general good (participatory governance).

Several writers claim that there has been a move from hierarchical governance to market governance in fisheries during the last 20 years, following the neo-liberal trend towards deregulation and privatisation (Kooiman (1999b:142). One reason for market governance's popularity during the 1980s and 1990s, according to Jentoft and McCay (2003:296), is that economists were held in much higher esteem by fisheries managers than were social scientists, who advocated the participatory mode. ITQs are now in operation in Iceland, New Zealand, and parts of Australia, Canada (Murray *et al*, this volume), Chile, Namibia, the USA, and Europe (in Denmark and the Netherlands) (Sissenwine and Mace 2001:13). However, in the EU as a whole, despite the trend towards deregulation in other policy areas, in fisheries, the trend has been in the opposite direction – towards greater regulation.

In critically appraising the market mode of fisheries governance, the first point to make is that it rests upon an over-simplified view of human motivation - "a one-dimensional *homo economicus*". (Kooiman 1999b:143). But fishing is more than a pecuniary activity; at least for some fishers it is a way of life, a form of self-expression, self-identification, and self-determination. Also, the market mode's extreme individualism ignores social and cultural influences on fishers' behaviour, such as concern for the marine environment (Jentoft and McCay 2003:297). Moreover, the market mode of governance does not entail the *elimination* of state involvement in fisheries management. Far from it: market governance depends on the state for several functions, including the tasks of establishing the terms of the market (for example, deciding the overall quotas, for shares of which fishers will compete); of monitoring the functioning of the market to ensure that fair competition is maintained; of ensuring that public goods such as adequate fish stock levels and the health of the marine ecosystem are not damaged; and of guaranteeing that private property rights are not violated. The fact is, that the market mode of fisheries governance is a highly regulated market – it does not entail letting everything rip.

Furthermore, where the market mode has been introduced, there are mixed messages about its success. For example, its advocates are highly positive about its beneficial effects on the fisheries in New Zealand (Clark *et al* 1998) and Iceland (Arnason 1996), pointing out that where ITQs are introduced, there is a marked reduction in overcapacity (Jentoft and McCay 2003:296) . But critics point out that a market system of ITQs was abandoned in the Faeroes, and replaced by a days-at-sea scheme (hierarchical governance), and that it is causing severe social problems in both New Zealand (*Fishing News* 7/11/03:6) and Iceland (*Fishing News* 20/8/04:6).

Another criticism is that market governance has a damagingly differential impact upon fishers (van Ginkel, this volume). As Jim Portus (Chief Executive of the English South Western Fish Producers' Organisation (SWFPO)) put it, "we do not need…monstrous market forces experiments with quotas which will benefit the few and impoverish the many" (quoted in Wood and Ritchie 2001:21). For instance, an ITQ system rewards those who are already in a market, but penalises those who are trying to get in (van Ginkel 1999:55-56). Moreover, market governance favours the offshore sector, which is highly capitalised, at the expense of the inshore sector, which is more artisanal. As a result, it has an adverse effect on local fishing communities, which rely heavily upon the inshore sector. Steps have to be taken by the state to protect these often remote local communities from being wiped out by globalising forces, because they may have little alternative employment prospects (Collet 1999:124).

Finally, it is important to note that market governance in itself will not necessarily maintain the level of fish stocks, still less look after the health of the marine ecosystem (Wilson, this volume). Indeed, market logic might dictate to capitalists a strategy of exploitation of stocks to the point of economic extinction, to gain a short term high return which can be "reinvested elsewhere" (Collet 1999:123). However, this criticism may be partly met from within the market mode of governance – for example, by an eco-labelling system, whereby consumers can choose to buy fish products solely from sources that are independently certified as sustainable (Jentoft and McCay 2003:296-7). Such a system is already in place, in the shape of the Marine Stewardship Council (MSC) (Long 1999), though after eight years of existence, the MSC has only managed to certify a fraction of the world's fisheries (*Fishing News* 27/2/04:7) and, with its limited funding, it is doubtful whether its scheme will ever have much impact on consumer choices.

Notwithstanding these criticisms, the market mode of governance has one significant value – it serves as an important corrective to the hierarchical mode in that it demonstrates that regulators should not try to 'buck the market', because rules that prevent fishers from making a living will be ignored. In other words, fishing regulations must be economically literate.

1.3.3 PARTICIPATORY GOVERNANCE

We now come to the third and last mode of fisheries governance – the participatory mode – on which this book is focused. The participatory mode is more variegated than are the two previous modes, in that it contains four distinct sub-types: industry self-regulation; co-management; community partnership; and environmental stewardship. Before examining these four sub-types in detail, however, there are some generic features of the participatory mode to be explained. First, its concept of the person is very different from those held by the other two modes. By contrast to the hierarchical concept of the master/subject relationship between regulators and regulated, and the market concept of producers and consumers, the participatory concept is that of citizens and stakeholders. Also, the participatory mode operates at the meso (civil society) level, that is, mid-way between the macro (state) level of the hierarchical mode, and the micro (individual) level of the market mode (van Vliet and Dubbink (1999:22). The four types of participatory governance are made up of four different cohorts or segments of civil society: industry; industry plus regulators; local communities; and environmentalists. Moreover, whereas for hierarchical governance, legitimacy lies in the formal system of parliamentary elections (van Vliet and Dubbink 1999:26), the essence of legitimacy in the participatory mode lies in the involvement of stakeholders in decision-making (see Hatchard, this volume), though the nature and extent of that involvement will vary from one type of participatory mode to another (Dryzek 1997:86). Furthermore, the characteristic style of the participatory mode is one of consensus-seeking negotiation, rather than either the hierarchical style of command, or the market style of exchange.

By contrast to the administrative rationalism of hierarchical governance (leave it to the experts), and the economic rationalism of market governance (leave it to the forces of supply and demand), participatory governance is depicted by Dryzek (1997:chapter 5) in terms of "democratic pragmatism" ("leave it to the people") (cf Kooiman 1999b:142). As Dryzek (1997:92) explains, this means putting politics back into governance in place of administration, which is characteristic of hierarchical governance, and economics, which is characteristic of market governance. Although both market governance and participatory governance employ against hierarchical governance the argument that fisheries, ecosystems and regulations are too complex for government to manage alone, they draw different conclusions. Market governance argues that only the market can provide solutions, whereas participatory governance argues that only the collective knowledge of all affected parties can deliver answers (van Vliet and Dubbink 1999:15). Two heads are better than one; collective wisdom outweighs individual wisdom.

Turning now to the roots of the participatory mode of fisheries governance, one root is post-materialism, a theory which Ronald Inglehart (1990) developed to account for the fact that in post-war Western countries, many citizens have reached the point where they are less concerned with the accumulation of material goods, and more concerned with their quality of life – that is, post-material values. These post-material values include environmental goods and greater self-determination, which in turn entail a demand for more public participation in political decisions. This leads us to another, closely related root, which is the appearance of new social movements (NSMs). NSMs, unlike old social movements such as trade unions, which demanded redistributive economic and social policies, have arisen to push for causes such as environmental protection, feminism, and community values. NSMs have spawned an explosion of non-governmental organisations (NGOs) demanding inclusion in decision-making forums.

A further root is loss of faith in experts. People are less inclined nowadays to defer to claims to superior knowledge held by bureaucrats and government scientists, and the value of experiential knowledge is becoming increasingly recognised. Also, there is increasing recognition that many features of decision-making in fisheries governance are value-laden, not value-free, and that the value judgements of the public should prevail over the value judgements of the experts (Sissenwine and Mace 2001:13). Another root is the spirit of devolution, particularly in the UK, where in recent years we have seen the creation of the Scottish Parliament, the Welsh Assembly, and moves towards regional assemblies in England. This is a response to the need to devolve decision-making to its lowest possible levels, which is formally endorsed by the EU's principle of subsidiarity.

Communicative rationality is a further root. This is a concept derived by Jurgen Habermas (1984) to denote the contemporary aspiration of civil society to engage in dialogue on the important political issues of the day in order to reach more reasoned decisions (Wilson, this volume). It differs from the administrative rationality of hierarchical governance, because it opens up the lines of dialogue to all citizens, not just the experts, and it differs from the economic rationality of market governance, in that it strives to reach universalistic, not individualistic, conclusions (Kooiman 1999b:164). For Habermas (1984:19), dialogue is a collective search for truth. Van der Schans (1999:115) makes the important point that this does not necessarily rule out all hierarchical regulations or economic drivers, but it does mean that they must pass the dialogic test of good reasons. And this test entails a process of interactive communication (Kooiman *et al* 1999:262; Hatchard, this volume).

The final root of the participatory mode is the failure of the other two fisheries governance modes (Hanna 2003:311). Crises in the fisheries drive managers to seek the help of stakeholders (Hall-Arber; Dunn, both this volume). Sen and Nielsen (1996:416) point out that in nearly all of the 22 cases of co-management that they studied, the rationale for setting up the co-management regimes was because the fishery was near, or at, the stage of over-exploitation.

There are several generic criticisms of the participatory mode of fisheries governance, which I will consider at more length in the final chapter. They can be summarised here as follows: right wing critics argue that participation is unnecessary, because experts have all the knowledge that they need; damaging, because it inhibits flexibility and slows down rapid responses to emergency; costly, because it absorbs considerable time and energy to organise; and subversive of representative democracy. Left wing critics argue that participation is a charade, cynically used by regulators to mask their domination, and to co-opt, and therefore neutralise, stakeholders.

Let us now turn to the four different types of the participatory mode of fisheries governance – industry self-regulation; co-management; community partnership; and environmental stewardship.

1.3.3.1 *Industry Self-Regulation*

The industry self-regulation version of participatory governance is the assumption by the fishing industry of sole responsibility for running the fishery (Sutinen and Soboil 2001:16; Symes and Phillipson 1999:63). Essentially, industry self-regulation is about fishers' organisations taking charge of their own destinies. This is why the terms 'autonomous self-management' and 'self-determination' have been used to characterise industry self-regulation. However, this does not mean entire independence: industry self-regulation is autonomous only within certain limits. For instance, safety rules laid down at national/international level could not be set aside by a fisheries organisation.

Examples of industry self-regulation are common in developing countries, as Johannes (2003:15) points out: "in indigenous fisheries…management is…often largely in the hands of the fishers". A partial example of industry self-regulation in a developed country is the large-scale offshore fisheries in the USA, where, in 1976, the government conferred on eight Regional Fisheries Management Councils (in which the majority of members are from the commercial and recreational fishing industry (Symes, this volume)) most of the responsibility for managing fisheries in federal waters (USCOP 2004:231). A clearer example is in New Zealand, where the Challenger Scallop Enhancement Company has entire responsibility forthe Southern scallop fishery (PMSU 2004:Annex D,para 4.4). A further example is in Normandy, France, where a fishers' organisation (CRPMEM), headquartered at Cherbourg, representing over 2000 fishers in 640 over-25 metre vessels, manages 85 per cent of the species within its allocated area (*Fishing News* 14/11/03:18). Within the UK, the best example of industry self-regulation is in inshore shellfisheries which have been subject to a Regulating Order, whereby exclusive fishing rights are vested in an organisation largely composed of fishers and charged with the responsibility of running the fishery – for instance, the Shetland Islands RO 1999, where management is in the hands of a limited company called the Shetlands Shellfish Management Organisation (Symes and Ridgway 2003:42). Also, as Stead (this volume) shows, the UK aquaculture industry is characterised by a high degree of voluntary self-regulation, whereby trade associations bodies set down detailed codes of conduct, to which they require their members to adhere.

Advocates of industry self-regulation argue that it has many benefits. For one thing, industries are very good at protecting their fisheries from external vessels which threaten to wipe out the stocks (van der Schans 1999:113-114). Also, it is claimed that the industries have first hand knowledge of fisheries, and that this experiential knowledge is invaluable for accurate management measures. Moreover, industry self-regulation shifts the responsibility for decision-making to those whose livelihoods depend on the measures being taken, so there is a considerable incentive for the industry to take steps that will protect the stocks, thereby safeguard their own long-term economic prospects. Furthermore, the fact that the industry is self-regulating suggests that fishers are likely to look favourably on the resulting regulations, and so compliance rates will be high. Finally, it is claimed that because the industry is in charge, it will bring peer pressure to bear on those fishers who continue to violate the rules.

However, critics of industry self-regulation argue that it has serious weaknesses. One weakness is that if its codes of practice remain voluntary, sanctions against their violation may not be strong enough to ensure compliance (Stead, this volume). Another weakness is that the industry's self-interest may not coincide with the public interest. For example, the industry may take the view that the protection of small cetaceans such as dolphins and porpoises is not a high priority, and, therefore, that their members are not required to take energetic steps to reduce their cetacean by-catch, if such steps would significantly reduce their profit margins. A further weakness is that there is no guarantee that members of the fishing industry will be able to agree on management decisions: self-regulation does not guarantee that everyone will be happy with fisheries management decisions. As Sissenwine and Mace (2001:14) note, "At present, the US National Marine Fisheries Service is coping with more than one hundred legal actions attempting to overturn fisheries management decisions. In almost all cases, the litigating parties actively participated in the debates leading up to the decision, but they disagreed with the outcome."

1.3.3.2 *Co-Management*

In a co-management system, management is generally shared between government regulators and representatives from the fishing industry (van der Schans 1999:119). Kooiman (1999b:163-164) points out that this does not mean mere consultation of the industry by the government, but genuine partnership in decision-making: "power sharing is a must" (Jentoft 2003: 4). For Symes and Phillipson (1999:64), the role of user groups in management is not "passive or reactive", but "active or proactive". Moreover, for Van Vliet and Dubbink (1999:23-24), co-management does not entail an adversarial relationship between the two sides (regulators and industry), but a genuine endeavour on both parties to reach the common good. This is not to say that co-management comes naturally to either side: as Langstraat (1999:78) notes, each side must have some incentive to co-operate.

Jentoft (2003:1) points out that while the concept of co-management is only about 25 years old, co-management regimes have existed in some parts of the world for centuries. There are now many examples of co-management. For instance, Hara and Nielsen (2003) describe co-management systems in Africa, while Nsiku (2003) focuses on the case of Malawi; Pomeroy and Viswanathan (2003) discuss co-management approaches in Southeast Asia and Bangladesh, focusing particularly on coastal fisheries, and case-studying co-management in the Phillipines, while Baird (2003) case studies Southern Laos; Begossi and Brown (2003) explain co-management regimes in Latin America and the Caribbean; Loucks *et al* (2003) outline co-management arrangements in North America, comparing power sharing in the USA and Canada; while Hall-Arber, this volume, focuses on the New England's groundfish fishery co-management scheme; and Metzner *et al* (2003) analyse the diverse set of co-management structures in Australia (cf Baelde 2003) and New Zealand. In Europe, there are several examples of co-management, the purest form being in Norway, which operates a centralised co-management system at national level (Hernes *et al*, this volume). Within the EU, which does not have a co-management structure at the intergovernmental level of the CFP, the Netherlands (van Ginkel, this volume) has the strongest co-management system at the national level (Symes *et al* 2003:124), while the UK has some features of a co-management system in its sectoral quota management by the Producer Organisations (POs) (Symes *et al* 2003:126) and in its regulation of inshore fisheries in England and Wales (the Sea Fisheries Committees (SFCs)) (Knapman, this volume; Symes and Phillipson 1999:81). Many writers, including, Symes (1999b:41), have argued for an extension of the principle of co-management to the regional level in Europe, and the recently established RACs may be a step in that direction (Symes, this volume).

There are many advantages claimed for co-management systems. For example, Symes and Phillipson (1999:65) list the following benefits of co-management: increased transparency; a wider source of knowledge; more rational regulations; greater legitimisation and compliance (cf van Vliet and Dubbink 1999:24); and reduced costs of surveillance. Kooiman *et al* (1999:264-5) argue that co-management is essential to get to grips with the "diversity" of fisheries. Given these major advantages, it is hardly surprising that such writers claim that co-management is essential for modern fisheries. Indeed, Kooiman *et al* (1999:260) state that "This model is more than an option: it is a necessity."

On the other hand, Pomeroy (2003:248) claims that "There are only two well documented cases of long-standing marine fishery co-management arrangements that work, in Norway and Japan". Also, Symes and Phillipson (1999) warn that co-management cannot be forced upon an unwilling industry and/or government. Both sides must be able and willing to make it work, and this cannot be taken for granted. For example, the industry side may lack the professional skills or the financial resources to handle important negotiations; or it may be too fragmented to organise itself into a coherent body; or it may be reluctant to shoulder responsibility for decision-making, fearing loss of its autonomy; or it may worry that it would have to share power with other stakeholders, such as environmentalists. For its part, the government may be psychologically unable to share power with the industry. The major problem lies in building trust between the two sides.

Also, co-management may mask, rather than eliminate, local tensions (Jentoft and McCay 2003:302; Singleton 2000:18; Hernes *et al*, this volume). Moreover, co-management still entails a significant, even dominant role for government (Pinkerton 2003:65; van Ginkel, this volume; Symes and Phillipson 1999:64). Indeed, Pierre and Peters (2000:49) argue that the state might have "co-opted social interests that might otherwise oppose its actions" (cf Singleton 2000:2). Also, co-management raises the thorny questions of who are the stakeholders, and how will they be selected to be members of the decision-making body? The usual answer to the first part of this question is 'regulators and fishers', but there is a growing feeling in certain quarters that other groups of stakeholders should also be included, representing anglers, fish processors, conservation agencies, environmental NGOs, consumers, and recreational interests (Mikalsen and Jentoft 2001). And what about scientists? Should fish biologists and marine ecologists be represented in co-management councils, rather than be assigned an advisory role as experts? This raises the deeper question of whether members of such councils are recruited because they represent particular interest groups, or because they possess particular expertise (Jentoft *et al* 2003; Rice, this volume). In theory, the list of potentially eligible representatives is almost endless, embracing, as Hemmati (2002:2) suggests, anyone who has "an interest in a particular decision" (cf DfID 1995). However, extending membership beyond fishers and regulators also risks diluting the principle of co-management.

Equally difficult problems arise in trying to answer the second part of the above question – how will the representatives be chosen, and how are they to be selected? For example, which regulators will take part – only bureaucrats, or also ministers? Will the fishers be represented exclusively by their national organisations? If so, might that not exclude certain kinds of fisheries? Another issue is whether all stakeholders are to be regarded as of equal weight in the co-management deliberations. Moreover, some writers argue that the elaborate process of deliberation in co-management regimes makes it difficult either to reach decisions at all, or to do so speedily or decisively (Symes and Phillipson 1999:83,92).

Nevertheless, despite these drawbacks, co-management remains the favourite form of governance in the view of many writers. Yet, as Symes (1999b:32) points out, it is not widespread in practice. Co-management is easier to establish at local levels, where it may resemble community partnership, but it is more difficult to organise on a larger scale, because of the greater diversity of fisheries interests. However, its advocates hold that it is at the higher levels that it is needed most.

1.3.3.3 *Community Partnership*

Turning to the third type of participatory fisheries governance – community partnership – we find a much more inclusive structure. Here, the emphasis is less on the industry *per se* (industry self-regulation), or even on the industry's co-decision making with the regulators (co-management), than on the industry sharing management responsibilities with the whole range of local stakeholders who have an interest in the marine resource (not including the government) (Sen and Nielsen 1996:406). The focus of community partnership is on local fisheries, on the assumption that, like direct democracy, it is only practicable on a small, face-to-face scale.

Examples of community partnership are common in developing countries, typified in artisanal inshore fisheries, often based on complex systems of 'sea tenure' (Jentoft and McCay 2003:299). For instance, Bird *et al* (2003:178) describe a successful community partnership in Mexico. They are less common in developed countries, though Vodden *et* al (this volume) case study a community partnership in Newfoundland; Kooiman (1999b:162) finds them in Vigo and Shetland; and Symes and Phillipson (1999:64) locate "historical fragments" of them in "Spain (*cofradia*) and Mediterranean France (*prud'homie*)." Moreover, several local initiatives in the UK are developing new community partnerships for managing fisheries: for example, Loch Torridon in Scotland; 'Invest in Fish' in the south west of England; and the Solway Firth Partnership (*Fishing News* 22/10/04:6). Another form of community partnership is based on community quotas – that is, the purchase of quota by local communities (such as local authorities) in order to lease it to local fishers and to prevent quotas being bought by companies outside the area. Three community quota schemes have been operating in Shetland, Orkney and Cornwall: though the Shetland and Orkney schemes were ruled illegal by the European Commission because they breached EU competition rules; while the Cornwall scheme is now privately funded (*Fishing News* 17/9/04:3). We can also find examples of community partnership in UK aquaculture – in the many local forums and Integrated Coastal Zone Management (ICZM) schemes operating especially in Scotland (Stead, this volume).

Advocates of community partnership argue that it has many benefits. They claim that local communities have more extensive knowledge of their fisheries and their economic and social impact than anyone else, and that local communities more naturally reach consensus over fisheries decisions (van der Schans (1999:114). Also, local communities have a huge incentive to safeguard their fish stocks – the very life of their community may depend upon it. Moreover, communal partnership is held to be the best safeguard for small-scale fishing, which is described by the Scotland-based Fishermen's' Association Ltd (FAL) as "the most efficient in creating employment and ensuring environmental and ecological benefits" (*Fishing News* 8/12/00:18). Furthermore, Sissenwine and Mace (2001:14) assert that there is a higher rate of compliance with locally made rules.

However, other writers are sceptical of these claims. Symes and Phillipson (1999:63), for example, argue that community partnership was more appropriate in the past, when there was not so much pressure on stocks, and therefore the main task of management regimes was simply to ensure that every fisher got a fair share of the abundant fishing opportunities available. Modern fisheries management, however, facing declining stocks, requires capacities that are not available in local fisheries, such as research skills to provide scientific assessments of the state of the stocks and the health of the ecosystem as a whole. Stead, this volume, notes that voluntary community partnerships such as ICZM suffer from not being legally compulsory. Also, van der Schans (1999:114) disputes the claim that, nowadays, local communities are particularly homogeneous and consensual in their values: the fact is that the extent of fisheries dependence is now much less than in the past, and so the views of the community on resource use are more varied. However, van der Schans (1999:115-117) is not claiming that, nowadays, local communities are incapable of participatory governance; all he is arguing is that they must base that governance on a process of Habermasian dialogue that does not depend on a pre-formed consensus of values. Nevertheless, critics argue that community partnership's much-vaunted principle of interactivity may not lead to the enunciation of the public interest. In other words, contrary to the predictions of Habermas's theory of communicative rationality, the general will may not prevail over particular wills, as each sub-group may pursue only its own self-interest (Jentoft *et al* 1999:252).

1.3.3.4 *Environmental Stewardship*

The fourth type of the participatory mode of fisheries governance reflects the growing power of environmentalism in fisheries policy. As the Royal Society of Edinburgh (RSE) put it in a recent report (2004:56), environmental integration of fishing is "the new culture". Symes and Ridgway (2003:9) state that "There is no escaping the inevitability of environmental integration; the question is not whether but how to do it."

Essentially, environmental integration entails subjecting the fishing industry to the same environmental requirements that every other user of marine resources has had to comply with in many countries for the past twenty years. If other industries are required to carry out satisfactory environmental impact assessments before they are allowed to set foot in the marine environment, "it is not clear why the fishing industry should be exempt from such procedures" (RSE 2004:56). The most common way of characterising this requirement is by using the concept of the ecosystem-based approach (EBA), which entails managing a fishery as part of the marine ecosystem within which it is situated. Instead of directing action at individual species and habitats that are at risk, which is a fire-fighting exercise that may be merely shutting the stable door after the horse has bolted, EBA concentrates on protecting the health of the whole marine environment, thereby taking care of potential threats to all individual species and habitats. With the ratification of the 1992 Convention on Bio-diversity (CBD), the EBA has become a legal necessity (Frid, this volume).

However, it is not as an idea, but as a structure, that makes environmental integration into environmental stewardship, and thereby into a type of the participatory mode of fisheries governance – a structure that is steadily tightening its grip on fisheries policy. This structure is forged by the link between environmentalism and participation – a link that is assumed in so much of the literature that it has become something of a received wisdom or even a necessary truth (I examine the basis of this assumption in chapter 20). This link between environmentalism and participation serves several purposes. The main purpose is to ensure that the aims of the EBA are determined by the society at large. A recent UK working group on marine nature conservation argued that public participation is necessary to the EBA in order to set its priorities, because the objectives of marine resource management are "a matter for societal choices" (DEFRA 2004:89). Additional reasons for the link between environmentalism and participation are to capitalise on the knowledge of a wide range of stakeholders, and to commit them to identification with the EBA.

This leads us to the variety of participative structural forms that environmental stewardship can take. The most direct form of environmental stewardship is where environmentalists are in complete charge of a fishery. A good example of this form is the Great Barrier Reef Marine Park (GBRMP), where "day-to-day management" is undertaken mainly by officers of the Queensland Parks and Wildlife Service (Day 2002:140). For the UK, SFCs provide a partial example of this term: in fact, as Knapman, this volume points out, since the 1990s they have been given environmental duties. The next most direct form of environmental stewardship is where environmentalists share responsibility for running a fishery. An example of this form is the Shetland sandeel fishery, which is jointly managed by the Royal Society for the Protection of Birds (RSPB – a leading UK environmental NGO (Dunn, this volume)); Scottish Natural Heritage (SNH – Scotland's statutory conservation agency); the Shetland Fisherman's Association (the local fishers' organisation); and the Scottish Executive Environment and Rural Affairs Department (SEERAD – the regulatory authority).

A more indirect form of environmental stewardship is exercised by nature conservation agencies (NCAs), where they have a statutory right and duty to designate marine sites and habitats for special protection (Eno and Gray, this volume). For example, in the UK, there are three NCAs (English Nature; Scottish Natural Heritage; and Countryside Council for Wales) with this authority. They have no power of direct regulation, still less of enforcement, but there is a legal obligation on the regulators to implement the advice of these statutory bodies, at risk of administrative penalties. So the NCAs wield significant effective control over fisheries, to the point where their recommendations could result in the closure of a fishery if it were judged to pose a serious threat to the marine environment.

The least direct form of environmental stewardship is that manifested by pressure exerted by environmental NGOs (ENGOs) (Todd and Ritchie 2000; Dunn, this volume). Such pressure may be exerted in five different ways:

1. **Confrontation** – This is a familiar technique employed by Greenpeace, an example of which was its campaign against industrial fishing in 1996, taking action against Danish trawlers to end sandeel fishing off the Firth of Forth by proclaiming a 30 mile exclusion zone off the east coast of Scotland, and preventing the Danish vessels from fishing in that zone by attaching buoys to their nets. Jim Slater (a Scottish fishers' leader) declared that: "Greenpeace have set themselves up as the governing body of the North Sea" (Gray *et al* 1999:124). Greenpeace is currently taking similar action against bass pair-trawling in the English Channel in protest against the cetacean bycatch of this fishery (*Fishing* News 18/03/05:3). The success of such confrontation critically depends upon sympathetic media coverage (Oliver, this volume).
2. **Legal action** – NGOs take advantage of environmental legislation which has third party appeal rights, to institute legal proceeding against regulators for failing to comply with their environmental duties. This litigious pressure is a familiar part of US fisheries culture (Hall-Arber; Symes, both this volume), and it is becoming an increasing feature of European (Eno and Gray; van Ginkel, both this volume) and Australian (Metzner *et al* 2003:181-2) fisheries practice. Greenpeace has recently issued a legal challenge in the UK High Court against the British government for failing to honour its obligations under the EU Habitats Directive to protect the dolphin population from pair trawling for bass in the Channel.
3. **Negotiation** – ENGO presence at the Esjberg North Sea Conference in 1995 was crucial in persuading the Conference to adopt two environmental criteria for fisheries, an outcome which Richard Banks (then chief executive of the National Federation of Fishermen's Organisations (NFFO)) declared "was a disaster for the industry... Environmentalists were allowed to influence decisions for the first time: the 'precautionary principle' and the closed areas principle were endorsed" (Gray *et al* 1999:127).
4. **Collaboration** – The best example of collaboration is over eco-labelling schemes (where environmental stewardship coincides with the market mode). For example, the San Francisco-based ENGO, Earth Island, awards its Dolphin Safe certificates to tuna fisheries across the world which adopt its approved method of fishing. According to Struan Stevenson, then chair of the European Parliament's Fisheries Committee, Earth Island has become "the all-powerful *de facto* regulator of the $2 billion international tuna industry" (*Fishing News International* 2003 November:6)
5. **Advice** – Many ENGOs produce advice by employing environmental scientists (in both the natural and the social sciences) who present reports to regulators, which are based on expert knowledge of the environmental impact of fisheries. Such scientists form an 'epistemic community', composed of like-minded advocates of the ecosystem-based approach.

This is not to say that the notion of environmental stewardship as a type of the participatory mode of fisheries governance is without difficulties. On the contrary, it bristles with controversy. One criticism is that environmentalism is not a form of governance in itself, but a normative principle which may be (and indeed has been) adopted by any of the existing forms of governance. However, such a criticism misses the point, because if we look at the reasons why these other modes of fisheries governance have adopted environmental perspectives, we will find that they are responding to pressure applied by environmentalists, and this is what governance is. As an editorial in *Fishing News* (4/3/05:2) pointed out, "Environmentalists are involved at all stages of the fishing regulation process and wield enormous power."

Another criticism is about the lack of clarity in the mission of environmental stewardship. This criticism relates to the ambiguity of the environmental concept of EBA. Different players in the environmental stewardship type of fisheries governance interpret EBA in different ways. The main contrast is between the **preservationists** and the **sustainable developers** (Eno and Gray, this volume). The **preservationists** seek to preserve the marine ecosystem in aspic – that is, to maintain it in, or return it to, its original, pristine condition before human activity damaged it. By contrast, the **sustainable developers** seek not to prioritise, but to integrate environmental considerations into fisheries management, to ensure that fisheries do not damage the marine ecosystem beyond repair. Here there is a recognition that humans are part of the ecosystem, and that therefore their economic and social well-being must be taken into account and balanced against the well-being of other parts of the ecosystem. Perhaps a practical way of bridging the gap between these two contrasting, but equally legitimate, interpretations of EBA, is zoning – that is, dividing the sea into some areas where fishing is forbidden, and other areas where it is allowed (albeit under certain conditions) (Symes, this volume).

A further criticism of environmental stewardship centres on its participatory credentials. Who are the stewards? In some examples of environmental stewardship, the stewards are activists in ENGOs; in other examples, the stewards are officials in NCAs; in yet other examples, the stewards are the general public (Hernes *et al*; Coffey, both this volume). There are even examples of fishers as environmental stewards – indeed, for some researchers, the future lies with fishers adopting the mantle of stewards of the sea (EFEP 2004). This leads us to another criticism of environmental stewardship – that it appears to marginalise, or even demonise, the fishing industry: fishers see the environmental movement as a threat. However, if we take the view that the environmental stewards are the general public, then fishers are included in their ranks, along with all other stakeholders. Moreover, fishers themselves may be regarded as having an unrivalled practical knowledge of the sea, which should be utilised to improve the health of the marine ecosystem. The final criticism of the environmental stewardship type of the participative mode of fisheries governance comes from the opposite direction – namely that, far from being a powerful force in fisheries management, it is strong only on rhetoric, not on action. I will return to some of these issues in chapter 20.

1.4 The wider implications of different modes of fisheries governance

We turn, finally, to a discussion of some of the broader issues which arise out of this analysis of modes of fisheries governance. There are three questions which warrant our attention. First, do any of the three modes of governance exist in pure, unmixed forms? The answer to this question is 'No': the three modes are 'ideal types' rather than actual regimes; actual regimes are different mixtures of the three modes (Kooiman 1999a:8). In practice, therefore, the difference between fisheries management regimes is whether the balance is struck more in favour of one mode rather than the others. For example, the CFP is a mixed regime where the balance is struck in favour of the hierarchical mode, because of the dominant role played by the central organs of the European Union – the European Commission, the Council of Ministers, and, to a much lesser extent, the European Parliament (EP). In some Member States such as the UK, this hierarchical mode is reinforced. However, there is also a strong element of market governance in the CFP; its markets policy includes neo-liberal principles such as (conditionally) equal access to common resources, and non-discriminatory treatment between Member States, together with rules about marketing standards, stabilising of market prices, support for producers' incomes, and safeguards of consumer interests, and it permits Member States to introduce ITQ schemes (as in Denmark and the Netherlands).

There are also some elements of participatory governance in the CFP (Coffey, this volume). For instance, the Fisheries Council formally exemplifies representative democratic decision-making (though its vexed relationship with the European Commission prompted the following poignant question posed by Alex Smith (Scottish Fishermen's Federation President): "Who runs Europe – the elected representatives or the non-elected Commission?" after the Commission invoked its emergency powers to impose a days-at-sea scheme (*Fishing News* 24/1/03:2)). Also, there is an Advisory Committee on Fisheries and Aquaculture (ACFA) made up of stakeholders from the Member States who represent widely different interests, which comments on proposals sent to it by the European Commission, though it has no decision-making role (Hawkins, this volume). Further evidence of the participatory mode in the CFP includes the facts that the 2002 CFP reform process entailed an extensive form of stakeholder consultation; that most major initiatives of the European Commission are now preceded by consultative exercises; and that the CFP permits Member States to make use of co-management at both national level (as in Denmark and the Netherlands) and sub-national level (as in SFCs in England and Wales).

Significantly, the then EU Fisheries Commissioner, Dr Franz Fischler, in introducing a three-day debate on the Commission's Green Paper on *The Future of the Common Fisheries Policy* (EC 2001), touched on all three modes of fisheries governance. On the hierarchical mode, he referred to "implementing more effective technical measures and strengthening and harmonising control and enforcement"; on the market mode, he stated that "in the longer term, market forces…could play a greater role in the CFP"; and on the participatory mode, he said that he was "particularly anxious to engage stakeholders in the review process" (*Fishing News* 15/6/01:7).

By contrast to the CFP, in New Zealand and in Iceland, the balance is struck more in favour of market governance, with a system of fully tradeable ITQs as private property rights, but administered by a strong, centralised, hierarchical state structure. In other countries, the balance is struck more in favour of participatory governance, such as in North America, where there are fisheries regimes that can almost be characterised as industry self-governance; in many other countries where there are co-management structures; in developing countries and in local areas in Europe (including Vigo and Shetland), where there are examples of fisheries regimes that are principally based on community partnership; and in environmental stewardship regimes such as the GBRMP (Day 2002). But all of these participatory fisheries are firmly nested into wider systems of both hierarchical and market governance.

The second broad question to be considered is 'is there a process of evolution or progression, from one mode of fisheries governance to another?' At first sight, the answer is probably 'No', because there is evidence that all sorts of shifts have taken place: for instance, from hierarchical to market governance in Iceland and New Zealand; from hierarchical to participatory [co-management] (and market) governance in Denmark, the Netherlands and the USA; from market to hierarchical governance in the Faeroes and in the EU (Jentoft *et al* 1999:257); and from participatory [community partnership] to hierarchical and/or market governance in developing countries. However, we may be able to discern a long-term trend whereby the balance has broadly shifted between the three modes – from participatory governance [community partnership] (from pre-history to the nineteenth century); to market governance (from the nineteenth century to the 1930s); to hierarchical governance (from the 1930s to the 1980s); back to market governance (during the 1980s); and on to participatory governance [co-management and environmental stewardship] (during the 1990s and from 2000 onwards, respectively). I would add a simultaneous contemporary shift towards community partnership, at least in localised inshore areas, and towards industry self-regulation in North American fisheries (both offshore and inshore). But the most important recent development, in my view, has been the inexorable rise of the fourth sub-type of participatory governance – environmental stewardship – elements of which are now present in nearly every fisheries regime in the developed world.

The third broad question is 'does the mode of fisheries governance in a country reflect its political culture?' There is some evidence to suggest that the answer to this question may be 'Yes'. For instance, the co-management type of participatory governance found in the Netherlands and Denmark may be related to their proportional representation electoral systems, and their traditions of coalition governments and corporatism in national politics. Symes and Phillipson (1999:60) draw the conclusion that this is why it is difficult for a country to change its mode of fisheries governance. According to Kooiman (1999b:160), there is also a relationship between the predominant mode of fisheries governance in a country and the structure of its fishing industry. For instance, where there is co-management, the fishers' organisations are strengthened and united, but where the industry is excluded from decision-making, it remains fragmented and divided (Symes and Phillipson 1999:71).

As we shall see in the chapters which follow, many of the issues raised in this chapter are taken up, exploring particularly the implications of the co-management and environmental stewardship types of the participatory mode of fisheries governance.

1.5 Synopsis of the chapters

There are four broad sections to the book. The first four chapters focus on questions of participation in the EU's CFP. The next four chapters discuss issues of co-management in Norway, the Netherlands, the USA and the UK. Then, after a chapter on industry self-regulation and community partnership in the aquaculture sector, three chapters explore different aspects of the environmental stewardship type of participatory governance. Finally, five chapters concentrate on the complex issue of integrating fishers' knowledge and expertise with fisheries science and management. The concluding chapter draws out the three main themes that run through these chapters – the value of participation; the relationship between participation and the ecosystem-based approach (EBA) to fisheries governance; and the role of fishers' knowledge.

In chapter 2, **Clare Coffey** examines several important generic questions relating to public participation – including 'Who are the public?; 'What does public participation entail?'; 'Why is public participation regarded as valuable?'; and 'How can we evaluate its effectiveness?' – before case-studying the EU's CFP. She shows how there is already an extensive amount of public participation in the CFP, especially after the 2002 reform process, and that the new RACs offer further opportunities for the public to get involved in decision-making. But she warns that public participation comes at a price, because it absorbs a considerable amount of time, energy and funding.

In chapter 3, **Jenny Hatchard** continues the theme of participation in the CFP by highlighting its three main democratic deficiencies – centralisation, politicisation, and externalisation – and showing how a distinction should be made between representative democracy (which exists in the CFP) and deliberative democracy (which does not). She explains how a recent interdisciplinary project into North Sea fisheries governance attempted to overcome these three deficiencies of representative democracy by using a process for obtaining stakeholder preferences called 'iterative stakeholder engagement' – which is a form of deliberative democracy – tied into an ecosystem-based approach to fisheries management. But whether the CFP will embrace deliberative democracy is unclear.

In chapter 4, **Tony Hawkins** also takes up the theme of the participatory deficit of the CFP, focusing particularly on the lack of stakeholders' involvement in the process of obtaining expert advice on fish stock assessments. He argues that fishers' contribution to this advice would be especially valuable, and he explains how the North Sea Commission Fisheries Partnership (NSCFP) was established in 2000 to provide a forum for fishers, scientists and others to develop a more collaborative method of fish stock assessment. The NSCFP was the prototype for the North Sea RAC, set up by the European Commission in 2004, but this is only a first step towards the goal of a more participatory CFP.

In chapter 5, **David Symes** interprets the RACs from the perspective of regionalisation, importantly linked to the concepts of ecosystem-based management and spatial planning. He shows how the thinking behind the RACs is bound up with the EU's commitment to good governance and the 2002 CFP reform process. However, Symes points out some of the difficulties faced by the RACs – including ensuring the representativeness of stakeholders; arriving at consensus; and delivering environmental integration of fisheries policy – and he concludes that the jury is out on their likely effectiveness.

The sixth chapter, by **Hans-Kristian Hernes**, **Svein Jentoft** and **Knut Mikalsen**, shifts the focus of stakeholder participation in fisheries governance from regionalisation to social justice. They explore this angle by using a case study of the so-called 'quota ladder', a unique allocative scheme in Norwegian fisheries, whereby TACs are shared between the different inshore and offshore sectors of the fleet in such a way that the larger the TAC, the greater the proportionate share allocated to the offshore sector. Hernes *et al* argue that this quota ladder distribution system was the key to securing the consent (by a social contracting process) of fishers to the fisheries regime, responsibility for which the government was then able to delegate to a co-management structure. However, Hernes *et al* claim that the quota ladder is too narrow in both focus and representation to solve all the distributive justice issues involved.

In chapter 7, **Rob van Ginkel** examines another kind of co-management system, introduced in the Netherlands in the early 1990s in order to deal with the compliance and environmental problems thrown up by the individual transferable quota (ITQ) system, which was given legal status in 1985. This is the 'Biesheuvel' regime, under which fishers have to organise themselves into eight management groups, supervised by the Dutch Fish Product Board (PVIS), which enforce their own management plans on their members, largely by employing peer pressure backed up by legal penalties. However, van Ginkel notes that the extent of participation is still very limited, and the Dutch fisheries governance system retains much of its old command-and-control character.

In chapter 8, **Madeleine Hall-Arber** analyses the way in which a co-management regime was introduced in the New England groundfish fishery in 2003. She outlines the previous constraints on participation in this fishery, and explains how the New England Fisheries Management Council (NEFMC) was driven, by harsh criticism of its proposed options for regulating the fishery, to invite the fishers to suggest their own management tools. Three fishing organisations offered plans, and the NEFMC chose the one submitted by the New England Seafood Coalition, because it was more flexible, more accommodating to science, and involved a wider range of stakeholders than its rivals. This case demonstrates that a co-management system does not have to be place-based, but can be set up by a process of competitive bidding between organised groups of stakeholders, though whether compliance rates will improve as a result, remains to be seen.

In chapter 9, **Paul Knapman** describes a fourth sort of co-management system – that of the 12 Sea Fisheries Committees (SFCs), which regulate inshore fisheries in England and Wales. Half of the members of the SFCs are elected councillors from local authorities (which fund the SFCs), and the other half are largely fishers' representatives. Knapman evaluates their contribution to good governance, including their environmental credentials, and concludes that ways have to be found of preventing some industry representatives from unduly influencing SFC decisions in their own interests; of increasing SFCs' level of funding; and of giving SFCs more legal flexibility.

The tenth chapter, by **Selina Stead**, continues the theme of participation in inshore fisheries, but switches attention from capture fisheries to aquaculture. After noting that European aquacultural management contains elements of all three modes of fisheries governance, she focuses on two types of the participatory mode – self-regulation, where the participants are largely members of the industry; and integrated coastal zone management (ICZM), which is a form of community partnership, where the participants are all the stakeholders in the area. Stead concludes that elements of both types are needed for effective aquaculture governance, but currently there is a greater need for more ICZM than for more self-regulation.

With chapter 11, by **Clare Eno** and **Mark Gray**, we turn to the first of three analyses of the environmental stewardship type of the participatory mode of fisheries governance. Eno and Gray rehearse the role of statutory nature conservation agencies (NCAs) in the management of fisheries. This role includes the designation of Special Areas of Conservation (SACs) to protect marine habitats and species; Special Protected Areas (SPAs) to protect seabirds; and Marine Nature Reserves (MNRs) to protect ecosystems, under powers derived from EU Directives. Such designations often entail restrictions on fishing opportunities. In addition, NCAs have an advisory/advocacy role, promoting the ecosystem-based approach in forums such as RACs, local partnerships, and official reviews of fisheries policy. However, the NCAs' effectiveness is sometimes blunted by their varied interpretations of what environmental integration entails.

In chapter 12, **Euan Dunn's** angle on environmental stewardship is the role played by environmental NGOs in fisheries governance. In his chapter, he shows how this role has shifted from a strategy of 'problem identification', on the fringes of fisheries management, to a strategy of 'problem solving', at the centre of fisheries management, and he provides five illustrations of generally successful ENGO problem solving efforts However, Dunn points out that there has been a vast expansion of the demands on ENGOs to engage in these exercises, and that this is stretching their resources to the limit and making them reflect whether they should alter the balance between lobbying and stakeholder involvement in governance.

In chapter 13, from a very different perspective, **Tim Oliver** continues the story of environmental stewardship by examining the role of the media in fisheries governance. He points out that the national media contributes significantly to the environmental pressure placed on the fishing industry, by providing space for gloomy green reports on the condition of the sea. However, he acknowledges that regional newspapers sometimes take the side of the fishing industry, and that the fishing trade press (in which he occupies a leading position) performs a valuable role in representing the opinions of fishers and their communities.

With **Chris Frid's** chapter 14, the environmental theme is situated within the last major theme of the book – that of the role played by marine science in the participatory mode of fisheries governance. Frid considers how science has been used and misused in the hierarchical mode of fisheries governance for over 100 years, and he rehearses the management failures that have occurred during that period. He argues that the way to avoid such failures in the future is to incorporate ecosystem objectives into marine science, and to engage in closer dialogue with fishers and other stakeholders. This strategy entails a significant challenge for scientists, both in adjusting to a new holistic approach to the marine environment, and in taking on new advisory and educational responsibilities.

In chapter 15, for **Jake Rice,** the central issue in this changing role of marine science is how best to integrate fishers' experiential knowledge into fisheries science advisory meetings, an issue raised earlier by Hawkins. Drawing on his own extensive experience of such meetings, Rice discusses five different scenarios of how this integration has been attempted in Canada. He concludes that, although it takes time for both fisher and scientific participants to become accustomed to such arrangements, if everyone involved tries to make the meetings work as advisory, rather than as adversarial, exercises, there is no reason why such inclusive approaches should not become an accepted part of the practice of fisheries governance.

In chapter 16, **Grant Murray**, **Dean Bavington** and **Barbara Neis** follow directly on from Rice's chapter, providing detailed empirical evidence of the utility of fishers' ecological knowledge (FEK) to fisheries governance in Newfoundland and Labrador. They contrast the hierarchical mode of governance and marine science which was in place in Canada during the 1970s and 1980s, with the participatory mode of governance, making use of fishers' knowledge, which grew up during the 1990s. Two case studies of the participatory mode – Atlantic cod and American lobster – are analysed, and the conclusion is reached that there are great benefits to be gained from such inclusive approaches, but only if they are properly designed, developed and funded.

In chapter 17, **Kelly Vodden**, **Rosemary Ommer** and **David Schneider** reinforce the conclusions of Rice and Murray *et al*, by reporting on three different ways of using collaborative learning in fisheries governance – hierarchy; networks; and community – all of which have been tried in the major Canadian *Coasts Under Stress Project.* They contextualise this comparative analysis by reference to the problem of 'scale' – that is, the need to choose a scale for fisheries governance that satisfies both the ecosystem-based approach and the human need to feel 'at home'. Their finding is that collaborative learning is of immense value in improving the quality of fisheries governance, and that the wider the extent of such learning processes, the better.

In chapter 18, **Jim Wilson** picks up on the problem raised by Vodden *et al* of the appropriate scale of the marine ecosystem for fisheries governance purposes. The problem is where to draw the line between an area that is too large (and therefore too complex) to be adequately understood, and an area that is too small to include important interactions with factors outside it. Wilson points out that this problem is ignored by conventional fisheries science, which concentrates on single species populations which are essentially scale-less. But the ecosystem-based approach meets the problem head-on by scaling the ecosystem to the maximum size that can be adequately understood, and, therefore, properly managed by a decentralised co-management regime.

In chapter 19, **Douglas Wilson** and **Alyne Delaney** also refer to the problem of the scale of the fisheries that are being managed, in a detailed analysis of the production of scientific knowledge for the EU's CFP. Their aim is to evaluate the way in which stakeholder participation is having an impact on the generation of the formal scientific advice that informs the governance of EU fisheries. Their conclusion is that conventional fisheries science should recognise its limitations; accept the fact that the boundary between objectivity and subjectivity is blurred; and engage in a co-operative exercise with fishers to find "serviceable truths" leading to "more flexible fisheries governance as well as better science".

My concluding chapter sums up the book's findings, and discusses the three main themes that emerge from the chapters: 1) the benefits and deficiencies of stakeholder participation in fisheries governance; 2) the relationship between the ecosystem-based approach and stakeholder participation; and 3) the role played by fishers' knowledge in fisheries governance.

References

Arnason, R (1996) 'On the ITQ fisheries management system in Iceland' *Reviews in Fish Biology and Fisheries* 6:63-90

Baelde, P (2003) 'Using fishers' knowledge goes beyond filling gaps in scientific knowledge – analysis of Australian experiences' *Fisheries Center Research Reports* 11(1):78-86, Vancouver BC, University of British Columbia Press

Baird, I (2003) 'Local ecological knowledge and small-scale freshwater fisheries management in the Mekong river in Southern Laos' *Fisheries Center Research Reports* 11(1):87-99, Vancouver BC, University of British Columbia Press

Begossi, A and Brown, D (2003) 'Experiences with fisheries co-management in Latin America and the Caribbean' in Wilson, DC, Nielsen, JK and Degnbol, P (eds) *The Fisheries Co-management Experience*, Dordrecht, Kluwer

Bird, KE, Nichols, WJ and Tambiah, CR (2003) 'The value of local knowledge in sea turtle conservation: a case from Baja California, Mexico' *Fisheries Center Research Reports* 11(1):178-183, Vancouver BC, University of British Columbia Press

Blyth, RE, Kaiser, MJ, Hart, PB and Edward-Jones, G (2003) 'An example of conservation and exploitation achieved through a voluntary fishery management system' *Fisheries Center Research Reports* 11(1):409-425, Vancouver BC, University of British Columbia Press

Clark, IN, Major, PN and Mollett, N (1998) 'Development and implementation of New Zealand's ITQ management system' *Marine Resource Economics 5:325-349*

Collet, S (1999) 'From sustainable resource use to governance of marine ecosystems: function and role of the ethic of the sea' in D Symes (ed.) *Alternative Management Systems for Fisheries*, Oxford, Blackwell Science Ltd

Day, JC (2002) 'Zoning – lessons from the Great Barrier Reef Marine Park' *Ocean and Coastal Management* 45:139-156

DEFRA [Department for Environment, Food and Rural Affairs] (2004) *Review of Marine Nature Conservation – Summary of the Working Group Report to Government*, London, DEFRA

DfID [Department for International Development] (1995) *Stakeholder Participation and Analysis*, London, DfID, Social Development Division

Dryzek, JS (1997) *The Politics of the Earth: Environmental Discourses*, Oxford, Oxford University Press

EC [European Community] (2001) *Green Paper: the Future of the Common Fisheries Policy*, vol 1 COM (2001) 135 final, European Commission, Brussels, 20 March

EC [European Community] (2002) *Communication from the Commission on the Reform of the Common Fisheries Policy* ("Roadmap"), COM (2002) 181 final, European Commission, Brussels, 28 May

EC [European Community] (2004) *Treaty Establishing A Constitution for Europe*, Official Journal of the European Union 47, 16 December

EFEP [European Fisheries Ecosystem Plan] (2004) *The North Sea Fisheries Ecosystem Plan*, Newcastle upon Tyne, Dove Marine Laboratory, Newcastle University

Fishing News (2000-2005) various articles, London, Informa

Fishing News International (2003) article, London, Informa

Gray, TS, Gray MJ and Hague, RA (1999) 'Sandeels, sailors, sandals and suits – the strategy of the environmental movement in relation to the fishing industry' *Environmental Politics* 8(3):119-139

Habermas, J (1984) *The Theory of Communicative Action, vol 1: Reason and the Rationality of Society*, Cambridge, Polity Press

Hanna, S (2003) 'The future of fisheries co-management' in DC Wilson, JK Nielsen and P Degnbol (eds) *The Fisheries Co-management Experience*, Dordrecht, Kluwer

Hara, M and Nielsen, JR (2003) 'Experiences with fisheries co-management in Africa' in DC Wilson, JK Nielsen and P Degnbol (eds) *The Fisheries Co-management Experience*, Dordrecht, Kluwer

Hardin, G (1968) 'The tragedy of the commons' *Science* 162:1243-8

Hayek, FA von (1944) *The Road to Serfdom*, London, Routledge

Hemmati, M (2002) *Multi-Stakeholder Processes for Governance and Sustainability*, London, Earthscan

Holden, M (1994) *The Common Fisheries Policy: Origin, Evaluation and Future*, Oxford, Blackwell Scientific Publications

Inglehart, R (1990) *Culture Shift in Advanced Industrial Society*, Princeton, NJ, Princeton University Press

International Fishing News (2003) article, London, Informa

Jentoft, S (2003) 'Co-management – the way forward' in DC Wilson, JK Nielsen and P Degnbol (eds) *The Fisheries Co-management Experience*, Dordrecht, Kluwer

Jentoft, S, Friis, P, Kooiman , J and van der Schans, JW (1999) 'Knowledge-based fisheries: opportunities for learning' in J Kooiman, M van Vliet and S Jentoft (eds) *Creative Governance: Opportunities for Fisheries in Europe*, Aldershot, Ashgate

Jentoft , S and McCay, BJ (2003) 'The place of civil society in fisheries management: a research agenda for fisheries co-mananegement' in DC Wilson, JK Nielsen and P Degnbol (eds) *The Fisheries Co-management Experience*, Dordrecht, Kluwer

Jentoft, S, Mikalsen, KH and Hernes, H-K (2003) 'Representation in fisheries co-management' in DC Wilson, JK Nielsen and P Degnbol (eds) *The Fisheries Co-management Experience*, Dordrecht, Kluwer

Johannes, RE (2003) 'Fishers' knowledge and management: differing fundamentals in artisanal and industrial fisheries' *Fisheries Center Research Reports* 11(1):15-19, Vancouver BC, University of British Columbia Press

Kooiman, J (1999a) 'Rethinking the governance of fisheries' in in J Kooiman, M van Vliet and S Jentoft (eds) *Creative Governance: Opportunities for Fisheries in Europe*, Aldershot, Ashgate

Kooiman, J (1999b) 'Experiences and opportunities: a governance analysis of Europe's fisheries' in J Kooiman, M van Vliet and S Jentoft (eds) *Creative Governance: Opportunities for Fisheries in Europe*, Aldershot, Ashgate

Kooiman, J (2003) *Governing as Governance*, London, Sage

Kooiman, J, van Vliet, M and Jentoft, S (1999) 'Creating opportunities for action' in J Kooiman, M van Vliet and S Jentoft (eds) *Creative Governance: Opportunities for Fisheries in Europe*, Aldershot, Ashgate

Langstraat, D (1999) 'The Dutch co-management system for sea fisheries' in Symes, D (ed) *Alternative Management Systems for Fisheries*, Oxford, Blackwell Science Ltd

Long, A (1999) 'The Marine Stewardship Council initiative: the development of a market incentive approach to achieving sustainable fisheries' in D Symes (ed.) *Alternative Management Systems for Fisheries*, Oxford, Blackwell Science Ltd

Loucks, L, Wilson, JA and Ginter, JJC (2003) 'Experiences with fisheries co-management in North America' in DC Wilson, JK Nielsen and P Degnbol (eds) *The Fisheries Co-management Experience*, Dordrecht, Kluwer

Metzner, R, Harte, M and Leadbitter, D (2003) 'Experiences with fisheries co-management in Australia and New Zealand' in DC Wilson, JK Nielsen and P Degnbol (eds) *The Fisheries Co-management Experience*, Dordrecht, Kluwer

Mikalsen, KH and Jentoft, S (2001) 'From user-groups to stakeholders: the public interest in fisheries management' *Marine Policy* 25:281-292

Nsiku, E (2003) 'The use of fishers' knowledge in the management of fish resources in Malawi' *Fisheries Center Research Reports* 11(1):148-162, Vancouver BC, University of British Columbia Press

Pierre, J and Peters, BG (2000) *Governance, Politics and the State*, London, Macmillan

Pinkerton, E (2003) 'Towards specificity in complexity: understanding co-management from a social science perspective' in DC Wilson, JK Nielsen and P Degnbol (eds) *The Fisheries Co-management Experience*, Dordrecht, Kluwer

PMSU [Prime Minister's Strategy Unit] (2004) *Net Benefits: a Sustainable and Profitable Future for the UK Industry*, London, Cabinet Office

Pomeroy, RS (2003) 'The government as a partner in co-management' in DC Wilson, JK Nielsen and P Degnbol (eds) *The Fisheries Co-management Experience*, Dordrecht, Kluwer

Pomeroy, RS and Viswanathan, KK (20030 'Experiences with fisheries co-management in Southeast Asia and Bangladesh' in DC Wilson, JK Nielsen and P Degnbol (eds) *The Fisheries Co-management Experience*, Dordrecht, Kluwer

RCEP [Royal Commission on Environmental Pollution] (2004) *Turning the Tide: Addressing the Impact of Fisheries on the Marine Environment*, 25th Report, London, RCEP

Rhodes, RAW (1996) 'The new governance: governing without government' *Political Studies* 64:652-667

Rosenau, JN (1992) 'Governance, order, and change in world politics' in Rosenau, JN and Czempiel, E-O (eds) *Governance without Government: Order and Change in World Politics*, Cambridge, Cambridge University Press

RSE [Royal Society of Edinburgh] (2004) *Inquiry into the Future of the Scottish Fishing Industry*, Edinburgh, RSE

Sen, S and Nielsen, JR (1996) 'Fisheries co-management: a comparative analysis' *Marine Policy* 20(5):405-418

Singleton, S (2000) 'Co-operation or capture? The paradox of co-management and community participation in natural resource management and environmental policy-making' *Environmental Politics* 9(2):1-21

Sissenwine, MP and Mace, PM (2001) 'Governance for responsible fisheries: an ecosystem approach', *Reykjavik Conference on Responsible Fisheries in the Marine Ecosystem*, Reykjavik, Iceland, 1-4 October

Sutinen, JG and Soboil, M (2001) 'The performance of fisheries management systems and the ecosystem challenge', *Reykjavik Conference on Responsible Fisheries in the Marine Ecosystem*, Reykjavik, Iceland, 1-4 October

Symes, D (1999a) 'Alternative management systems: a basic agenda for reform' in D Symes (ed.) *Alternative Management Systems for Fisheries*, Oxford, Blackwell Science Ltd

Symes, D (1999b) 'Regionalisation of the Common Fisheries Policy' in D Symes (ed.) *Alternative Management Systems for Fisheries*, Oxford, Blackwell Science Ltd

Symes, D and Phillipson, J (1999) 'Co-governance in EU fisheries: the complexity and diversity of fishermen's organisations in Denmark, Spain and the UK' in J Kooiman, M van Vliet and S Jentoft (eds) *Creative Governance: Opportunities for Fisheries in Europe*, Aldershot, Ashgate

Symes, D and Ridgway, S (2003) *Inshore Fisheries Regulation and Management in Scotland: Meeting the Challenge of Environmental Integration*, Scottish Natural Heritage Commissioned Report No. FO2AA405, Edinburgh, SNH

Symes, D, Steins, N and Allgret, J-L (2003) 'Experiences with fisheries co-management in Europe' in DC Wilson, JK Nielsen and P Degnbol (eds) *The Fisheries Co-management Experience*, Dordrecht, Kluwer

Todd, E and Ritchie, E (2000) 'Environmental non-governmental organisations and the Common Fisheries Policy' *Aquatic Conservation: Marine and Freshwater Ecosystems* 10:141-149

USCOP [US Commission on Ocean Policy] (2004) *Preliminary Report of the US Commission on Ocean Policy: Governors' Draft*, Washington, DC <http://www.oceancommission.gov>

van der Schans, JW (1999) 'Governing aquaculture: dynamics and diversity in introducing salmon farming in Scotland' in Kooiman, J, van Vliet, M and Jentoft, S (eds) *Creative Governance: Opportunities for Fisheries in Europe*, Aldershot, Ashgate

van Ginkel, R (1999) 'Capturing and culturing the commons: public-private dynamics in the Dutch oyster and mussel industry' in J Kooiman, M van Vliet and S Jentoft (eds) *Creative Governance: Opportunities for Fisheries in Europe*, Aldershot, Ashgate

van Vliet, M, and Dubbink, W (1999) 'Evaluating governance: state, market and participation compared' in J Kooiman, M van Vliet and S Jentoft (eds) *Creative Governance: Opportunities for Fisheries in Europe*, Aldershot, Ashgate

Wood, D and Ritchie, E (2001) *Future Governance: Lessons from Comparative Public Policy*, Unpublished report on the Public Hearing on the CFP Green Paper, Brussels, 5-7 June

CHAPTER 2
WHAT ROLE FOR PUBLIC PARTICIPATION IN FISHERIES GOVERNANCE?

CLARE COFFEY
Senior Fellow, Institute for European Environmental Policy, Ave des Gaulois 18, B 1040 Brussels, Belgium

Abstract

Public participation is a key ingredient of good governance, aimed at a) engaging the public, b) resolving conflicts and supporting implementation, and c) improving decision quality. It is useful to consider ways of analysing the effectiveness of public participation by reference to these three aims: the level of public satisfaction through ongoing engagement of sections of the public; the degree to which conflicts are resolved and compliance with decisions is achieved; and the extent to which public views are reflected in the outputs and outcomes of decisions. I examine the 2002 Common Fisheries Policy (CFP) reform process for insights into the European Commission's use of public participation to achieve these aims. I note how the Commission's public participation efforts were reinforced during this process, but I conclude that more innovative efforts could be used in future, particularly to engage new 'publics', and that finding a balance between the three aims within the newly formed Regional Advisory Councils will be particularly important.

2.1 Introduction

During the last two decades, public participation has become very fashionable, and is now routinely promoted by national and local authorities, civil society groups, social scientists and scientific experts alike (Maurer *et al* 2003). Although practice in this area is certainly uneven, public participation is no longer seen as a luxury for the western middle classes but as an essential element of good governance, including the governance of natural resources (Pring 2001).

The public participation agenda has emerged as a result of developments on various fronts, including international civil and political human rights discussions, as well as the sustainable development discourse. The European Union (EU) has engaged – in one way or another – in these discussions and gradually incorporated some provisions in EU primary and secondary legislation, which is applicable to all EU Member States and increasingly also to the EU institutions themselves. In doing so, the EU has also sought to respond to the growing public dissatisfaction with the EU, in an attempt to (re)kindle public support for the European project (Coffey 2001).

Neither the human rights agenda nor the EU's specific interests in participation have focused explicitly on fisheries. They have however contributed – directly or indirectly – to changes in the way Europe's fisheries are being governed. Certainly a combination of these factors, plus the particular difficulties surrounding the legitimacy and effectiveness of European fisheries policies, appear to have helped to force public participation high up the 2002 Common Fisheries Policy (CFP) reform agenda. The result was not only to

T. S. Gray (ed.), Participation in Fisheries Governance, 27–44.

open up the 2002 CFP reform process itself; the reforms also opened the way for more substantial and lasting changes to the CFP's governance structure, in support of greater participation.

This paper outlines the concept of and issues surrounding public participation. As will become evident, a comprehensive overview, let alone analysis, of public participation in relation to European fisheries governance is better suited to a book than a single chapter. Therefore, having explored general public participation in decision-making and some of the issues associated with this, the chapter focuses on experience in relation to the European Commission's organisation of the EU Common Fisheries Policy reforms of 2002, with a view to providing an initial assessment of practice.

2.2 Public participation: A vague and multi-faceted concept

2.2.1 DEFINING PUBLIC PARTICIPATION

Despite its apparent ubiquity in public and non-governmental organisation (NGO) documents, 'public participation' is an ill-defined concept, meaning different things to different people. This may be helpful for some but it can also lead to much confusion and frustration, and can even contribute to a loss in confidence in management bodies and processes. Before proceeding with an analysis of public participation in relation to the CFP reform it is, therefore, useful to identify what public participation means, at least in the context of this chapter, and why it is being promoted so widely and vigorously.

2.2.1.1 *Who Is The 'Public'?*

While at first glance a relatively simple term, the word 'public' is often used loosely in the context of public participation and is often interchanged with the terms 'stakeholders', 'users', 'interests' or 'civil society'. While each of these groupings is significant to the discussion on public participation, none of them is synonymous with the 'public'. The 'public' or 'publics' potentially include everyone. Public participation should arguably include more than just organised representative groups and non-governmental organisations but also a mass of other individuals. In practice, the 'public' will normally be limited to some sub-sections of society: for example, those more affected by an issue, those who decision-makers determine to be the public, and/or those who are able to make their voices heard (Lazarow undated). For the sake of this chapter, 'public' is taken to mean all those affected, recognising that this is still a rather 'elastic' definition.

2.2.1.2 *What Does Public Participation Involve?*

Participation can loosely be defined as the process through which the public participates in, influences, and shares control over decision-making, be it priority setting, resource allocation, or access to goods and services (World Bank 1996). Participation can occur during any or all stages and levels of decision-making, including decisions relating to the initial identification of a policy need, to subsequent policy formulation, execution, monitoring, enforcement and review (House of Lords 2000). And, although participation is most frequently associated with the development and implementation of public policies, it is equally and increasingly relevant to private sector activities as well.

There is a whole spectrum of ways in which the public has been involved or engaged in public decision-making, ranging from relatively weak but large scale participation through mainstream information provision and general elections, to very strong and interactive participation where stakeholders share decision-making with government, as is the case under co-management systems.[1] Between these two extremes lie a number of other relatively conventional participation methods, such as issuing consultation papers and organising stakeholder meetings. The emphasis on public participation has additionally spawned new approaches to generate discussion and deliberations amongst the wider public: that is, reaching beyond stakeholder groups. This has involved focus groups, citizens' juries, consensus conferences, community panels and deliberative polls, methods that are considered to be at an early stage in their development (Weale 2002). Other options include internet dialogue, standing consultative panels and foresight programmes (House of Lords 2000), as well as film/video, radio/TV ads and electronic information provision.

The chosen method of participation should reflect the size and make-up of the 'public', which in turn should reflect the subject of discussion and the chosen objectives of the exercise. More limited but ongoing stakeholder exercises may be suited to participation on matters of technical detail, whereas the general public may be engaged less regularly in discussions on more strategic, political or moral issues. Stronger or more intensive participation will normally be associated with engaging smaller numbers of people, but this does not mean that more active dialogue need necessarily be restricted to those most affected by decisions.

2.2.1.3 *Getting The Public To Engage*

The chosen method for participation should correspond to the public's level of motivation for getting engaged. A number of factors have been identified that influence the scale and nature of participation. At a rather intangible level, these include the existence of a sense of civic duty, a desire to get engaged or at least not to be excluded, and the perception of the organisation seeking engagement. More concretely, perhaps, people will be influenced by costs of participation and non-participation, direct benefits to be gained and the likelihood of influencing decisions (see Webler 1998). The level of awareness of the issue and decision-making processes, and the distribution of costs and benefits of engagement are also important (Rydin and Pennington 2000). In the case of narrowly targeted programmes, such as aid to the fisheries sector, benefits are concentrated while the costs are thinly diffused over society at large, a situation that would most likely encourage beneficiaries of existing aid to seek engagement, whereas the general public would be rather disinterested. Securing participation will arguably be more resource intensive as motivation decreases.

2.2.2 THREE KEY REASONS FOR PUBLIC PARTICIPATION

While public participation can potentially involve everyone, and can certainly take many different forms, the definition of the public and methods of engagement should reflect the purpose of participation. Although this too may vary considerably, depending on the

[1] Public participation is often understood, incorrectly, as being the same as co-management. Conversely, Phillipson (2002:56) notes that "Much too often…the term 'co-management' is used rather loosely to embrace almost any occasion of user group participation…"

institution or 'public' concerned, most would agree that public participation *could* be sought for the following three reasons:

- **Public engagement** – the public's right to take part in its own governance is now seen as a basic human right, to be defended and strengthened for its own sake. This is to be achieved by opening up decision-making processes and, ideally, supporting a dialogue with the public. As Rydin and Pennington (2000) noted, a policy that has been developed with the involvement of a wider range of parties is seen, in itself, as being more desirable. The implication here is that a basic level of participation should involve as many people as possible, or as appropriate given the issue at stake.
- **Conflict resolution** – public participation also provides an opportunity to discuss, manage and even resolve differences in views and positions, ideally smoothing the way to agreement on particular issues, as well as supporting implementation and therefore policy outcomes further down the line. Conflicts can exist between authorities and the public; among different sectors of the public; and between individuals and between whole sectors of society, all of which participation can help to identify, address and accommodate.
- **Decision quality** – aside from the 'feel good' factor, there are practical reasons why participation is being sought by policy-makers. Public engagement should widen the pool of views, values and ideas taken into account in policy design, going beyond those captured by officials or expert consultants. This should help to rectify particularly strong imbalances in the political process, flush out competing perspectives, and identify the most appropriate policy for addressing the problem, taking fuller account of costs and benefits of decisions for society at large. The suggestion is that a range of experts and stakeholders are valuable, because the result of their involvement should be improved policy outputs.

The benefits of public participation are widely seen to be self-reinforcing: policies are more likely to succeed if they involve public participation and public participation is an important element in securing effective policy-making.

These arguments for public participation strategies are compelling, and can be readily applied to EU fisheries policy and fisheries management. Parts of the fisheries sector and associated coastal communities feel alienated as they become exposed to increasing regulation and other governmental interference, at the same time as struggling to cope with financial hardship and social change. The general public, even though it is largely removed from marine issues, which are rather out of sight, evidently regards the marine environment as serving important social and moral ends. Fishing and fish farming interests are subject to conflict among themselves and/or with environmental interests, which multi-stakeholder participation can help to resolve or mitigate. The importance of getting stakeholder and consumer buy-in is very great, notwithstanding the general difficulty in overseeing fishing and enforcing rules using conventional policing methods. There is also a high level of scientific and technical complexity and uncertainty surrounding the fisheries management issue, and, while decision-making must be informed by scientific and expert advice, the public can contribute further knowledge, either because it is directly involved (such as fishermen), or because it has an interest in the problem at hand (such as civil society groups or the public at large) (López and González 1996). For a variety of reasons, therefore, public participation in fisheries governance promises to be very beneficial.

2.2.3 EVALUATING PUBLIC PARTICIPATION

The arguments for public participation appear to be uncontroversial, and public participation is all but universally encouraged in academic literature, policy documents and NGO lobbying materials dealing with governance. "Everyone seems to agree that participation is a good thing and that non-participation is bad" (Chandler 2001:1). However, evidence of participation strategies actually contributing to the three objective identified above is less clear-cut. A part from supporting democracy for democracy's sake, public participation does appear to add value simply by channelling more information to the public (the first main reason for participation identified above). However, evidence of public participation actually resulting in reduced conflict or improved policy outputs and outcomes is patchy (Maurer *et al* 2003). Wilson (2003) concluded from a review on participation in environmental management that participation does increase both compliance with, and the quality of, the resulting policies, but only under three conditions: if participation is well structured and transparent; if at least some real decision-making is participated in; and if a bargaining space exists in which the participants can reach a compromise. Others suggest that public participation can – if handled badly – actually be counterproductive. Rydin and Pennington note accounts of the policy process which "highlight a propensity towards special interest capture and bureaucratisation as the reality of participation in practice" (2000:153).

With this in mind, and before examining public participation in relation to the CFP, it is useful to consider whether it is possible to analyse public participation in a systematic fashion, with a view to evaluating its effectiveness:

- Participation to strengthen public engagement – the quality or meaningfulness of public participation in itself will depend on the way in which participation is arranged and how opportunities are used: that is, the process and extent of engagement. A generous definition of the 'public' will also be important, bearing in mind that a basic level of participation should encompass or reflect a relevant but very wide audience. The quality of participation will depend on the match between the situation, the definition of 'public', the methods used and the resources applied. Evaluating participation processes will be difficult but could be judged by whether the public feels satisfied with participation. The level of **ongoing or institutionalised engagement by different sections of the public** could perhaps be taken as a proxy indicator for satisfaction.
- Participation as a means to resolve conflict and smooth the way for implementation – attempts to evaluate success in relation to this heading will encounter problems in trying to establish cause and effect. It may be possible, by analysing statements and dialogue, to ascertain **whether and to what extent conflicts remain**. Here it will be important to recognise that participation processes may initially lead to new conflicts being aired, which should not necessarily be seen as negative. As regards the effects of participation on implementation and outcomes, there will need to be a sufficient time lag to allow the effects to work themselves through to the stage when behavioural change can be expected. Improved compliance with new policies could be a useful indicator here, although participation should support implementation in ways that go beyond compliance alone. Evaluation of progress would be heavily dependent on qualitative research, as implementation behaviour will be affected by many

factors, including confidence, buy-in or goodwill generated by the decision-making process.

- Participation as a means to improve the quality of decisions – it will also be difficult to establish cause and effect between participation and decisions subsequently taken, at least in the absence of detailed and qualitative research. However, it should be possible to ascertain the **extent to which public views are reflected in decisions** even if this does not necessarily confirm a connection between the two. If a link between participation and policy outputs can be established, judgements have to be made about what is a 'good' and what is a 'bad' decision, and here both the decision itself (output) and the eventual outcome of that decision is relevant. Finally, it should also be borne in mind that an evaluation of practice and resulting decisions and outcomes could find that practice is relatively poor but that participation has nevertheless been influential and beneficial.

In Figure 2.1, the strength of participation (weak, medium, strong) is distinguished from the extent of participation (narrow, medium, wide) in order to demonstrate the above conclusions diagrammatically.

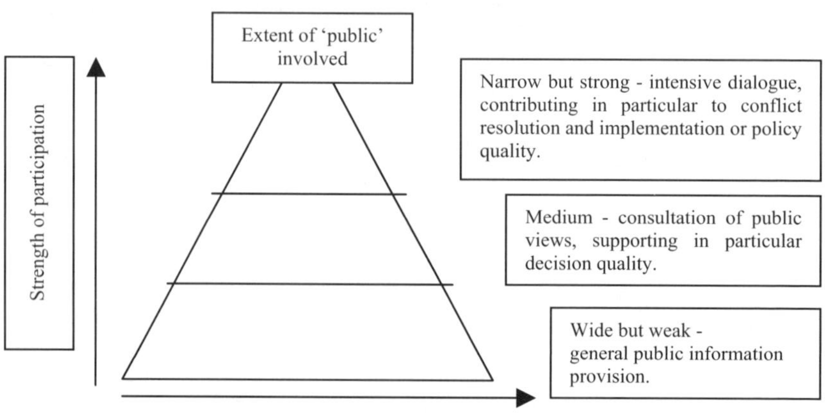

Fig. 2.1 Strength and extent of public participation

- Weak – simply supplying information widely to the mainstream public as well as groups within that. This is likely to mean conveying relatively simple messages or general messages. It contributes in an important way to the delivery of the first objectives, to a lesser extent the second and not really the third.
- Medium – participation is two-way, involves a medium sized public and potentially more complex issues. It could potentially deliver most on the third objectives for participation, since it is likely to be too narrow to engage everyone directly and too wide to tackle and resolve conflicts.
- Strong – participation that involves intensive dialogue and even co-management. Discussions could focus on detailed issues with a narrow group of people. It would contribute most to the second objective, as well as substantially to the third one.

Given resource constraints, a growing emphasis on public service efficiency and cost-effectiveness, time constraints associated with decision-making, and a general interest in avoiding unnecessarily protracted discussions, there is clearly a trade-off to be made between the strength of participation and the extent of the 'public' involved. The result of that trade-off will depend on which of the three objectives is being prioritised.

In managing the 'strength' versus 'extent' trade-off, the following points appear to be particularly relevant:

1. Public participation strategies can be combined to secure narrow, medium and wide participation at the same time, so that all three objectives are delivered in a relatively efficient way;
2. Even weak participation efforts are important since they offer an opportunity for everyone or at least large sections of society to be better informed;
3. Medium level participation balances access to numbers with the ability to engage on more detailed issues. It may be particularly useful as a means of accessing a wide range of expert views, beyond the 'core' stakeholders. Although the immediate public engaged will be limited, information may be disseminated more widely by non-governmental organisations, local authorities, and industry groups;
4. The most intensive but narrowest participation may be organised in such a way that prioritises stakeholder conflict resolution whilst also seeking to access specific expertise and views. Again, other organisations may then step in to disseminate information more widely to the public;
5. If participation efforts seek to address conflicts and access expertise at the same time, however, there is a risk that other public views are not brought into the process, whilst the potential for capture by sectoral interests may be increased. The risk of sectoral capture can be reduced if membership is rebalanced but this will demand greater investment (human and financial) in the process, as the level of motivation is likely to be lower among non-sectoral interests. It may also create, rather than resolve, conflicts, as the core stakeholders feel challenged by the new publics brought into the dialogue.

Accordingly, decision-makers should aim to establish participation that covers the range of public(s) to some degree, and that targets stakeholders and representatives of other publics for more intensive dialogue. The challenge is to select the right people and to engage them in the right way, rather than necessarily trying to get everyone involved in detailed decisions, without giving sufficient attention to how this is done.

2.3 Public participation in practice: EU fisheries policy

2.3.1 GOVERNANCE ON THE EU AGENDA

The idea that all individuals have a right to participate in decision-making has emerged through international human rights discussions, with participation in government and access to information among the universal civil and political rights established in international law (Anderson 1998). The sustainable development discourse has also emphasised public participation, including in relation to various major groups[2], and participation in decision-making on **environmental** matters is now a right enshrined in international law under the Århus Convention (UNECE 1998).[3] The European

[2] Public participation is referred to in Principle 10 of the Rio Declaration signed at the 1992 UN Conference on Environment and Development (UNEP 1992)

[3] Adopted in 1998, the Århus Convention entered into force in 2001.

Community is committed to these international participation provisions,[4] which are gradually also being reflected in EU legislation applicable to the 25 Member States and to the EU institutions themselves.

Environmental rights aside, the 'governance' question also steadily rose up the EU's agenda during the late 1990s. A serious governance debate had been precipitated by the crisis of the European Commission's forced resignation in March 1999, the prospect of enlargement, and the changing global demands being placed on the EU. A significant additional, though largely inseparable, driver for reform has been the level of public dissatisfaction about how the EU is being governed and how effective policies are in practice. Successive editions of the Eurobarometer catalogue pointed to the demise in public confidence in the EU, but the extent of the problem was brought into sharp focus by demonstrations against EU meetings of Heads of State and Government, including rioting at the Gothenburg Summit in 2001, which resulted in the death of one demonstrator.

In recognition of these problems, the Commission launched a debate on 'European governance' in autumn 2000 and, following a year-long exercise in reflection and consultation, the White Paper on European Governance (EC 2001a) emerged in July 2001. This focused primarily on improving governance within the confines of the existing European Treaties, and included 'participation' among its five principles of good governance.[5] A month earlier, at Gothenburg, EU leaders had agreed to elements of a European Commission (EC 2001b) proposal for an EU Sustainable Development Strategy that, *inter alia,* sought to extend public participation by promising earlier and more systematic dialogue with stakeholders, and routine public hearings on Commission proposals. The governance and sustainable development agendas were eventually linked to efforts to improve regulation by introducing prior impact assessment procedures that included public consultation elements (EC 2002a). In 2001, EU leaders also embarked on a more fundamental process to review and revise the European Treaties themselves, and established a 105-strong Convention to start this process. Part of the reason for this exercise was to simplify the Treaties and make the EU more transparent and accessible to the public. The result was a new draft constitution that was presented at Thessalonica in June 2003.

2.3.2 PUBLIC PARTICIPATION AND THE CFP REFORM

Although governance and participation are fashionable words in the EU vocabulary, a link between rhetoric and reality is, of course, never guaranteed. The question for this chapter is whether the public participation discourse has been or is starting to be embedded in the reality of EU fisheries or whether fisheries (and other sectors, potentially) has been left behind in the governance debate.

---

[4] The European Community signed up to the Rio Declaration and the Århus Convention. A proposal to ratify the Convention is before the Council of Ministers (European Commission 2003a). Directive 2003/35/EC (EC 2003b) provides the main instrument for the Community's implementation of the public participation provisions in the EU Member States. There is also a European Commission (2003c) proposal regarding the application of Arhus provisions to the EU institutions themselves.

[5] The White Paper suggests that participation is desirable in order to improve the quality, relevance and effectiveness of policies, as well as creating more confidence in the end result and the institutions. There is no reference to participation as being an important principle in its own right, and in furtherance of democracy.

2.3.2.1 *The CFP – Plentiful Opportunities for Participation*

The CFP – like other EU common policies – consists of a large number of individual items of legislation, all of which are developed at the EU level by the Council, Commission and (to a lesser extent) Parliament. Once adopted at the EU level, measures normally have to be implemented by the individual Member States. In some cases, provisions are directly binding on individual actors (that is, fishermen) in which case policy-related decision points will occur at EU level only. But in most cases, legislation requires some activity by the Member States and even the sub-national regions. Participation 'opportunities' can therefore arise at numerous times during any single policy cycle and at various levels of governance; and there can be hundreds of policy cycles working in parallel at any one time.

EU aid to the fisheries sector under the Financial Instrument for Fisheries Guidance (FIFG) demonstrates the complex nature of EU policy and the extensive participation opportunities that can arise in relation to this single item of legislation. The basic FIFG rules are set at EU level before being translated into national or regional funding programmes. Actual funding is decided on a project-by-project basis, with the help of a programme monitoring committee and supporting bodies. The project plans and results are evaluated *ex-ante* and *ex-post* at programme and EU level. As a result, the devolved nature of this EU policy means that participation can occur at the policy, programme and project stages, which are all designed, administered and evaluated at EU, national and regional levels.

Before the 2002 CFP reform, a number of systems or mechanisms have been in place to support participation in relation to some of the many decision points that exist. For example,

- Commission proposals have for some time been informed by 'official' input from a range of national, scientific, technical, industry and environmental interests. The formal forum for participation by industry and environmental NGOs is the Advisory Committee on Fisheries and Aquaculture (ACFA), where opportunities are provided to comment and contribute views on CFP related issues. Further opportunities for industry (though not environmental) participation were provided by the series of regional fisheries management meetings convened by the Commission before the 2002 reform agreement.
- Compared to the Commission, the Council has traditionally been rather poor at securing public participation, with no known attempts to engage the public or stakeholders in fisheries working group or Council deliberations. It is up to individual Member States to decide whether or not to consult nationally on proposals in advance of Council discussions, and practice varies, particularly in terms of participation by environmental or wider public interests.
- The European Parliament and its committees provide most access to the public, and its members are the only EU politicians directly elected. Committees also organise public hearings from time to time as a means of consulting stakeholders in the process of drafting specific reports or opinions. This is additional to direct lobbying which industry and environmental interests engage in routinely.

Overall, and despite frequent suggestions to the contrary, public participation arrangements were in place before the CFP reforms in 2002. Nevertheless, there was a widespread feeling that public participation in the CFP was inadequate (Coffey 2001), and it is true that participation was targeted predominantly at industry and, to a lesser extent, environmental stakeholder groups, rather than seeking to engage other sections of the public.

2.3.2.2 *Participation in the Commission's Preparations for the 2002 CFP Reform*
Opportunities and mechanisms for public participation in EU level fisheries governance have existed for some time but they took on a new dimension during the 2002 CFP reform debate. The increase in participation emerged during a time of widespread disenchantment with the resource management aspects of the CFP, from many fisheries managers as well as industry, NGOs and sections of the public. Coffey noted (1998:3) that "[i]t is perhaps a sign over the extent of feeling over the CFP, that work on the 2002 review has already commenced". The result was that governance and public participation shaped the very form of the discussions, as well as providing an objective for the reforms themselves.[6] According to Gray, the reform process signified to some extent a shift from "the discourse of authoritarianism to the discourse of democracy" (2003:16). The result was a new basic CFP agreed by the Ministers in December 2002 which included "broad involvement of stakeholders at all stages of the policy from conception to implementation" among a set of principles of good governance to guide the future development of the CFP (EC 2002b: Article 2 (2) (c)).

During the CFP review process, the Commission (DG Fisheries) employed new and, for it, novel methods to strengthen engagement with the sector, other interest groups and, arguably, even the general public or interested sections thereof. The main methods used included the following.

Questionnaire
In March 1998, a detailed nine-page questionnaire covering most aspects of the CFP was sent to approximately 350 organisations representing a wide range of interests from across Europe. Approximately 175 responses were received (EC 2000). The receiving organisations included some who had not previously had direct contact with DG Fisheries. This was a rather targeted approach to engage both the sector and 'new' stakeholders who had traditionally received little attention in the development of EU fisheries policy. It was apparently intended to generate greater confidence in the institutions, the CFP and the reform process as a whole. By asking for responses to the questionnaire, the Commission also initiated a longer-term dialogue over the best way to reform the policy.

Series of Regional Consultation Meetings
Thirty meetings were organised by the Commission and attracted in the region of 1,500 actors (EC 2000) from fishermen's organisations and environmental NGOs, national and environmental authorities, as well as trade unions, institutes and academia. The meetings brought Commission officials in direct contact with interested individuals, going beyond those engaged by the questionnaire and raising issues not covered by the questionnaire (Hatchard 2003). The meetings ran for nearly one year (Sept 1998-June 1999). According to the Commission, the purpose of the meetings was: first, to give representatives of the fisheries sector and other interested groups an opportunity to engage in a direct dialogue with the Commission's services on the future of the CFP, elaborating on answers given to the questionnaire, expressing national/regional/local views on the review of the CFP in 2002, as well as on any other issue of particular

[6] For an overview of the reform process, see *El Anzuelo* (various issues)

interest to them; and second, allow Commission officials to have direct contact with the various actors and their demands in all the Member States, so that the broadest possible range of views and ideas could be gathered (EC 2000). The consultation meetings built on the questionnaires in three key ways: extending the range of people actually engaged; allowing direct contact to be made with policy-makers and thus helping to address the 'remoteness' of the CFP; and supporting two-way discussion. By touring the EU, the Commission also managed to overcome some of the financial and time-related motivational barriers relating to participation.

Green Paper on the Reform of the CFP

In March 2001, the Commission produced a Green Paper which was effectively a discussion document intended to precede more concrete communications or proposals. The document was well publicised and made accessible on the Commission's web pages. Interested parties, that is, stakeholders and the public at large, were invited to comment on it. Although publication of the Green Paper was a rather passive and weak method to support participation, it did give people an opportunity to find out about the Commission's thinking and express to them their views on an important document about future European fisheries governance

Public hearing

The Commission held a Hearing on the CFP Reform in June 2001, which provided an opportunity for more than 400 individuals to hear at first hand about the Commission's reform ideas, and for individuals to express their views both during the formal sessions and between sessions. More than 106 oral representations were made (Hatchard 2003; Ritchie and Wood 2001), with environmental views featuring heavily amongst these. However, the format of the hearing meant that the Commission gathered information, rather than encouraged real debate on any given issue. This consequently provided a valuable means of promoting openness and transparency, engaging a broader public, and providing a channel for expressing views, but it did not offer much in terms of meaningful two-way dialogue. The nature of the event, and the costs of attending it, also meant that it attracted the most interested public groups that also had resources to travel or otherwise were well placed to attend.

Roadmap

Communication efforts were stepped up during the CFP reform process, with the Commission issuing a 'Roadmap' (EC 2002c) that set out key steps to be taken as part of the reform, including production of a number of Commission proposals and more general communications. Rather like the Green Paper, this Roadmap is best seen in the context of the wider reform process, including also the media activities that accompanied its release and subsequent discussions. Nevertheless, it too was a rather weak participation tool, designed to inform stakeholders of the path that was being taken by the Commission, and how this would affect the fisheries sector in particular. In this way, it was a confidence-building measure, to support participation *per se* and reduce resistance to the reforms.

Other Communication Methods

Considerable information was produced and made accessible via the DG Fisheries website, in order to inform the press and others interested in understanding the reforms and the arguments for them. CFP reform issues were subsequently covered by the mainstream press, including the *Financial Times* and the *Economist* magazine, although some of this was linked to the replacement of the Danish Director-General during 2002. There were also various informal meetings, with public groups entering into dialogue with DG Fisheries staff, particularly involving fishermen's representatives and European NGOs, such as the World Wide Fund for Nature (WWF). The Fisheries Commissioner and staff also embarked on a series of missions to present the reform package and to provide opportunities for questions to be raised. This tour marked the end of the Commission's activities and supported the start of debates on national positions to be taken on the reform proposals.

Taken in addition to the general mechanisms for participation in the Commission's work, the CFP reform process clearly employed a range of different methods in an attempt to extend engagement beyond the 'usual suspects', engaging stakeholders in discussions on what were relatively complex reform issues, and using stakeholder views throughout the reform process, to build up a reform package. In rolling out its participation activities, the Commission clearly also committed significant amounts of resources in terms of time and funding, suggesting a real desire to invest in communication and dialogue.

Despite the approaches used, however, the Commission relied heavily on more conventional participation methods, notably consultation papers and stakeholder meetings. More innovative methods such as focus groups, consensus conferences, citizens' juries or community panels were not used, nor was funding made available to ensure participation in meetings by the less motivated 'publics'. The Commission's approach reflects a tendency to see more intensive public participation exercises as a way of engaging 'core' stakeholders and experts, rather than exploring options to engage other 'publics' in the process.

2.3.2.3 *Assessing the Effectiveness of the Commission's Participation Efforts*

Recognising the limits of the types of participation methods applied by the Commission, it is nevertheless useful to assess the effectiveness of the Commission's efforts in relation to the three main objectives of participation that were outlined above.

Level of ongoing engagement by different sections of the public

Throughout the reform process, the Commission managed to engage a far wider audience than had previously been the practice in this area. From the launch of the questionnaire through to the 2001 hearing, attempts were made to widen stakeholder engagement. Nevertheless, the methods used did not allow the Commission to reach much beyond user groups and representative organisations. In some cases, notably the regional meetings, stronger engagement with affected communities was possible, but wider engagement with the general public was not really sought, despite the 2002 reforms providing a suitably strategic opportunity to generate greater public interest and awareness of marine environmental issues. Nor were the Commission's methods able to effectively gauge the public values in this area: for example, on issues such as having a healthy marine environment; the morality of intensive aquaculture; bycatch; or EU fishing agreements with African countries. Instead, the Commission appeared to have relied heavily on NGOs to communicate messages to the public and also to provide 'public' feedback on the Commission's activities.

The main 2002 reform process was concluded at the end of 2002, with the focus moving to more detailed discussions on implementation of the 'new' CFP. Since that time, the level of engagement with stakeholders has changed and, to some extent, deepened.

NGOs have remained involved through the main advisory committee structure, and have secured a place in the Regional Advisory Councils (RACs) that are being established by the Commission. Fishermen's groups have used much the same mechanisms to stay engaged, although also benefiting from participation in discussions on new EU stock recovery plans. EU-level NGO media communication activities have reduced somewhat, however, with implications for wider public engagement. The fact that stakeholders have remained engaged may have been more of a reflection of the importance of the discussions than a particular happiness with the process. At the very least, however, it does suggest that stakeholders were sufficiently happy so as not to walk away from the discussion table during or after the reforms.

Mitigation of Conflicts Between Groups and Supporting Compliance

At the beginning of the reform process there were many areas of disagreement between the fisheries sector and other interests, and within the fisheries sector. The questionnaire responses and subsequent discussions in the regional meetings highlighted conflicting interests between: fishermen and environmental actors; deep-water versus coastal sectors; and northern versus southern interests (Hatchard 2003). The final reform agreement sought to reconcile these different interests, in particular by making the environment more central to the CFP. The fisheries sector appears to have accepted the inevitability of this shift and the reforms could therefore be deemed to have succeeded in terms of managing and reconciling at least some conflicting interests. There were even examples of sector and environmental interests actively collaborating in the interests of long term environmental and social sustainability of the industry, for example, involving WWF UK and the Scottish Fishing Industry.

However, the nature of discussions surrounding the detailed implementation of the CFP reform, particularly in relation to the development and adoption of EU stock recovery plans, suggests that the reconciliation was to some extent superficial. This is, perhaps, not surprising, as the 2002 reforms were largely strategic in nature, leaving most of the important details of implementation to subsequent negotiations. Overall, however, the reform process did help to bring together the different interests and engage them in a more mature dialogue with one another, as well as with the Commission services.

The reform process itself, including the questionnaire and Roadmap, also allowed stakeholders to see how they would be affected by a 'new' CFP, and the costs and benefits of different approaches. This should have built confidence that reforms were in the interests of the fisheries sector and associated communities, as well as the public more widely. For example, the Roadmap and some of the Commission documents that followed, sought to emphasise that restructuring would be undertaken in a way that would minimise negative impacts and ease the inevitably difficult transition to long-term sustainability. In particular, the Commission emphasised recent socio-economic trends, such as the persistent decline in employment in the sector, and the fact that action was needed if this trend was to be reversed. In this respect the process should have improved general confidence and buy-in into the process which was in everyone's interests; clarified the overall long-term direction of reform; and reassured fishermen that nothing unexpected was going to happen. In turn, this should have strengthened considerably the likelihood of 'compliance' with the 'new' CFP. However, there is little evidence of such increased harmony and compliance, especially after the unpopular Fisheries Council decisions in December 2003 and December 2004 on TACs, days-at-sea, and the North Sea cod recovery plan.

Improved Quality of Decisions

The 2002 CFP reform proposals did reflect a wider range of issues than had previously been tackled by the CFP. The new CFP basic regulation and the Commission proposal that preceded it, referred in particular to ecosystem-based management, precautionary principle, and good governance, which were all new to the main CFP framework. As such, public participation by environmental and other public interest groups did appear – at least at first glance - to influence to a significant degree, the shape and content of the new and improved CFP agreement.

Of course, in the absence of detailed analysis of a number of official documents and interviews, it is impossible to establish with any certainty the extent to which the 'new' CFP was actually influenced by the public participation that did occur, given the possibility of there being a number of intervening variables. The Commission had a particular interest in coming forward with a new policy that was more effective and also more cost-effective. Additional stakeholders could have been brought into the process in order to bolster the Commission's negotiating position in relation to certain Member States and certain elements of the fisheries sector. Nevertheless, the Commission's approach to the reforms was certainly a great improvement on previous participation efforts and undoubtedly supported all three public participation objectives identified in this paper. Table 2.1 contains a summary of the above points.

Table 2.1. Public participation in the 2002 CFP reforms: strengths and effectiveness

Methods used	*Strength of participation engagement*	*Extent of public engaged*	*Conflicts resolved and implementation supported*	*Policy design improvements*
Questionnaire	Medium – involved two-way communication but with little facilitation or support.	Limited to 350 recipients, representing a range of stakeholder groups, including environmental NGOs.	Helped to identify conflicts and interests	Resulting in an extension of the initial menu of issues to be tackled, and thus informed the Commission's agenda.
Regional meetings	Medium – stronger than the questionnaire, with the Commission actively going out to talk to stakeholders, including NGOs. Attendance depended on access to resources.	Approx 1,500 delegates, representing a range of stakeholder groups, including environmental NGOs.	Helped to focus on specific issues and thus support conflict resolution, and general trust in CFP and institutions	Allowed additional information to be accessed from more people, including informal exchanges. This is likely to have highlighted potential costs and benefits of different actions.
Green Paper	Weak – a report was produced, made available on the internet and responses welcomed.	Potentially the general public, although coverage mainly in industry/specialist press	By setting out options, the Green Paper allowed stakeholders to see how their interests were being promoted	Responses to the Green Paper should have allowed some modifications to be made, to support a more effectively delivery even if the overall approach remained unchanged.
Hearing	Medium – delegates were invited to present their views, but there was no real dialogue. Participation depended on ability to attend a Brussels event.	400 delegates, mainly representatives from different industry and stakeholder groups.	Venue for raising issues and discussion, informally, particular issues. Provided an opportunity to promote environmental arguments to the Commission and other stakeholders.	Provided an opportunity to hear different views, although perhaps few new ones.
'Roadmap' (EC 2002c)	Weak – this consisted of information and communication, rather than two-way discussion.	Potentially the general public, although coverage mainly in industry/specialist press	The Roadmap allowed stakeholders to see how they would be affected and what their role would be in future. It should have built additional confidence that nothing unexpected was going to happen.	This was intended to be a tool to communicate to others, not to inform the Commission's policy-making.
Other media based activities, including press packs & speeches	Weak - this consisted of information and communication, rather than two-way discussion.	Accessible to the general public, particularly where picked up by NGOs and mainstream media organisations, and thus disseminated more widely.	Other activities should have helped to support better understanding of different proposals and impacts, as well as generally cementing commitment to the policy.	Other media and communications activities were also aimed at engaging the public's interest rather than gathering views.

2.4 Conclusions and future prospects

In the discussion on fisheries governance and public participation, it is important to understand the potentially wide-ranging implications of the latter concept. In particular, public participation means more than simply reinforcing engagement of existing sectoral and even environmental public interest groups, but also calls for the involvement of other publics, at least if the potential of public participation is to be maximised. Poorly constructed efforts to enhance public participation may not in fact engage different or relevant publics, and may instead reinforce existing patterns of participation involving relatively powerful and vocal interest groups.

The CFP provides a useful focus for those trying to evaluate public participation strategies and practice in the EU. Not only are there numerous CFP measures in place and in the pipeline at any one time, but the nature of individual measures also varies widely, some using very top-down command-and-control type approaches whilst others leave a great deal to be decided at the national and sub-national levels. At the EU level alone, there are several institutions that can engage the public. The result is that the CFP provides a rich seam for those wishing to examine practice in public participation. Critics who demand greater public participation should reflect that efforts to ensure it could potentially become completely overwhelming. A trade-off has consequently to be made so that limited resources are tailored as much as possible to the specific objectives being pursued.

This chapter has examined practice in relation to the Commission's handling of the 2002 CFP reform process – just one institution's role in one of many policy processes – but a particularly major and strategic one that attracted widespread interest from a range of actors and organisations. Although it only provides a very limited view of public participation in relation to a rather atypical policy formulation process, the CFP reform process does offer an opportunity to examine 'best practice' within the Commission's DG Fisheries and the extent to which this is keeping up with broader thinking and approaches on public participation. The conclusion from the analysis is that the Commission stepped up efforts in this area, significantly enhancing the more permanent participation structures and mechanisms that existed before and during the 2002 process. However, the Commission's good practice appears to compare less favourably with state of the art methods that are being developed and promoted elsewhere, particularly as regards engaging new groups or 'publics' in policy discussions.

The 2002 CFP reform process has led to a new emphasis on governance and stakeholder participation. This is most obviously demonstrated by the RACs that are being established, building on an implementing Regulation adopted in 2004 (Regulation 2004/585). RACs are to bring the CFP closer to individuals and communities most affected by fisheries. While not having formal decision-making powers, their views will in certain circumstances have to be taken into account by the Commission and the Member States.

RACs offer the prospect of stronger engagement by certain groups in the EU policy formulation process, creating the space for stakeholders to participate in two-way communication and ongoing dialogue, with the quality of discussions hopefully maturing with time. This should allow conflicting views to be aired and worked through, with the overall legitimacy of decisions potentially being enhanced, to some extent at least. RACs should also ensure additional knowledge is brought to the table, although participation is to be limited to a relatively small group of individuals. Environment and consumer groups are to be represented on RACs, but the fishing industry will dominate. Other 'experts' may be invited to meetings but only if members so desire.

The make up of RACs reflects perhaps a desire above all to resolve conflicts and build confidence in the CFP – the second objective of public participation; while improvements in the quality of decisions and outcomes – the third objective – appear to be served less well by RACs. Rectifying this imbalance could mean changing the RAC membership, but also using alternative participation methods to engage other 'publics'. Consideration would also need to be given to ways of getting other less 'motivated' groups effectively involved. This includes environmental groups that currently have access to RACs but lack the necessary resources to participate fully in their work. It is also important to remember that RACs will not meet broader objectives aimed at informing society at large about fisheries policy. Overall, therefore, RACs should be seen as very welcome, but only another step toward true public participation in EU fisheries policy-making.

Acknowledgements

I would like to thank Dirk Reyntjens and Dr Joan O'Mahony for their helpful comments on an earlier draft.

References

Anderson, MR (1998) 'Human rights approaches to environmental protection: an overview' in AE Boyle and MR Anderson (eds) *Human Rights Approaches to Environmental Protection*, New York, Oxford University Press

Chandler, D (2001) 'We know the answer is public participation...but what's the question?' unpublished conference paper, presented at *Developments in Public Participation*, Barcelona, June

Coffey, C (1998) 'Countdown to 2002: the CFP review – unravelling the process' *El Anzuelo.* Vol 1:2-3, London, Institute for European Environmental Policy

Coffey, C (2001) *Good Governance and the Common Fisheries Policy: an Environmental Perspective*, Peterborough, IEEP/English Nature

EC [European Community] (2000) *Report from the Commission on the regional meetings arranged by the Commission in 1998-1999 on the Common Fisheries Policy after 2002*, COM(2000)14, Brussels, European Commission, 24th January

EC [European Community] (2001a) *European Governance: a White Paper*, COM(2001)428, Brussels, European Commission, 25th July

EC [European Community] (2001b) 'A Sustainable Europe for a Better World: a European Union Strategy for Sustainable Development' *Communication from the Commission*, COM(2001)264, Brussels, European Commission, 15th May

EC [European Community] (2002a) *Consultation document: Towards a reinforced culture of consultation and dialogue - Proposal for general principles and minimum standards for consultation of interested parties by the Commission*, COM(2002)277, Brussels, European Commission, 5th June

EC [European Community] (2002b) *Council Regulation (EC) No 2371/2002 of 20 December 2002 on the conservation and sustainable exploitation of fisheries resources under the Common Fisheries Policy*, Brussels, European Council of Ministers

EC [European Community] (2002c) *Communication from the Commission on the reform of the CFP ("Roadmap")*, COM(2002)181final, Brussels, European Commission

EC [European Community] (2003a) *Proposal for a directive of the European Parliament and of the Council on access to justice in environmental matters*, COM(2003)624, Brussels, European Commission

EC [European Community] (2003b) 'Directive 2003/35/EC of the European Parliament and of the Council of 26 May 2003 providing for public participation in respect of the drawing up of certain plans and programmes relating to the environment and amending with regard to public participation and access to justice' *Official Journal of the European Communities* L 156, 25/06/2003 P. 0017-0025, European Parliament

EC [European Community] (2003c) *Proposal for a regulation of the European Parliament and of the Council on the application of the provisions of the Århus Convention on Access to Information, Public Participation in Decision-making and Access to Justice in Environmental Matters to EC institutions and bodies*, COM(2003)622, Brussels, European Commission

El Anzuelo (various issues) European Newsletter on Fisheries and the Environment, London, IEEP

Gray, TS (2003) 'CFP reform - plus ca change?' Paper presented at the Political Studies Association Annual Conference, Leicester University, April

Hatchard, JL (2003) 'Stakeholder consultation and the European Common Fisheries Policy' *Political Studies Association Annual Conference*, Leicester University

House of Lords (2000) *Science and Society*, House of Lords Select Committee on Science and Technology Third Report, HL Paper 38, Session 1999/2000, London, HMSO

Lazarow, NS (undated) *Demystifying Public Participation: a Case for Developing Environmental Indicators for Public Participation*, Australia, Co-operative Research Centre for Coastal Zone Estuary and Waterway Management, Griffith University

López, C and González, G (1996) 'Lay knowledge and public participation in technological and environmental policy' *Philosophy and Technology* 2(1 Fall):53-72

Maurer, C, Ehlers, S and Buchman, A (2003) *Aligning Commitments: Public Participation, International Decision-making, and the Environment*, Washington DC, WRI Issue Brief, World Resources Institute

Phillipson, J (2002) *Widening the Net: Prospects for Fisheries Co-management*, Newcastle University upon Tyne, Centre for Rural Economy

Pring, G (2001) *The Law of Public Participation in Mining and Resources Development*, Washington, DC, IIED/WBCSD

Ritchie, E and Wood, D (2001) *An Analysis of the Public Hearing on the CFP Green Paper - Future Governance: Lessons from Comparative Policy,* Unpublished Report

Rydin, Y and Pennington, M (2000) 'Public participation and local environmental planning: the collective action problem and the potential of social capital' *Local Environment* 5(2):153-169

UNECE [United Nations Economic Commission for Europe] (1998) *Århus Convention on Access to Information, Public Participation in Decision-making and Access to Justice in Environmental Matters*, Århus Denmark 25th June

UNEP [United Nations Environment Programme] (1992), 'Rio Declaration' resulting from *United Nations Conference on Environment and Development*, Rio de Janeiro 13-14 June

Weale, A (2002) 'Public consultation and public policy on risk' in A Weale (ed) *Risk, Democratic Citizenship and Public Policy*, Oxford, Oxford University Press

Webler, T (1999) 'The craft-and theory dialectic of public participation' *Journal of Risk Research* 2(1): 55-71.

Wilson, DC (2003) 'The community development tradition and fisheries co-management', in DC Wilson, JR Nielsen and P Degnbol (eds) *The Fisheries Co-management Experience: Accomplishments, Challenges and Prospects*, Dordrecht, Kluwer Academic Publishers

World Bank (1996) *World Bank Participation Sourcebook*, Washington DC, World Bank

CHAPTER 3
ENGAGING STAKEHOLDER PREFERENCES THROUGH DELIBERATIVE DEMOCRACY IN NORTH SEA FISHERIES GOVERNANCE

JENNY HATCHARD
School of Geography, Politics & Sociology, Newcastle University, Newcastle upon Tyne, NE1 7RU, UK

Abstract

The question of how the preferences of a broad range of stakeholders can be effectively brought into the process of fisheries governance is one that has yet to be resolved in the North Sea context. To date, a top-down style of fisheries governance, exemplified by the Common Fisheries Policy (CFP), has failed to meet the expectations of all those involved in North Sea fisheries – from the science community to the fishing industry and from environmentalists to politicians. Part of this failure has been attributed, by stakeholders, to three democratic deficiencies of the CFP – its centralisation, its politicisation and its externalisation – which have collectively caused the exclusion of the majority of stakeholders from the process of fisheries governance This chapter considers what prospects two models of democracy – representative (currently in operation in North Sea fisheries governance) and deliberative (unexplored in North Sea fisheries governance) – offer for successfully engaging a broad range of stakeholders. I argue that the current governance framework is characterised by both 'thin' (electoral) and 'thick' (corporatist) types of representative democracy, but that knowledge of stakeholder preferences obtained by a process of deliberative democracy offers a better way of strengthening the legitimacy and effectiveness of North Sea fisheries governance. Research conducted using iterative stakeholder engagement (ISE) – derived from the deliberative model – to develop a framework for ecosystem-based fisheries management in the North Sea is employed to support this claim.

3.1 Introduction

The notion of including stakeholders in fisheries governance is gaining currency in the European Union (EU) (World Commission on Environment and Development 1987; EC 2002a, 2002b). However, there is no commonly agreed understanding of how this might best be achieved. The North Sea – one of the most important of the EU's regional seas – manifests two types of stakeholder participation in fisheries governance: a 'thin' or electoral type of representative democracy (for example, the EU Fisheries Council); and a 'thick' or corporatist type of representative democracy (for example, co-management in the Netherlands and Norway). However, the thin type is a very diluted style of stakeholder representation – few UK stakeholders feel that the UK Fisheries Minister is representing them at the EU Fisheries Council. Similarly, the thick type is also a restricted style of stakeholder representation as co-management has historically only included resource users in the governance process, thereby effectively excluding all other stakeholders. Additionally, co-management tends to limit participation to a select few representatives of the fishing industry, leaving many resource users with no voice, particularly if they are not affiliated to the dominant national associations. Thus, while

T. S. Gray (ed.), Participation in Fisheries Governance, 45–64.

the thick type of representative democracy has achieved a greater degree of stakeholder participation than has the thin type, it is still an incomplete solution to the problem of increasing the level of stakeholder participation in fisheries governance.

This chapter considers an alternative to the representative democratic model of stakeholder participation in fisheries governance (whether thin or thick). Iterative stakeholder engagement (ISE) is drawn from the deliberative model of democracy and consists of a systematic and sustained dialogue with a broad range of fisheries stakeholders about their preferences regarding the structures, styles, instruments and controls of fisheries governance. This deliberative democratic model of fisheries governance offers a means by which more stakeholders can obtain more effective access to the fisheries governance process. The viability of this alternative model of stakeholder participation in fisheries governance is supported by empirical evidence from a recent European project – European Fisheries Ecosystem Plan (EFEP) – which employed ISE to consult with a wide variety of stakeholders from several countries around the North Sea.

3.2 Fisheries stakeholders in the EU

In many parts of the world, the involvement of stakeholders has come to be seen as a necessary component of marine fisheries governance. The advantages of such involvement are well rehearsed in the literature (Jentoft & McCay 1995; Nielsen 2003; Hatchard *et al* 2004). For example, it is claimed that stakeholder participation in the governance process will improve the perceived legitimacy of the governance system in general, and of specific regulations in particular, in the minds of resource users and the wider stakeholder community, which will in turn increase compliance (Pinkerton 1989; Hatcher 2000). It is also claimed that incorporating stakeholders' views into the decision-making process will take advantage of their wide experience and knowledge, which will produce better policies, not least because stakeholders will be able to shed light on the likely implications and consequences of different management measures for both ecological and societal systems (EFEP 2004; McCay and Finlayson 1995).

However, acceptance of the concept of stakeholder participation, observable in both the developing world and developed countries, has only recently begun to spread in the actual governance of European seas, where fisheries governance has traditionally been characterised by a centralised, politicised and coercive style of management, orchestrated by the Directorate-General for Fish (DG Fish) in Brussels, and legitimised by a thin type of representative democracy in the shape of the Fisheries Council, the members of which are the elected Fisheries ministers of the Member States. Any tendencies towards a thick type of representative democracy have been restricted to the internal fisheries policies of individual Member States within the Union (of which the Netherlands' co-management system is a good example).

However, in the not too distant future, if we are to believe the rhetoric of European fisheries' political masters, it is likely that stakeholder participation will be established as a permanent element of EU fisheries governance, alongside, if not in place of, what has hitherto been a wholly top-down affair. This is a significant change in the governance process, the importance of which for the fishing industry, and for environmental interests, should not be underestimated, although it is yet to be established what form this participation will take – representative, deliberative or otherwise.

This probable shift comes after two damaging decades of centralised governance of Europe's 'common pond' under the Common Fisheries Policy (CFP), which was introduced in 1983 (Holden 1994; Symes 1997; Cooper 1999). The impact of this period of governance, which has been described by many stakeholders (including the majority of those who participated in the EFEP consultation) as divorced, via many layers of bureaucracy, from the reality of fishing and the environment on the ground, is particularly clear in the North Sea. During these decades, there has been a significant decline in many demersal commercial fish stocks – most notably cod, but also other species such as plaice – and the "decimation" (in the words of fishers) of many fishing communities – few of which have alternative means of income provision.

Those critics who subscribe to the view that the manner of European fisheries governance has contributed to (if it is not the sole cause of) the current 'crisis in fisheries', cite three negative features of the EU's fisheries governance system: **centralisation**, **politicisation** and **externalisation**.[1] First, the EU's *centralisation* of authority excludes most resource users from the decision-making process. The only legitimate way in which members of the North Sea's fishing industry can influence this process is through lobbying of their national governments, the European Commission and the Council of Ministers, via their national fishers' associations. Given that these associations do not necessarily represent all individual fishing operations that are impacted by policy decisions, the centralised governance system has effectively excluded a significant proportion of the fishing industry from influencing the decision-making process. The failure of the CFP's traditional governance framework to give due consideration to the impact of management policies on the socio-economic fabric of fishing communities, may be attributed to this disconnection between the elite policy-making sphere and the experiences of fishers, their families and their communities.

This centralisation of fisheries governance within the EU also affects the opportunities of non-fishing industry interests to influence the policy process. Environmentalists, for example, like the fishing industry, have no direct channel through which they can participate in decision-making processes. Instead, they, too, are left with the sole option of lobbying politicians on issues that they consider significant – such as the protection of the North Sea's cetacean population, and the impact of bottom trawlers on the habitats and benthos of the seabed. There are, however, two key distinctions that must be made between the lobbying by these interest groups and by resource users: their relative capacity and opportunity to pressure governments and influence policy. In terms of capacity, in general, environmental groups are better funded and more experienced than the fishing industry, which gives them an advantage in their efforts to lobby the political system. In terms of opportunity, the potential of different groups to (successfully) lobby political institutions lies in the existence of a favourable political climate. In the case of North Sea fisheries, environmental interests seem to have synchronised more closely in recent years with the climate of government policy than have fisheries interests.

This brings us to the second issue, that of *politicisation*, which indicates the degree to

[1] The criticisms of the CFP's historical failure to effectively govern fisheries, outlined in this section, were expressed by EFEP's stakeholders during the ISE, but they also reflect opinion from the wider stakeholder community, as documented by both the media and academic research.

which fishers are vulnerable to the discretion of their national governments in negotiating on their behalf at the European table or in implementing European policies. Many fishers regard the EU's governance system as heavily politicised. The CFP's *modus operandi* is the Total Allowable Catch (TAC) system, which sets annual quotas to regulate how much of each commercial species a national fleet is legally allowed to catch. Following a process of scientific assessment and analysis, which is charted elsewhere in this volume, the ultimate decision regarding the composition of national quotas is taken by the Council of Ministers. The Fisheries Council bases this decision, officially, on scientific advice from the International Council for the Exploration of the Sea (ICES), and on economic advice from the Science, Technical and Economic Committee for Fisheries (STECF), communicated by the European Commission. However, fishers who observe the process believe that, in some cases, decisions are based not on these scientific and economic factors, but on far wider political considerations (such as the Commission's wish to support Spain's fledgling democracy). If so, this means that fishers' interests have been played off against other national interests, to the detriment of both fish stocks and the fishing industry.

The disenfranchisement of the fishing industry, and, to a lesser extent, other interested parties, caused by the centralised and politicised nature of fisheries governance, has created feelings of alienation from those who govern, on the part of those who are governed. According to critics, this feeling of alienation highlights the third significant characteristic of European fisheries governance – **externalisation**. Being excluded from the policy process means that fishers have been required to comply with instructions from the centralised authority of the EU over which they have had no say, and which therefore appear as external to them. The feeling of helplessness that this externalisation has engendered is compounded by the EU's policy of employing a big stick – mainly in the form of substantial fines – to ensure that fishers comply with regulations. These coercive tactics have done little to address the problem of the illegitimacy with which fishers regard the governance structure. That this in turn reduces the likelihood of compliance by fishers with regulations (over which they have not been consulted), has been well documented (Hatcher *et al* 2000).

Issues relating to governance are regarded as important factors in the course that the North Sea's fisheries have taken ever since the inception of the CFP. But the emphasis that has been placed on the governance issue as a cause of the problems faced by North Sea demersal fisheries, increased sharply with the prospect of the reform of the CFP in 2002, when there was significant pressure for a change in the governance structure as a possible solution to the problems facing both fish stocks and the fishing industry. Decentralisation, regionalisation and national control were all considered as possible solutions. In the end, the European Commission advocated a weak form of regionalisation by establishing advisory councils for regional seas (see Symes, and also Hawkins, in this volume). As a result, the North Sea now has its own Regional Advisory Council (RAC), consisting of a range of stakeholders, which will advise the Commission on issues specific to its geographical area.

What was common to all the calls for the devolution of power away from the European centre was the insistence that fisheries governance must include some level of what is called 'stakeholder participation'. The fact that the European Commission and the Member States had come to accept the notion of stakeholder participation, is indicated by the way in which CFP reform was tackled. Unlike previous reforms, the 2002 reform process featured a long drawn out series of discussion and consultations with a wide variety of non-governmental stakeholder groups, in addition to the usual contributions from national governments and lobbying pressure from fishers' organisations (Ritchie and Wood 2001; Coffey, this volume). These consultations took in as wide a scope as possible of interests, including the fishing industry, the marine environmental movement, sea anglers' groups, and academic experts.

Since the 2002 reform, European fisheries have witnessed increased pressure on the European Commission to institute a framework through which stakeholders can have a more participative role in fisheries governance. The RAC structure is one mechanism for achieving this goal of better, more inclusive governance. However, this institutional solution may not satisfy the growing calls from fishers for an increased sense of control over their own livelihoods and industry, nor does it give environmental groups the level of influence over marine policy that they would like. Equally, it remains to be seen whether it will act successfully as an arbiter between these different interests. In fact, in the North Sea context, many stakeholders, from across the spectrum, remain sceptical of the RAC's capacity to deliver significant changes in governance. If they are right, regionalisation may well result in only a slight shift of power from one political elite to another.

In my view, what is crucial to the success of fisheries governance in the future is not this "additional layer of bureaucracy" (as some EFEP stakeholders described the North Sea RAC) – although it will be helpful in focusing more closely on the specific problems of the North Sea, as opposed to other European seas – but rather a mechanism in place which ensures the systematic identification of stakeholder preferences and an equally systematic procedure for feeding those preferences into the governance process. This is the argument that is advanced in this chapter, based on my experiences with ISE in the EFEP project. But first, the meaning of the term 'stakeholder' must be clarified.

3.3 Meaning of 'stakeholder'

A key revelation that came from approaching stakeholders for the EFEP project was less their views on the central subject of the consultation – ecosystem-based fisheries management for the North Sea – than their reactions to being considered 'stakeholders'. Some, particularly those familiar with policy environments, had long been aware of the term, and, for the most part, merely questioned which groups of people the project categorised as stakeholders. However, others, particularly from the fishing industry, had only come across the term as a result of the 'stakeholder consultation' that was conducted as a part of the CFP reform process. For many of them, this EU consultation process had not delivered what they hoped it would. For example, some felt that their preferred option for reform of European fisheries management – national control – had been sidelined, or ignored. These individuals, who had effectively been deceived by what they perceived to be an illusion of stakeholder involvement in the policy process, were much more wary of the prospect of being 'stakeholders' once more. One fisher dropped out of the process halfway through, having come to the conclusion that he no longer wanted to be a stakeholder in Europe.

Other respondents were openly derisory about the concept of 'stakeholding'. These individuals regarded the 'stakeholder' concept as flawed, partly because of their perception of who would count as a 'stakeholder'. For example, some fishers perceived the concept to be a tool by which non-fishing interests could be incorporated into the governance process, thereby reducing their own role and capacity to influence policy. To them, the term 'stakeholder' was a threat to their existing (admittedly limited) position in the decision-making process. This kind of scepticism and concern was not confined, however, to members of the fishing industry. There was also some anxiety from environmental groups that the 'stakeholder' concept was a means of bringing autonomy to resource users and excluding interest groups. These concerns reflected ambiguity in the way the term 'stakeholder' has been used – sometimes referring to a broad church of interests; at other times limited to user groups.

Fear of not being designated a stakeholder is, therefore, one anxiety. Another anxiety is fear of not being regarded as an **equal** stakeholder. Those with a fishing interest in the North Sea, for example, suspect that their political masters would argue that, while both the fishing industry and the environmental movement are stakeholders, the environmental groups are the better guardians of the marine ecosystem. In this scenario, some stakeholders are more important than others. Being designated a stakeholder, therefore, does not guarantee that one has equal status to all other stakeholders. This uncertainty undermines the value of the title of 'stakeholder', and it makes those who hold this title unsure whether or not their views will be taken into account by decision-makers. This ambiguity surrounding the term 'stakeholder' has cast something of a shadow over its potential value in fisheries governance. The failure of the EU to pin down its meaning has fed the perception that it is a term that can be manipulated for reasons of political expedience, thereby undermining the legitimacy of the concept.

What was the meaning of the term 'stakeholder' in the EFEP project? Simply put, for the purposes of the project, stakeholders were those who had a stake in North Sea fisheries and related ecosystem features. Working from the literature on stakeholders (Wijnberg 2000; Mikalsen & Jentoft 2001), an understanding of the term 'stakeholder' was developed that was tied to two key criteria: stakeholders are first, those individuals/groups who are *dependent* upon the North Sea; and second, those individuals/groups who have an *interest* in the North Sea (Hatchard *et al* 2003). Commentators tend to impose a hierarchy on these criteria – ranking dependent stakeholders as primary, and interested stakeholders as secondary (Wijnberg 2000).

In selecting the stakeholders to include in the EFEP consultations, I adopted this hierarchical assessment by including more stakeholders from the dependent group (the fishing industry) than from any interest group. This is a justifiable choice, given that those directly dependent on North Sea fish resources for their weekly income, their social interactions and/or for their cultural identity, are likely to experience more significant direct impacts from changes in fisheries policy than are individuals or groups who are simply interested in the resource and its environment. This is a form of proximity analysis, analogous to the ripple effect of dropping a stone into a pool of water – creatures closest to the impact will be more directly affected than will those further away.

The result of this categorisation of stakeholders for the project was the inclusion of a large number of dependent stakeholders from the fishing industry, and a roughly equal total number of stakeholders from a variety of different interest groups (including regulators, policy-makers, environmentalists and scientists) in the research sample. In this way, the question of the definition of 'stakeholders' was resolved within the EFEP project.

3.4 Two models of stakeholder participation in fisheries governance

The way in which stakeholders were involved in the EFEP project exemplifies the second of two models for involving stakeholders in fisheries governance. These two participatory models are tied to two different models of democracy (Held 1996; Cunningham 2002). The first model, **stakeholder representation**, is the product of representative democracy. Both 'thin' (electoral) and 'thick' (corporate) types of this model can be found within the North Sea context. The limited representational systems operating in the UK and at the EU level are examples of the thin or electoral type of representative democracy; while the co-management systems of the Netherlands and Norway are examples of the thick or corporate type of representative democracy. The second model, **stakeholder preferences**, is the product of deliberative democracy. There is no institutional example of such a system within the North Sea context, but the ISE approach adopted for EFEP's dialogue with North Sea stakeholders is an illustration of how such a system might operate. In this section, the practical advantages and disadvantages of each of these two democratic models for involving stakeholders in fisheries governance are explored.

3.4.1 MODEL ONE: STAKEHOLDER REPRESENTATION – REPRESENTATIVE DEMOCRACY

The concept of representation of stakeholders within fisheries governance and its decision-making processes incorporates a wide range of institutional arrangements, from the thin types such as parliamentary democracy, to the thick types such as co-management. In the North Sea, there are examples of both thin and thick types of stakeholder representation. However, before I discuss them, it is necessary to clarify the meaning of representative democracy.

In John Stuart Mill's (1861/1991) seminal treatise, *Considerations on Representative Government*, representative democracy is advocated as the best practicable way to meet both the democratic requirement of public participation and the practical necessity of administrating what Held (1996:107) denotes a "complex mass society". However, Mill's *ideal* polity was what he termed "popular democracy" or "completely popular government" – the classical style of direct democracy in operation in the city-states of the Ancient World. He put forward two justifications for direct democracy: first, the rights and interests of all people "are only secure from being disregarded when the person interested is himself able, and habitually disposed, to stand up for them. It is humiliating to be excluded from the political pale...It is a great discouragement to an individual, and a still greater one to a class, to be left out of the constitution; to be reduced to plead from outside the door to the arbiters of their destiny, not taken into consultation within" (Mill (1861/1991:254). Second, such a system would bring greater prosperity. That these two arguments apply to fisheries governance is evident from the demand, particularly from stakeholders within the fishing industry, for more notice to be taken of their views; for power to be devolved away from Brussels and closer to home; and for self-governance, which, they claim, would bring greater prosperity and more chance of achieving sustainability.

However, Mill recognised that direct democracy would be untenable in any society bigger than a small town, and, as far as Britain as a whole was concerned, he looked to the national political changes taking place – particularly the series of Reform Acts from 1832 onwards which gradually extended the franchise – for a model that would balance the twin aims of popular participation and effective administration. Effective administration, he conceded, could be achieved by a well-educated and experienced bureaucracy; but unless it was combined with some element of popular participation, he believed that the political system would lack vitality and innovation. Thus, only through popular participation (even to a limited extent, such as by electoral voting), combined with an organised bureaucracy, would it be possible to establish a stable and progressive government:

> …the only government which can fully satisfy all the exigencies of the social state, is one in which the whole people participate; that any participation, even in the smallest public function, is useful; that the participation should everywhere be as great as the general degree of improvement of the community will allow; and that nothing less can be ultimately desirable, than the admission of all to share in the sovereign power of the state. But since all cannot, in a community exceeding a small town, participate personally in any but some very minor portions of the public business, it follows that the ideal type of a perfect government must be representative. (Mill 1861/1991:255-6)

Mill's theory of representative government has stood the test of time, exemplifying the most successful and accepted political model in the contemporary western world. Importantly for this chapter, the influence of this model of representative democracy has not been limited to national political systems, but has infiltrated into all levels of society to the point where representation is the accepted means by which the general populace influences policy of all kinds. However, Mill deals with only one of two types type of representative democracy – which we have termed the 'thin' or electoral type – characterised by local and national governmental elections open to all adult citizens, but in which the extent of stakeholder participation in decision-making is low. The other type of representative democracy – which we have termed the 'thick' or corporate type – is characterised by elections restricted to members of a particular industry, where the extent of stakeholder participation is higher.

We can see both these types of representative democracy at work in North Sea fisheries governance. At the thin end of the spectrum of representative democracy in North Sea fisheries governance is the EU. The emphasis of European fisheries governance, particularly in the North Sea, is on the bureaucratic aspect of representative democracy, rather than on its participative dimension. Indeed, a phrase that I often heard used to describe the EU's CFP during the EFEP project was "bureaucracy gone mad". The popular participation aspect of the CFP has largely been limited to national elections, which provide formal political authority to Ministers in the important Fisheries Council, and select representatives to serve in the European Parliament, which has some limited authority in fisheries issues. However, a powerful bureaucratic machine in the form of the unelected European Commission, is responsible for day-to-day fisheries decision-making, which may illustrate Mill's contention that bureaucracy without popular participation to complement it, is liable to lead to sclerosis.

Also occupying the thin end of the spectrum of representative democracy is the UK's national mode of applying the CFP to its fisheries, where stakeholder participation in fisheries governance is quite minimal. Apart from the formal opportunity of all citizens to vote in national elections by which governments are elected, which in turn select fisheries ministers, members of resource user and interest groups in the UK only have the opportunity to lobby government. Generally, representatives of sectors of society take on this role, and as a rule, it is the bigger groups that shout the loudest who are most heard. As far as fisheries governance is concerned, such groups include fishers' organisations at both the national level, in the form of the National Federation of Fishermen's Organisations (NFFO) (which covers England and Wales), and the Scottish Fishermen's Federation (SFF), as well as more regional and local groups, and interest groups such as the environmental lobby. Thus, at the thin or electoral end of the spectrum, it seems that the link between stakeholders (of all kinds) and their political representatives is nominal and, possibly, even fractured, with few attempts made to bridge the gap, and no system in place, beyond parliamentary procedures, to make even this weak process an accountable one.

However, at the local level in the UK, there are some examples of a thick or corporatist type of representative democracy. For instance, the 12 Sea Fisheries Committees (SFCs) in England and Wales, which manage local inshore fisheries, are made up of representatives from the fishing industry, as well as from local government (Knapman, this volume). Also, there are many Producers Organisations (POs) in the UK, which have been given the power to allocate quota to fishers, and there are several examples, such as Invest in Fish South-West, of 'partnerships' comprising fishers, other resource users, local authorities, and environmentalists, with responsibility for managing local stocks. The possibilities for involving stakeholders in fisheries governance through these local institutions have been explored by Phillipson and Crean (1997). However, apart from POs, these 'thicker' participative regimes tend to be confined to inshore waters.

By contrast, some other North Sea states have instituted a thick or corporatist type of representative democracy for offshore fisheries governance to supplement the limited electoral foundation of their democratic representation, and to support (with the exception of Norway which is outside the EU) their interaction with the limited system of representative democracy inherent in the CFP. In Denmark, for example, fisheries governance is based on cooperation between the government ministry (the Ministry of Agriculture, Fisheries and Food) and industry. This collaboration has been formalised by the creation of several advisory, or consultative, boards, which provide user-group participation in management policy-making. Two boards are especially important: the EU Advisory Board and the Advisory Board for Commercial Fisheries. A variety of organisations participate in policy-making through the advisory board system, and, although it is fairly centralised and top-down, local government institutions and environmental groups do succeed in influencing decisions.

Less broadly-based, but more participatory, are the co-management systems in operation in Norway and the Netherlands (Van Ginkel; Hernes *et al*, both this volume; Hersoug and Rånes 1997). These corporatist-style systems enable formal interaction and cooperation between government and representatives of fisheries interests to develop in order to agree national fisheries policy (Sen & Nielsen 1996:406; Pomeroy and Berkes 1997). It is important to note, however, that in both countries the participation is limited to resource users – dependent stakeholders – and excludes interest groups, such as environmentalists. For example, in Norway, the Norwegian Fishers' Association works with the Norwegian Ministry of Fisheries to establish their quota allocation mechanism (Hernes *et al*, this volume), while in the Netherlands, the fishing industry, in the guise of two key organisations – the Nederlandse Vissersbond and the Federatie van Visserijverenigingen – works in close conjunction with government to ensure that the group system – the Biesheuvel groups – operates effectively to maximise quota uptake and ensure compliance with regulations (van Ginkel, this volume). The key difference between the two systems is that, unlike Norway, the Netherlands' scope for decision-making within this co-management framework is limited by the strictures of the CFP.

What is clear from the examples of Norway and the Netherlands is the central role played in co-management by the national fishermen's organisations. It seems that the industrial representative arrangements in North Sea states for dealing with fisheries governance are targeted at ensuring the representation of resource users (dependent stakeholders) rather than other interests (interested stakeholders), and only those resource users who are represented by the main fishers' organisations. So co-management seems to be an extension of the privileged lobbying position that industry elites hold in the more limited representative systems of the UK and the EU, rather than, as Noble (2000:76) hopes, facilitating "participatory democracy" in fisheries governance.

There is thus a range of types of stakeholder representation in countries around the North Sea, from basic citizen rights to vote for, and to lobby, their representatives, to corporatist arrangements for the representation of resource users in management decision-making via their industry representatives. The latter arrangements strengthen the links between individuals and their political representatives by including industry representatives in governance decision-making. However, even in these co-governance systems, many stakeholders are frustrated, either because they are excluded altogether from the decision-making structure (like the environmentalists), or because their representatives are out of touch with their problems (like some fishers).

I contend that the representative route (whether thin or thick) – as illustrated at EU level and within the key North Sea fisheries states – is not the only democratic route, nor is it, as has been demonstrated by the dissatisfaction expressed by stakeholders, sufficient to meet the requirements of modern "good governance" (Kröger 2001). Accordingly, it is important to search for other ways of providing opportunities for popular democracy. One such possibility – that of stakeholder preferences determined by a deliberative democratic process – is explored in the following section.

3.4.2 MODEL TWO: STAKEHOLDER PREFERENCES – DELIBERATIVE DEMOCRACY

As we have seen, fisheries governance mechanisms based on the theory of representative democracy are an accepted feature of North Sea fisheries governance. However, we must not let this established system preclude a search for alternatives and/or supplementary mechanisms drawing on another dimension of the democratic principle – its deliberative dimension. The purpose of this chapter is not to examine all the possibilities that exist within the deliberative model of democracy – that objective is, in any case, being pursued elsewhere: for example, in the analysis of the value of deliberative democracy for environmental political agendas (Dryzek 1997, 1998; Smith

2003). Instead, my aim is to look at a particular exemplification of deliberative democracy – that relating to stakeholder preferences – and to consider its potential application to North Sea fisheries governance. In this section, I will, first, examine the concept of 'stakeholder preference' and its relationship to deliberative democracy; and, second, discuss the way in which EFEP employed iterative stakeholder engagement (ISE) to initiate a dialogue with a broad range of North Sea stakeholders to ascertain and understand their preferences.

3.4.2.1 *'Stakeholder Preference' and Deliberative Democracy*

It could be argued that representative democracy reveals 'stakeholder preferences'. Indeed, liberal democratic theory uses the concept of 'preference aggregation' to mean a process of decision-making by majority voting. Members or citizens come together and express their preferences through a vote (Held 1996; Cunningham 2002; Elster 1998:6; Smith 2003:5). Preference aggregation operates on the basis of several assumptions: that all who are entitled to vote are equal; that the majority vote provides the 'agreed' solution; and that the system operates within a liberal democratic state whose citizenry will, by and large, accept those outcomes. As Cunningham (2002:163) points out, this 'liberal' approach "pictures citizens entering a democratic political process with fixed preferences that they aim to further by use of democratic institutions and rules." In the fisheries world, we see preference aggregation being conducted, at the highest level, by the Council of Fisheries Ministers in the EU.

However, this formal representative system is a poor way of identifying stakeholder preferences, because, in such a system, there is only a remote and fractured connection between voters' intentions and government policy. As such, representative preference aggregation amounts to what Goodin (2003:48ff) describes as "recording", rather than "respecting", preferences and does little to address any conflict between different groups of stakeholders and between stakeholders and regulators regarding "what public policies should be or how they should be arrived at and enforced" (Cunningham 2002:164). The unsatisfactory outcome of the CFP reform consultation process illustrates the limitations of this approach.

In contrast, the deliberative model of democracy offers a more "respectful" approach to 'preference', one that forges a closer link between stakeholder preferences and the governance system (Goodin 2003:48ff) in two ways. First, deliberation of preferences entails discussion of the reasoning behind preferences in order to reach a common consensus. Cunningham (2002:164) describes this process: "...those engaged in deliberative-democratic practices must be prepared to question and to change their own preferences and values. In such practices each gives reasons for his or her initially favoured views aiming thereby to persuade others to adopt them." This possibility of having a genuine influence on the policy process through collective discussion and decision-making is likely to make stakeholders feel they are part of the governance process. Second, according to Fiorino (1996:194), deliberative democracy located at "the lowest possible level" (1996:209), promises to close the participation gap that has been created by the centralisation of policy and by top-down decision-making processes (characteristic of European fisheries policy to date).

This deliberative model of democratic theory has been widely applied to environmental and natural resource management and decision-making in recent years (Dryzek 1997, 1998; Elster 1998; Fiorino 1996; Smith 2003). A key objective of this endeavour has been to find a way for democracy, and democratic governance processes, to reflect the plurality of environmental values held by those involved in environmental politics (Smith 2003:5-6) and to enable individuals and groups to express the intensity of the preferences they hold (Fearon 1998:45) Additionally, unlike representative preference aggregation, preference deliberation offers some political space and opportunity for the transformation of preferences to take place (Elster 1998:6; Gambetta 1998:22-3; Cunningham 2002:164) This is particularly important in the environmental context where information is complex and sometimes sparsely available and which can affect stakeholder preferences if they are aware of it. Accordingly, deliberative democrats' interpretation of preferences offers a better prospect of informed consensus-building.

Graham Smith (2003) is one proponent of this deliberative approach to environmental governance. For him, the ascertaining of citizen preferences provides a practicable means of reflecting environmental value pluralism. Smith proposes three deliberative procedures for the inclusion of citizens' preferences in the policy process: **mediation**, which involves the participation of members of affected interests in conflict resolution or problem solving; **citizen forums**, which enable a cross-section of the population to hear evidence and discuss issues of public concern; and **citizen initiatives and referenda**, which allow a population to vote directly on policy issues (2003:80-99). Similar procedures are also recommended by Fiorino (1996:209).

This deliberative approach to environmental governance is directly applicable to participatory fisheries governance institutions. Deliberative analysts' understanding of the problems of environmental governance and their causes are very relevant to North Sea fisheries in which centralised institutions are paramount, in which views may be heard but not listened to, and in which the vast diversity of environmental values are not explored and reflected by the policy process. In addition, despite limits identified by Fearon (1998:47) and Elster (1998:101) regarding incentives for participants to misrepresent their position, the deliberative mode encourages the search for common ground and consensus between diverse interests.

However, the particular deliberative procedures that Smith (2003:80-99) proposes do not seem the most appropriate way to address two of the three key flaws of current North Sea fisheries management – centralisation and externalisation. First, although **mediation** could be a positive means of resolving specific policy conflicts between regulators and regulated, such an *ad hoc* procedure would not alter the top-down, centralised dynamics of the fisheries governance system. Second, **citizen forums** open the door to public debate, and this could have the effect of diluting, rather than increasing, the role of dependent stakeholders in fisheries governance, thereby perpetuating their externalisation. Finally, **citizen initiatives and referenda** do not provide much opportunity for genuine deliberation. However, on the positive side, all three of Smith's procedures, by reducing the role of explicitly political interests, could help to reduce the politicisation of fisheries governance and decision-making.

Let us now turn to an alternative deliberative procedure for preference deliberation, used by my colleagues and I, in our work with stakeholders, conducted as part of the EFEP project. I characterise this method as 'iterative stakeholder engagement' (ISE)

3.4.2.2 *Iterative Stakeholder Engagement (ISE)*

EFEP (2004) was a three-year project, funded by the European Union, which aimed to develop a framework for ecosystem-based fisheries management for the North Sea. Iterative stakeholder engagement (ISE) was incorporated into the project from the beginning, entailing a two-way process of discussion and information transfer which primarily involved two phases of (where possible) face-to-face semi-structured interviews with a variety of stakeholders from four North Sea states – Denmark, the Netherlands, Norway and the UK – and from both dependent and interested sectors (Hatchard *et al* 2003, 2004). This lengthy process of iterative dialogue between researchers and stakeholders sought to examine stakeholders' "deeper preferences" and detect changes in them over time, rather than the "superficial expressions" (Goodin 2003:49) that a one-off paper-based survey would have yielded. The ISE process also provided opportunities for stakeholders to reflect on their own preferences and, through some multi-participant consultations and consistent reporting of aggregated findings to stakeholders, those of others. Goodin describes this internal process as "democratic deliberation within".

The first ISE phase (Hatchard *et al* 2003), which was conducted in late 2002/early 2003 at the beginning of the project, sought to establish stakeholders' overall impressions of the health of North Sea commercial fish stocks and that of the North Sea ecosystem, and their preferences regarding four aspects of fisheries management: structures; styles; instruments; and controls. The general preferences of the EFEP stakeholders regarding these four aspects are presented in Table 3.1.

The views and preferences which stakeholders expressed during this primary ISE process were used by the project team first, to assist in identifying significant features of the North Sea ecosystem (Ragnarsson *et al* 2003), and second, to inform a process of ecological modelling of a variety of management scenarios aimed at the protection of particular ecosystem features and functions (Piet *et al* 2003; Silvert *et al* 2003).

The second phase of engagement (Hatchard *et al* 2004), took place during 2004. In this second phase, researchers employed a key ISE tool – a visual discussion facilitator – to visually convey to stakeholders the possible ecological impacts of the management preferences they had expressed in the first ISE phase. This computerised visual aid facilitated discussion of a range of management scenarios for meeting particular ecosystem goals and of their ecological and societal implications. These scenarios were: spatial effort management; effort redistribution; human-consumption and industrial fisheries effort reductions; gear regulations; fisheries protected areas; and conservation protected areas.

Table 3.1. Summary of overall management preferences of EFEP stakeholders, expressed in the first consultation (Hatchard *et al* 2003)

Aspect of management	*Overall stakeholder preference*	*Reasoning*
Structures	Regional North Sea management	The overall consensus between stakeholders was that it is not possible for any one fishing fleet to manage their role in the North Sea independently of other fleets. North Sea fishing impacts on stocks and the ecosystem are inter-connected and should be managed as such. However, stakeholders also shared the view that the EU's CFP has been unsuccessful in its pursuit of a sustainable policy. A regional form of management could achieve this as long as all the different interests – national and sectoral – work together for common aims. Such a development represents to stakeholders an opportunity to move away from the politicisation which has characterised the CFP, and to limit management decisions solely to North Sea actors. Importantly, while there was some support for regional North Sea management among Norwegian stakeholders, they did not advocate a loss of their own sovereignty to such a body.
Styles	Inclusiveness	Overall, stakeholders agreed that fisheries management at all levels needs to incorporate, in particular, the fishing industry into decision-making. The top-down coercion of the CFP, mirrored by a lack of inclusiveness within some states, was criticised by stakeholders from both inside and outside such management regimes. They argued that the exclusion of key stakeholders from decision-making and developing regulations has created a climate within which the fisheries management system is not accorded much legitimacy or respect by those it seeks to regulate, and some suggest that co-management within states offers one means by which this problem could be addressed.
Instruments	Long-term strategy	Stakeholders argued that the broad brush, reactive approach which has long characterised European fisheries management must be replaced by a proactive, carefully targeted and sensitive approach. Thus, stakeholders advocated a system in which management instruments, whose impacts – environmental and economic – have been assessed, should be combined in a targeted manner to address particular issues. The opinions of stakeholders regarding the form which these regulatory instruments should take did, however, vary. There was controversy over whether the status quo of the quota system (that many regarded as unfair) should be maintained, albeit modified, or whether it should be replaced wholesale by a system which regulates the amount of time vessels can spend fishing. However, stakeholders did agree that technical management must play a central role and that further research needs to be conducted in this area. There was also support for 'real-time' closures and for work to be done to reduce discards.
Controls	Clarity and equality	The emphasis given by stakeholders with regard to controls was on the clarity of regulations, which should be enforced in a consistent way, incurring the same penalties, throughout the North Sea.

Table 3.2. Acceptability of management scenarios for meeting particular ecological objectives expressed in the second consultation (Hatchard et al 2004)

Measure	*Ecological objective*	*Acceptability*	*Implementation*	*Data requirements*
General effort reduction	Reduce direct effects of fishing on target stocks	Maybe	Days-at-sea	Ecological and socio-economic impacts of effort reductions to date
High frequency effort restrictions	Reduce discards of undersized target stocks	Yes	Real-time closures in areas of high fishing intensity	Fleet distribution and catches details to be collected on a real-time basis
Low frequency effort restrictions	Reduce bycatch of non-target stocks	Maybe	Limited entry and/or gear exclusion in areas that are not intensively fished	
Fleet effort restrictions	Increase prey availability for other fish predators	No	N/a	N/a
	Reduce mortality of lower trophic levels to increase feed availability for high-value stocks	Yes	Reduction of industrial fishing effort	Impact of industrial fishing on prey availability
Minimum mesh size regulations	Reduce discards of under-sized fish	Yes	Possible increase in mesh size used by some fisheries	Long-term ecological and economic effects of an improved age structure in commercial stocks
Discard ban	Reduce discards	Maybe	Combine with days-at-sea; real-time closures; and a more flexible quota system or quota abolition	Fleet movements
Technical bycatch regulations	Reduce bycatch of charismatic species Reduce bycatch of unwanted fish	Yes	Measures to be applied to culpable fleets and to geographical areas where bycatch is a documented problem	Location of bycatch problems
Fisheries protected areas	Reduce direct fishing impacts on juvenile stocks	Yes	Real-time gear exclusion zones	Location of nursery grounds
	Reduce direct fishing impacts on spawning stocks	Maybe	Flexible seasonal gear exclusion zones	Spawning ground locations
Conservation protected areas	Protect unique habitats	Yes	Year-round gear exclusion zones	Biological evidence of the need for long-term protection
	Protect essential fish habitats	Maybe	Year-round gear exclusion zones	Links between spawning and juvenile stocks and habitat types
	Protect significant sessile species	Maybe	Year-round gear exclusion zones	Links between species and habitats
	Protect 20-30% habitat types	No	N/a	N/a

The purpose of this second process was to enable the research team to ascertain and understand stakeholder preferences about how ecosystem-based fisheries management, and its associated goals, of the North Sea might be most effectively and acceptably implemented. The preferences expressed during this second phase are presented in Table 3.2. These preferences covered the stated ecological objectives; the scenarios for meeting those objectives; the implementation of each scenario; and the data requirements which stakeholders saw as crucial to the success of scenarios in meeting their objectives.

As Tables 3.1 and 3.2 demonstrate, the outcomes of these two connected phases of dialogue with stakeholders, regarding their preferences for the implementation of ecosystem-based fisheries management, were analysed by using an approach that *engaged* with the preferences of those stakeholders and sought to identify common ground between their sometimes similar and sometimes disparate views to steer the course of EFEP's research. The continuous iteration between stakeholders and researchers and the multi-directional transfer of information provided many opportunities for deliberation and development of preferences.

The EFEP project was conducted over the course of three years and featured much interaction between researchers and stakeholders. This two-way and sustained process was an important feature of stakeholder involvement, which it would be necessary to replicate in any governance scenario featuring stakeholder preference engagement. One of the advantages of the face-to-face meetings between stakeholders and researchers was that, as Cunningham (2002:164) suggested, the reasoning that underpinned the aggregated views of stakeholders could be teased out. The direct connection between researchers and stakeholders meant that the meaning of their views, and stakeholders' own explanations for holding them, were not lost in a process that saw the data changing hands. Such a connection would need to be replicated if stakeholder preferences were to be brought into fisheries governance. Admittedly, this would be difficult and would come up against the same constraints that pushed Mill away from direct democracy towards representative democracy. However, the link between the expression of stakeholder preferences and the reading of those preferences by policy-makers needs to be as strong as possible.

Given the geographical scope of the North Sea and the significant commitments of all the different stakeholder groups, both personally and professionally, it would be unrealistic and costly to continuously replicate the detailed interviewing process that was adopted for the EFEP project. However, in the current age of advanced techniques of communication, it would not be unfeasible to suggest that stakeholder preferences could be expressed on a regular basis using an electronic forum structure (recommended by Rossiter 2002). An e-forum could invite new ideas for policies from all stakeholders; could be an opportunity for fishers to draw attention to environmental and economic problems that they have noticed, thereby contributing their hands-on knowledge for the benefit of all; and could provide a means by which policy-makers could consult instantaneously on policies they are developing. Although such a process would not have the face-to-face feature of the EFEP consultations, modern computing technology would enable it to be an interactive process.

What I am proposing here is not to replace, but to supplement, the elements of representative democracy present within North Sea fisheries governance by a deliberative democratic procedure for the systematic engagement of stakeholder preferences. My plan is for the provision of a consistent and regular opportunity for stakeholders of all persuasion to contribute their views, and for there to be a framework of accountability in place that would ensure that this process is transparent, and that stakeholders' preferences are taken into consideration. This continuous ISE would "respect" as well as "record" stakeholder preferences regarding the North Sea and its fisheries as well as providing a means of conflict-resolution and consensus-building.

3.5 Discussion

Let us now compare how the two models of stakeholder involvement in fisheries governance – stakeholder representation and ISE – would deal with the three deficiencies of the CFP mentioned earlier – centralisation; politicisation; and externalisation. In the case of centralisation, it is clear that the thin type of representative democracy does little to address the failure of decision-makers to consider the impacts of decisions on fishers and their communities – a failure which is caused by the distance that exists between policy-makers and fishers as a result of the bureaucratic nature of governance. In contrast, the thick type of representative democracy – particularly co-management – goes some way to bridging the gulf between governors and governed. Here, the representatives of resource users have the opportunity to put their case to government and to work co-operatively to develop and agree upon shared solutions. Co-management in the Netherlands, for example, has meant that the Dutch fishing industry has had a significant input into Dutch national fishery policy making. However, the benefits of co-management for reducing the centralising impacts of fisheries governance are limited by the fact that it is usually representatives of national fishers' organisations who have the opportunity to participate in co-management forums. This means that a considerable majority of fishers are excluded from direct participation, and that those who are represented are very much reliant on their national organisations to be effective and efficient in that representation. Co-management also excludes interested stakeholders.

If, alongside co-management arrangements, there were mechanisms for recording and considering dependent and interested stakeholders' preferences for fisheries governance, and their reasons for holding those preferences, the policy process would reflect stakeholder opinion much more accurately. This deliberative method of decentralising the process of fisheries governance, so that information is gathered from Fiorino's (1996:209) "lowest possible level consistent with policy-making", would help to make policy-makers much more aware of the reality of the impacts of their policies.

With regard to the problem of politicisation of fisheries governance, the thin type of stakeholder representation not only fails to deal with it, but exacerbates it. The distance created between decision-making and stakeholders by a system of political representation leaves the decision-making process vulnerable to manipulation. In contrast, the thick type of representation, in the form of co-management, helps to reduce opportunities to distort fisheries policy for political gain. This is because, within corporatist arrangements, industry representatives and political representatives at the national level agree on policy choices before the final political arbitration process at the European level within the Council of Ministers. It is true that, at the EU level, there is no co-management system in place, and political trade-offs regularly occur in both the Council and the Commission, but it can be hoped that the exposure of political decision-makers to the feelings and priorities of stakeholders (both dependent and interested) as a result of deliberative processes of stakeholder preference engagement, would make them rethink any schemes that barter with the future of fisheries and their environment.

Finally, the thin representation of stakeholders within North Sea fisheries governance at the European level also fails to deal with the third of the governance flaws – that of externalisation. Indeed, it has contributed to this deficiency, in that weak links between resource users, and a decision-making machinery dominated by bureaucracy on a day-to-day basis, have had an alienating effect which has precipitated resistance to regulations which are not regarded as legitimate – a major cause of non-compliance. The response to this apathy by those who govern has thus far been to adopt coercive measures. A striking contrast at national level is provided by the thick representation of the fishing industry Dutch co-management system, which has been successful in reducing the need for coercion and facilitating compliance with quota allocations. Regular deliberative canvassing of stakeholder preferences and interaction with them would also contribute to reducing this feeling of exclusion. This was evidenced to a limited degree by the gradual acceptance of (if not necessarily agreement with), the concept of ecosystem-based management among some of the stakeholders over the course of the EFEP project (Hatchard *et al* 2004).

It is thus clear that, although the thick representative model for involving stakeholders more closely in fisheries governance has contributed something in reducing the negative impacts on fisheries, communities and stocks, of the decision-making system in the North Sea, the deliberative model enables decision-makers in fisheries governance to learn much more about the wider socio-economic impacts of policy decisions and to understand the reasoning underpinning stakeholders' preferences. Moreover, since communication works both ways, greater stakeholder involvement via the deliberative process would facilitate a considerable improvement in the stakeholders' understanding of policies and regulations and the reasons why they are adopted. Finally, it is important to note that, unlike the thick representative model, ISE ensures the contribution and inclusion of both dependent and interested stakeholders.

3.6 Conclusion

The two democratic models – stakeholder representative democracy and stakeholder deliberative democracy – discussed in this chapter, are mechanisms designed to achieve the same things, but in different ways. For too long, representative mechanisms – electoral and corporatist – have been solely relied upon to deliver participation in fisheries governance. Deliberative forms of participation are becoming ever more important in our growing consultative climate which developed from stakeholder demands for more inclusive governance (Hatchard and Gray 2003). By supplementing stakeholder representation with a deliberative system of ISE, it is possible to overcome the three key problems that have been identified in relation to North Sea fisheries governance – centralisation, politicisation and externalisation. Equally importantly, the deliberative mode of aggregation of stakeholder preferences constitutes an achievable change in a system that has consistently resisted genuine reform.

Acknowledgements

This work was supported by the EC through study contract Q5RS-2001-1685 *European Fisheries Ecosystem Plans: Northern Seas*. My thanks go to Tim Gray and Knut Mikalsen and the rest of the EFEP team.

References

Cooper M-P (1999) 'The Common Fisheries Policy of the European Union – a lesson in how not to make policy' *Politics* 19(2):61-70

Cunningham, F (2002) *Theories of Democracy: A Critical Introduction*, London and New York, Routledge

Dryzek, JS (1997) *The Politics of the Earth: Environmental Discourses*, Oxford, Oxford University Press

Dryzek, JS (1998) 'Political and ecological communication' in JS Dryzek and D Schlosberg (eds) *Debating the Earth: The Environmental Politics Reader*, New York, Oxford University Press: 584-597

EC [European Community] (2002a) *Communication from the Commission on the Reform of the CFP ("Roadmap")*, COM(2002)181final, Brussels, European Commission

EC [European Community] (2002b) 'Council Regulation No. 2371/2002 on the conservation and sustainable exploitation of fisheries resources under the Common Fisheries Policy' *Official Journal of the European Communities* L358/59, Brussels, European Council of Ministers

EFEP (2004) *The North Sea Fisheries Ecosystem Plan*, EU Quality of Life and Management of Living Resources: QLRT-CT-2001-01685, University of Newcastle

Elster, J (1998) 'Introduction' in J Elster (ed) *Deliberative Democracy*, Cambridge, Cambridge University Press

Fearon, JD (1998) 'Deliberation as discussion' in J Elster (ed.) *Deliberative Democracy*, Cambridge, Cambridge University Press

Fiorino, DJ (1996) 'Environmental policy and the participation gap' in WM Lafferty and J Meadowcroft (eds) *Democracy and the Environment: Problems and Prospects*, Cheltenham, Edward Elgar

Gambetta, D (1998) '"Claro!": an essay on discursive machismo' in J Elster (ed) *Deliberative Democracy*, Cambridge, Cambridge University Press

Goodin, RE (2003) *Reflective Democracy*, Oxford, Oxford University Press

Hatchard, JL and Gray, TS (2003) 'Stakeholders and the reform of the European Union's Common Fisheries Policy: The institutionalisation of a consultation culture?' *Maritime Studies* 2(2):5-20

Hatchard, JL, Gray, TS and Mikalsen, K (2003) *European Fisheries Ecosystem Plan: Stakeholder Consultation*, EFEP, EU Quality of Life and Management of Living Resources: QLRT-CT-2001-01685, Newcastle upon Tyne, University of Newcastle

Hatchard, JL, Gray, TS, Mikalsen, K and Brookfield, K (2004) 'Second stakeholder consultation' Annex 1 in OAL Paramor, CL Scott and CLJ Frid (eds) *European Fisheries Ecosystem Plan: Producing a Fisheries Ecosystem Plan*, EFEP, EU Quality of Life and Management of Living Resources: QLRT-CT-2001-01685, Newcastle upon Tyne, University of Newcastle

Hatcher, A, Jaffrey, S, Thébaud, O & Bennett, E (2000) 'Normative and social influences affecting compliance with fishery regulations' *Land Economics* 76(3):449-461

Held, D (1996) *Models of Democracy*, 2nd Edition, Cambridge, Polity Press

Hersoug, B and Rånes, SA (1997) 'What is good for fishermen, is good for the nation: co-management in the Norwegian fishing industry in the 1990s' *Ocean and Coastal Management* 35(2-3):157-172

Holden, M (1994) *The Common Fisheries Policy: Origins, Evaluation and Future*, Oxford, Fishing News Books

Jentoft, S and McCay, B (1995) 'User participation in fisheries management' *Marine Policy* 19(3):227-246

Kröger, M (2001) 'Report of working group: consultation and participation of civil society' in the *White Paper on European Governance*, European Commission

McCay, BJ and Finlayson, AC (1995) 'The political ecology of crisis and institutional change: the case of the northern cod' Paper presented to the Annual Meeting of the American Anthropological Association, Washington, DC, November 15-19

Mill, JS (1861/1991) *On Liberty and Other Essays*, Oxford and New York, Oxford University Press

Mikalsen, K and Jentoft, S (2001) 'From user-groups to stakeholders? The public interest in fisheries management' *Marine Policy* 25:281-295

Nielsen, JR (2003) 'An analytical framework for studying: compliance and legitimacy in fisheries management' *Marine Policy* 27:425-432

Noble, BF (2000) 'Institutional criteria for co-management' *Marine Policy* 24:69-77

Phillipson, J. and Crean, K (1997) 'Alternative management systems for the UK fishing industry' *Ocean and Coastal Management* 35(2-3):185-199

Piet, GJ, Daan, N, Jaworski, A, Paramor, OAL, Ragnarsson, SA, Scott, CL and Star, B (2003) *Fisheries and the Removal of Ecosystem Components*, EFEP, EU Quality of Life and Management of Living Resources: QLRT-CT-2001-01685

Pomeroy, RS and Berkes, F (1997) 'Two to tango: the role of government in fisheries co-management' *Marine Policy* 21(5):465-480

Ragnarsson, SA, Hill, L, Jaworski, A, Piet, GJ, Paramor, OAL, and Scott, CL (2003) *The North Sea Significant Web*, EFEP, EU Quality of Life and Management of Living Resources: QLRT-CT-2001-01685

Ritchie, E and Wood, D (2001) 'An analysis of the public hearing on the CFP Green Paper' *Future Governance: Lessons from Comparative Public Policy*, unpublished

Rossiter, T (2002), Unpublished research with Scottish fishers, University of Aberdeen

Sen, S and Nielsen, JR (1996) 'Fisheries co-management: a comparative analysis' *Marine Policy* 20(5):405-418

Silvert, W (ed.) (2003) *Managing Uncertainty in Ecosystem Model Dynamics and the Implications and Feasibility of Specific Management Scenarios*, EFEP, EU Quality of Life and Management of Living Resources: QLRT-CT-2001-01685

Smith, G (2003) *Deliberative Democracy and the Environment*, London, Routledge

Symes, D (1997) 'The European Community's Common Fisheries Policy' *Ocean and Coastal Management* 35(2-3):137-155

Wijnberg, NM (2000) 'Normative stakeholder theory: the link between ethics and politics' *Journal of Business Ethics* 25:329-342

World Commission on Environment and Development (1987) *Our Common Future*, Oxford and New York, Oxford University Press

CHAPTER 4
THE ROLE OF PARTNERSHIPS IN THE GOVERNANCE OF FISHERIES WITHIN THE EUROPEAN UNION

TONY HAWKINS
Chairman of the North Sea Commission Fisheries Partnership
University of Aberdeen, Department of Geography, St Mary's, Kings College, Aberdeen, AB24 3UF

Abstract

There is a lack of confidence in the ability of the European Union (EU) to solve outstanding and urgent problems, and there is criticism of its institutions and distrust of the way power is exercised. The loss of faith is especially strong over the management of fisheries. After 30 years of the Common Fisheries Policy (CFP), those fisheries are in a state of crisis. Emergency measures have been imposed following years of management failure. One of the most obvious flaws has been the failure of the Commission to involve stakeholders in shaping fisheries policy and delivering fisheries management measures. Yet people often give of their best when they are brought together to resolve problems and take decisions themselves, with experts serving as advisers and facilitators. Within the North Sea Commission Fisheries Partnership (NSCFP), fishers have recently been working together with scientists and technical experts to resolve some of the difficulties in assessing the state of the North Sea fish stocks. Soon, a new organisation – the North Sea Regional Advisory Council (NSRAC) – will be formed to take this initiative further and provide advice on fisheries management directly to the Commission. The new Council will require a significant change in working culture on the part of all those involved, and especially by the European Commission.

4.1 Introduction

Within the European Union (EU), there is a perception that government is increasingly remote from the people and from democratic structures. A recent White Paper on governance within the EU (EC 2001a:3) has pointed out that "many people are losing confidence in a poorly understood and complex system to deliver the policies that they want. The Union is often seen as remote and at the same time too intrusive". The creation of the EU has involved the joining together of countries with very different political cultures. Inevitably, the system of governance is a compromise and has been arrived at through political expediency rather than administrative efficiency or democratic principle.

Administration of the EU's affairs has been largely delegated to civil servants within the European Commission, aided by technical experts. The Commission initiates legislation, implements community policy, manages the EU's budget and is responsible for the Union's relations with outside countries. Day to day management of the Union's affairs is essentially carried out by the Commission. Yet the Commission is made up of

T. S. Gray (ed.), Participation in Fisheries Governance, 65–83.

an un-elected college of Commissioners, appointed by the governments of Member States and led by a President adopted as a result of discussions behind closed doors between the Member States. Much of the work of the Commission is intended to be directed by Councils of Ministers on which all the Member States are represented. However, deep divisions between Member States often render the Councils ineffective, especially where far-reaching strategic decisions must be agreed and implemented. There is a tendency for the Councils to argue about points of detail and issues of equity in the treatment of individual Member States. As for the European Parliament, it lacks the legal power to hold the Commission accountable for its actions, and is able only to comment on proposals and develop overall strategy. The Parliament's Fisheries Committee has shown little willingness to grapple with the complex issues of fisheries management. The close scrutiny of public affairs by elected representatives, which takes place within many of the Member States themselves, is, therefore, seldom achieved within the EU.

The Commission has identified the reform of European governance as one of its strategic objectives, and leaders of the Member States have agreed a new Constitutional Treaty for the EU (EC 2004). This Treaty reforms the EU to make its institutions more transparent, more accountable, more efficient and better able to meet the challenges of the 21st century. It also spells out that the EU is a union of nation states, and has only those powers that Governments have chosen to confer upon it. However, the Treaty is still regarded as controversial and it has yet to be ratified by Member States.

Remarkably, the conservation of marine biological resources under the Common Fisheries Policy (CFP) is one of only five areas of exclusive competence within the proposed new Constitutional Treaty for the EU (EC 2004). In this respect, the CFP stands alongside the major areas of customs union, competition rules for the internal market, monetary policy and common commercial policy. This extraordinary elevation of marine conservation reflects the complexity of fisheries management within the EU. Symes *et al* (2003) have pointed out that not only is the European coastline highly fragmented and deeply indented, with distinctive regional seas, but responsibility for management is divided amongst large numbers of coastal states

However, loss of faith in governance – by which I mean the manner of governing or regulating – is especially apparent in the field of fisheries. After 30 years of the CFP, the fisheries of the EU are in a state of crisis. Strong emergency measures, with adverse consequences for fishers, and the industries and communities which depend upon them, have been imposed in the Baltic Sea, Irish Sea and North Sea, following years of management failure. Fishing effort is progressively being curtailed and catch limits reduced, at great cost to coastal communities in some Member States; so far without significant improvements to the fish stocks. At the same time, the fishing fleets of other Member States continue to receive subsidies which enable them to expand their fishing power. In these circumstances, the whole system of governance is in question, together with the competence to govern of the Commission, the Council of Ministers and their advisers.

The CFP and its current administration exemplify many of the wider deficiencies in governance that exist within the EU. The Policy itself lacks clear strategic objectives, contains many contradictions, and it is administered through a centralised command-

and-control system, where decisions are made at the top, by the Commission and Council of Ministers, and then imposed, not always effectively, upon the fishing industry. The system of governance is a hierarchical one. The measures adopted are devised behind closed doors by the Commission, aided by a narrow selection of technical experts, the majority of whom work for the governments of Member States. The decisions themselves are often draconian, with major effects upon coastal communities.

For any decision-taking body to produce sensitive and responsible decisions, it is necessary that it should be accountable to those affected. The Commission, in managing the fisheries of the European Commission, is accountable neither to voters nor to the fishing communities. Fishermen, fish buyers and sellers, boat builders and the institutions which support them play no part in management, although they have to live with the consequences of the decisions taken.

This chapter considers, first, the inherent flaws in the governance of fisheries within the EU, focusing particularly on the narrow basis of its scientific advice, and the top-down nature of its governing structure. Second, the chapter explains an attempt by stakeholders to address these flaws through the formation of a fisheries partnership – the North Sea Commission Fisheries Partnership – which is currently engaging with the European Commission and concerned Member States to establish a Regional Advisory Council for the North Sea, to provide advice on the management of fisheries. The chapter's conclusion is that, in the future, such Councils may facilitate greater participation in the governance of fisheries by those most affected.

4.2 The contradictions of the Common Fisheries Policy

The Commission has admitted the failure of the EU's policies on fisheries, most notably in its Green Paper on the Future of the CFP (EC 2001b). The Green Paper discusses the weaknesses and challenges of the Policy. It points out that the CFP has not delivered sustainable exploitation of fisheries resources and will need to be changed if it is to do so. The shortcomings of the CFP can be expressed in conservation, economic and political terms.

So far as conservation is concerned, the Green Paper stresses that many stocks are outside safe biological limits. They are too heavily exploited or have low quantities of mature fish, or both. The situation is particularly serious for demersal fish stocks such as cod, hake and whiting, and if current trends continue, many stocks will collapse. At the same time, the available fishing capacity of the Community fleets far exceeds that required to harvest fish in a sustainable manner. In economic terms, the fisheries sector is characterised by fragility resulting from over-investment, rapidly rising costs and a shrinking resource base. Politically, stakeholders do not feel sufficiently involved in the management of the policy and many believe that there is no level-playing field in terms of compliance and enforcement.

There are also inherent contradictions within the CFP. Its main objectives, derived mainly from the Common Agricultural Policy, are to:

- Increase productivity;

- Ensure a fair standard of living for fishermen;
- Stabilise markets;
- Assure the availability of supplies;
- Supply consumers at reasonable prices;
- Ensure there is no discrimination across Member States;
- Protect the environment, according to the precautionary approach;
- Provide for rational and responsible exploitation on a sustainable basis.

However, in a number of respects these objectives are contradictory. They aim to:

- Conserve fish stocks, but they promote fishing activities;
- Modernise the fleet, but they limit fishing effort;
- Implement measures centrally, but they require Member States to enforce them locally;
- Maintain employment, but they reduce fleet capacity;
- Ensure a good income for fishermen, but the supply of fish is declining;
- Prohibit discrimination, but they discriminate on access to fishing.

Although the Commission has diagnosed the deficiencies of the CFP in its Green Paper, significant reform has yet to be achieved. The Commission's 'Roadmap' on the reform of the CFP, published in 2002 (EC 2002a) following a period of consultation on the Green Paper, emphasises the need for stronger conservation measures and controls. It preaches the need for better governance, including the principles of openness, participation, accountability, effectiveness and coherence. However, so far, the Commission and Council of Ministers have done little to secure the necessary improvements.

4.3 The small cadre of expert advisors

At the heart of fisheries management is the need for expert advice. Scientific advice is required on the state of fish stocks and the impact of fishing upon them. In addition, ecological advice is needed on the state of marine ecosystems, the interactions between fish stocks and other key species, and the impact of fishing upon non-target species and the wider environment. The advice required is not limited to scientific questions, however. It is also important for fisheries managers to have advice available on the appropriateness and practicality of management measures, and, in addition, on the economic and social impact of those measures upon fishers and fishing communities.

In the management of the fisheries of the Community, advice comes from a relatively small cadre of experts. These are mainly biological scientists employed by governments within national fisheries laboratories. In addition to providing advice directly to the Member States, these scientists also come together within the International Council for the Exploration of the Sea (ICES), an old-established body which is contracted to provide advice to the European Commission. They also meet within the Scientific, Technical and Economic Committee on Fisheries (STECF), a body set up by the Commission itself to provide advice on scientific, technical and economic issues.

The pre-eminence of government scientists in providing advice on fisheries management, and the virtual exclusion of independent scientists and those with practical experience of fishing, may have originated in the requirement for expensive facilities and official backing to carry out fisheries science. Landing statistics and details of the fishing fleet must be obtained from the different ports in each country. Sampling of fish landings is necessary to determine the length, age and condition of the fish. Voyages on commercial fishing vessels by scientific observers must be carried out to monitor the fish that are caught (many of which may be discarded). Research vessels must conduct surveys of the abundance of fish. Most of these activities can only be carried out within large, well-equipped fisheries research institutes. Such facilities have been established within all the littoral Member States and are funded directly by their governments.

There is also the possibility that Member States prefer scientific advice to come from organisations under their control. The position of government scientists has always been ambiguous. As individuals, they may strive to be objective and independent, but the work that they engage in, the funding for their research and their ability to speak out on controversial issues, is constrained by the fact that they are government employees who cannot be seen to contradict their political masters.

Because fish resources are shared between Member States, and the fish themselves are often mobile, it is necessary to bring together results of research and monitoring from many countries in order to assess the state of fish stocks properly. ICES, which has had its headquarters in Copenhagen since 1902, long pre-dates the formation of the EU. ICES serves both as a repository of data and the organiser of working groups which carry out fish stock assessments, and it is the body responsible for providing scientific advice to the Commission and adjacent non-EU states like Norway, Iceland and the Faeroe Islands, through its Advisory Committee on Fishery Management (ACFM). It also serves as a forum in which scientists can discuss and debate the main issues of fisheries science, through its many study groups and symposia, its journal (the *ICES Journal of Marine Science*) and its Annual Science Conference. ICES operates in a wider context than fisheries science. It co-ordinates and promotes marine research across the whole of the North Atlantic, and acts as the prime source of advice on the marine ecosystem to the governments and international regulatory bodies within the North Atlantic Ocean and adjacent seas.

ICES has carried out a valuable role in providing independent scientific advice on fisheries for over a century. However, it has been slow to adapt to changing conditions and to the heavy demands placed upon it. It is symptomatic of the problems within ICES that the organisation shows a strong interest in its own history (Anderson 2002), a liking for ceremony and an aversion to change. The current wider debate within Europe on the role of experts, and the need for openness and transparency in rendering advice, has passed ICES by. There is an urgent need for the reform of the ICES structure and procedures. The skills required to handle the complex fish stock assessment models are in short supply. The scientific data on which fish stock assessments are based are largely incomplete and are inaccessible to independent experts. The scientific working groups which carry out assessments have generally remained closed both to independent observers and to stakeholders affected by the assessments. ACFM itself, which assembles the scientific advice and forwards this to the Commission, has been completely closed. ICES has underestimated in the past the capacity of fishers to participate in the collection of data on the fish stocks, and has neglected the wishes of fishers to understand and question the advice provided.

The manner in which ICES operates, with government scientists from different Member States expected to play an independent role, does provide it with some protection from political pressures. However, this does not necessarily guarantee the independence of the scientific advice offered up. The absence of any peer review and the lack of openness in its activities mean that internal pressures acting against independence can remain hidden and protected from scrutiny. The precautionary and limit reference points for the fish stocks, established under the precautionary approach adopted by ICES, and which greatly influence management advice, are set internally and are not subjected to wider discussion. There is little participation in the process by fishery managers, let alone fishers. Indeed, the participation of non-scientists is regarded within ICES as likely to compromise the integrity of the organisation.

Even more seriously, ICES is unable to provide advice on economic and social aspects of fisheries. It is constituted in such a way that it cannot look at the consequences of its advice, or weigh the benefits of conserving fish stocks against the impact of the conservation measures upon fishers and fishing communities. Indeed, this restriction is often seen as one of ICES' strengths, rather than a weakness. Moreover, ICES lacks expertise on the practicalities of applying management measures, whether these involve adjustments to fishing gears, reductions in fishing capacity, or restrictions in catches. Many of the measures introduced as a result of ICES advice have proved difficult to implement in practice, and have underestimated the ingenuity of fishers to circumvent them.

There is another source of expert advice to the Commission. The STECF is the European Commission's own scientific, technical and economic advisory body on fisheries. The committee is made up of experts chosen by the Commission from those nominated by Member States. It is composed of biologists, economists and fishing gear technologists. The STECF is asked by the Commission to examine the advice it receives from ICES and to comment on that advice. In addition, the Commission often convenes STECF sub-groups or independent expert groups to address specific issues. The deliberations of STECF are not carried out in public and cannot be subjected to query or debate before advice is provided to the Commission. Its membership tends to be drawn from the same institutes represented within the ICES committees, which limits its independence and its ability to peer review the ICES advice. Technical experts from the fishing and fish processing industries are absent from the committee. The Commission has been slow to encourage STECF to develop and extend its economic expertise. Social scientists do not play any role. As a consequence, it has not been possible for STECF to provide full and independent advice to the Commission on the validity of the scientific advice, the practicality of the management measures proposed, or the likely effects of those measures upon the fishing industry, associated industries, and fishing communities.

The systems for providing expert advice, which include not only ICES but also the national fisheries institutes and the Commission's own STECF, thus essentially involve the same narrowly-based group of mainly biological scientists with its own culture and attitude. The advice is centred on the assessment of fish stocks. More broadly-based advice on the state of the ecosystem, and the effects of fishing upon it, is largely lacking. Little use is made of the expertise of fishers in deciding upon management measures. Moreover, very little attention is paid to the provision of economic and social advice. It is important in managing fisheries to look at the costs, as well as the benefits of management measures. Those costs must include weighing up the adverse effects of any action, both in advance of the measure being adopted, and after the event. That has not been the practice of the Commission or of the Council of Ministers, which directs the Commission.

4.4 The narrow basis of the scientific advice

The main advice provided to fisheries managers is in the form of assessments of the state of individual fish stocks in different areas. These assessments involve examination of the catches, and of the changes that take place in the age composition of fish over time. The data are slow to be collated and analysed and the assessments require long data series. The assessments are inherently long term, and their ability to project the future state of fish stocks is limited. Although there is a progressive increase of confidence in the assessments after years have elapsed, there is always uncertainty about the current and future state of the stocks. This uncertainty poses particular problems in assessing the effects of management measures. Currently, the effects of major changes to the management regime cannot be assessed until three or four years have elapsed. Often, new measures are introduced before any assessment has been made of the efficacy of previous measures.

Stock assessments would benefit greatly from improvements in the capacity for collecting data from the fishery and analysing it immediately. The assessments would also be improved by the incorporation of up-to-date information from fishers themselves on their recent catches and on changes in the behaviour and distribution of the fishing fleets. There is a need for ICES to consider new paradigms for the more rapid and up-to-date assessment of fish stocks. New and independent assessments made using different methods would help to validate those obtained by the current methods. They would also enable more prompt evaluation of the efficacy and impact of management measures.

Currently, there is also a lack of attention paid to multi-species factors. Each fish stock is evaluated separately, although it is recognised that different species interact with one another in a complex way. For example, cod are predators of herring, sandeels and Norway lobsters. Conversely, sandeels are eaten by a wide range of other fish species. Changes in one fish stock do affect other stocks. Moreover, fish are part of a wider ecosystem containing predators (like seabirds and sea mammals) and also providing prey as food for fish. The wider ecosystem may be affected through changes to the fish stocks. For example, seabirds may be deprived of food through heavy fishing on their prey, and charismatic organisms may be affected directly by fishing activities, such as the destruction of deep-sea corals by trawls, or the incidental capture of cetaceans in drift nets. The growth of predator populations such as seals, may have an effect upon fish stocks. It is now widely accepted that there must be a move towards an ecosystem-based approach to fisheries management, but there has been a paucity of ideas on how this should be achieved, and little progress has been made towards achieving it.

Fish stocks are also affected by changes in the environment, although these changes are not taken into account in the assessments. The latter assume that the biological reference points for the stocks remain stable with time. Recent changes in ocean climate, and observed changes in species composition within an area like the North Sea, have drawn attention to the effects of environmental change upon fish stocks and the need to take this into account in setting biological reference points. There is little point in setting, say, a target for spawning stock biomass based on a past level, which is unlikely to be reached under the prevailing climatic conditions.

An array of management measures has been imposed upon the fisheries within the North Sea in recent years. Chief amongst them have been controls on fishing outputs, including catch limits and minimum landing sizes. Various technical measures have been introduced to improve the selectivity of fishing gears for particular species and sizes of fish. These have proved difficult to apply and enforce in the mixed fisheries of the North Sea where several species of differing size are being caught simultaneously. Discarding has been prevalent, where fish are caught and brought on to the deck but then thrown back into the sea without being declared as landings. This practice, which is seen by fishers as wasteful, has undermined support for the management measures. The imposition of closed areas and closed seasons has received some attention, but has not been applied consistently or efficaciously. Controls on fishing effort have been introduced, limiting the numbers of vessels, their size and fishing power. Most recently the numbers of days-at-sea have been regulated. There have been problems in enforcing all these management measures, and different Member States have differed in the attention they have paid to enforcement. One result of this lack of control has been that landings data cannot be trusted and do not provide a sound basis for the stock assessments carried out by scientists.

It is in the area of selecting, operating and enforcing management measures that advice from fishers would be especially valuable. Fishers are aware of the deficiencies in the different measures and know how the imposed controls may be circumvented. Fishers are themselves responsible for the large discrepancies between actual and recorded landings, and their estimates of the magnitude of these differences are needed. Before their aid can be enlisted in assessing the stocks more effectively, however, fishers need to be persuaded that there is a level playing field, and that their views are being listened to. They need to be full and trusted participants in the management of the fisheries, alongside other stakeholders.

There is growing distrust by fishers both of the stock assessments and the scientific advice given to management. This distrust may arise in part from the different perspectives of fishers and scientists on the abundance of fish and on the factors that control the abundance of fish. Fishers are very aware of natural fluctuations in the abundance and distribution of fish, although their own knowledge is local and personal. They distrust the simple models employed by scientists, and question the validity of the data collected. They tend to resist any suggestions that fish stocks are over-exploited and are reluctant to accede to further restrictions upon their abilities to fish. Fishers maintain that for some species and areas, the information available to them is better than the information available to scientists. The lack of any means for incorporating fishers' knowledge and experience into the traditional scientific advice is certainly a major disadvantage. Moreover, the unwillingness of scientists and fishery managers to address these concerns of fishers presents a serious barrier to future progress.

4.5 Dealing with uncertainty

There is often great uncertainty in assessing the state of fish stocks. Some key species like the monkfish are hardly assessed at all, and provisional or precautionary catch limits are set, often based on previous landings. The data on which the assessments of even the most important species are based are often known to be inaccurate because of undeclared and misreported landings. Moreover, scientists admit that their knowledge of the dynamics of fish populations is often very limited.

The role of uncertainty is now recognised as an important one in decision-taking. Increasingly, both experts and politicians have been persuaded to adopt a 'precautionary approach', especially with respect to environmental risks. There are differences of opinion over the exact meaning of the term 'precautionary approach', but essentially it proposes that potential risks should be dealt with by the imposition of constraints, even in the absence of scientific certainty. This approach is regarded by its proponents as a more sensitive and cautious way of dealing with potential threats than waiting for hard scientific proof as a prerequisite for decision taking. The problem is, of course, that it may lead to a decision being taken which is risk averse and which may subsequently prove to be inappropriate. There are already pressures, acting within ICES, which tend to make the scientific advice cautious and conservative. ICES and the scientists who participate within its working groups and committees will be severely criticised if fish stocks fail and they may be under pressure to place restrictions upon fishing.

It is, therefore, important that the precautionary approach should retain a dimension of reversibility, as the advance of knowledge could show a particular decision to be no longer justified. The action taken should be commensurate with the risk, and there should be consistency in the way the approach is applied. It also has to be recognised that evidence on a particular issue can be gathered from many sources. It is not only the evidence of scientists that is relevant. An uncertain scientific assessment may be validated by seeking information from other sources, including fishers. There is a need for more dialogue between scientists, politicians and the public over the precautionary approach and how it is applied.

4.6 The role of the European Commission

The European Commission plays a key role in both managing the fisheries of the EU and deciding upon future fisheries policy. It has a small number of experts of its own but, by and large, it is composed of non-specialist civil servants, either seeking a career within the Commission, or seconded from Member States. Its decisions are meant to be taken to meet the requirements of the Council of Ministers but it is often called upon to undertake management on a short time scale. Dissent within the Fisheries Council and an inability of Member States to agree on strategic issues, often gives the Commission control over decisions.

The Commission and Council of Ministers take advice from ACFM and STECF, but they have also established an Advisory Committee on Fisheries and Aquaculture (ACFA), made up of stakeholders, to comment upon proposals. The Commission tends to operate by producing a consultation paper or draft recommendation, fully worked-out, and then asks for comments from ACFA, often at short notice. ACFA itself is very large, and on it sit stakeholders from all the Member States, representing different interests and holding very different views. It is not a body that can reach a consensus on any issue, or participate in decision taking. All it can do is ensure that a diversity of comment is presented to the Commission on its proposals.

As we have seen, the top-down style of management exercised by the Commission has failed to meet important biological, economic and political objectives for the governance of fisheries. It is time to consider more open ways of proceeding. I believe that openness and accountability should be the guiding principle of all management, but especially management in a crisis. Stakeholders and the general public must be told in advance of any risks and of any steps needed to resolve the crisis. The Commission must take account of the views of an informed public. Stakeholders' concerns must be dealt with if management decisions and measures are to be complied with. Those concerns are exacerbated within the EU by the complexity of decision-making, which takes place between different countries with different interests. There is a fear by stakeholders that decisions will be taken as a result of compromise between Member States, or by trading one interest off against another within the Fisheries Council, rather than through a fair and objective assessment of the issues. Some countries place fisheries high on their list of priority interests, while others are more inclined to give way on fisheries issues in the expectation that they will gain in some other area. The danger is that management decisions may be taken which are perceived to be unfair by fishers. These decisions may then be difficult to implement. The suspicion of Commission actions, which has developed amongst fishers, has had severe consequences. Many of the management measures imposed have been subverted by fishers, while policing is uneven and not always able to prevent breaches of the regulations. Fish may be landed illegally; catches may be attributed to areas other than those where they were taken; and fishing gear regulations may be avoided. One unfortunate result of these practices is that the data available to scientists are then unreliable and unable to support the need for objective advice.

4.7 The need for wider participation

There is widespread and growing public distrust of experts and the role they play in public affairs. This distrust has been fuelled, especially within the United Kingdom, by some pronounced deficiencies in scientific advice, such as in the control of animal diseases, and the preparation of forensic evidence assessing the impact of biotechnology and promoting human health. Expertise in a narrow field of science does not necessarily convey an ability to take decisions in a wider context, especially where social and economic factors are important. As Mary Midgley (2001) has pointed out, the learned are often importantly foolish. There is also concern about the way expert advice is used to support political decisions. There is suspicion that political decisions come first, and that scientific advice from government employees is subsequently used to justify those decisions. It is, therefore, important in all areas of governance that expert advice is presented in an open and transparent way. The information on which decisions are based must be available to others, and any uncertainties must be pointed out. Peer-review of the science is essential, and expert advice must be subjected to close scrutiny and critical assessment by others. Even if the scientific advice is valid, other wider considerations have to be taken into account; scientific recommendations have to be balanced against other factors in deciding public issues. Practical aspects of implementation and economic and social factors are also important.

Currently, the Commission is accountable only to itself. There must be changes in the way the fisheries are managed. There is a particular need to change the way that advice on the fish stocks and management of the fisheries is provided. There must be wider participation in the process of assembling and then producing expert advice. Participation by outside experts, by stakeholders and by the public, should not be considered as unnecessary or inconvenient but rather as a way of extending and enriching the process of gathering information. Fishers and other stakeholders now want to know how decisions are taken. They want to be able to pose questions, hear answers directly, and present their own views, rather than simply accept advice given from behind closed doors.

In addressing the problems of fisheries governance, it is evident that the provision of advice to the Commission should not be confined to a narrow range of experts. New institutional arrangements are needed which will ensure a multidisciplinary approach; greater participation by those affected by Commission decisions; greater independence of the decisions; and less dominance by the bureaucrats responsible for imposing regulation. Assessment of potential threats should take account of all costs and benefits – direct and indirect, social and economic. Ignoring these other factors inevitably results in poor decision-taking.

Public inputs to policy debates are not merely 'opinions', but provide other forms of knowledge and experience, and other values, as well as raising questions that scientists and bureaucrats have neglected. There needs to be a long-term process of mutual learning between the public, important stakeholders, and those who formulate expert advice. This interaction will necessarily involve new institutional relationships and new, more participatory, forms of governance. People often give of their best when they are brought together to solve problems and take decisions themselves, with experts involved in a subsidiary position as advisers and facilitators. The principle of a partnership is to share risks and benefits. There is a need for such partnerships for the governance of European fisheries. I suggest that such partnerships would be able to promote sustainable fisheries through a common vision and consensus, with fishers and other stakeholders fully involved in policy-making, decision taking and operational management. This is the thinking behind the founding of the North Sea Commission Fisheries Partnership.

4.8 The North Sea Commission Fisheries Partnership

A conference at Haddo House Aberdeenshire in 1998, considered how best to deal with the differences of opinion which arose between fishers and scientists over the state of fish stocks. The main outcome was a proposal that a standing forum should be established for fishers, scientists and others to develop and implement improvements in the assessment of fish and shellfish stocks and in the regulation of fisheries. The forum would promote more sustainable management of fisheries by bringing together those with the strongest interest in fishing.

A partnership between fishers and scientists was established as a pilot project in 2000 by Aberdeenshire Council and the North Sea Commission. The North Sea Commission is an alliance of regional and local governments from around the North Sea. It includes Norway, as well as regions from the Member States of the EU that are heavily dependent upon fishing and fish processing. Funding was initially provided by the European Community PESCA Initiative. Subsequently, the North Sea Commission Fisheries Partnership (NSCFP) successfully applied to the Interreg IIIb North Sea Programme (a European Regional Development Fund) for funding to develop its work for a further three years.

The Partnership includes representatives of fishers and scientists from eight countries around the North Sea. Fishers are drawn from the national organisations representing fishers. Scientists are drawn from the main government fisheries research institutes. The objectives of the Partnership are to:

- Improve the exchange of views between fishers and fishery scientists;
- Promote dialogue between fishers from different countries around the North Sea;
- Further develop scientific advice on the state of fish stocks, making better use of information held by fishers;
- Evaluate the costs and benefits of different fisheries management measures and develop consensus views on these;
- Progressively involve other stakeholders, including fishery managers, fish merchants, processors and conservation groups in discussions of the management of the North Sea fisheries;
- Work for the implementation of new and more sustainable management measures, taking full account of biological, economic and social factors.

This Partnership has already gone a long way towards developing greater trust between fishers and scientists. Some of the initial difficulties in communicating have already been overcome. It has proved possible for the Partnership to discuss sensitive and controversial subjects, without the trust that is developing between the partners being breached. It has also achieved improvements in the process of assessing North Sea fish stocks.

The Partnership has made particular progress in reforming the system for obtaining and presenting scientific advice. In 2000 and 2001, ICES scientists made the results of their preliminary assessments available to the Partnership at an early stage, allowing comments from fishers to be fed back to scientists. In 2002, 2003 and 2004 this process was extended by the Partnership employing independent scientists from outside Europe to review the ICES Working Group assessments of demersal stocks in the North Sea. Discussions took place between the Chairman of the ICES Working Group, the independent scientists and members of the Partnership. Though these discussions did not lead to a consensus on the state of the fish stocks, they did lead to greater understanding of the problems of assessing fish stocks by all parties. Fishers were able to provide information on the way the fisheries were prosecuted, and in the case of one species, this resulted in a better understanding by scientists of the problems of assessing saithe, with adjustments being made to the assessment.

This process of reform has not yet gone far enough. In 2002 and 2003, the Partnership sought permission to attend, as observers, the meeting of ACFM, which produces the definitive advice on the fish stocks. Subsequently, at the ICES statutory meeting, some ICES delegates remained firmly opposed to the introduction of further transparency in the preparation of advice on the stocks. It was argued in support of the exclusion, that fishers would attempt to influence the assessments unfairly, and that fishers' representatives would be unable to keep silent about the conclusions of ACFM until the duly appointed date for releasing them. The European Commission and other regional management bodies advised by ICES have supported these arguments against the inclusion of fishers as observers. However, the Partnership has continued to seek observer status on ACFM, and in 2004 the ICES Secretariat permitted an observer from the Partnership to sit in on part of the ACFM deliberations. The exclusion of fishers from ACFM in the past has generated suspicion and distrust over the way the ACFM advice is arrived at. The future may well see a different approach.

Cooperation between fishers and scientists has also resulted in the development of an annual survey of fishers' views on the state of fish stocks in different parts of the North Sea. Since 2001, surveys have been jointly organised between the Partnership and Europêche, an organisation representing fishers' representative bodies. Questionnaires are sent out to fishers seeking their views on the state of the main North Sea stocks, and the results are then collated and presented to the appropriate ICES Working Group, to assist with preparation of the subsequent scientific advice.

In 2002, 2003 and 2004, the Partnership organised a joint Study Group on Fishers' Information with ICES. Discussions between fishers and scientists within the Study Group have been wide ranging, but several central themes and issues have emerged. Of particular importance and interest to all was agreement on the need to reform the management structure operating within the European Community. The fishing industry wished to be more closely involved in the decision-making process. The scientists wanted better and more accurate data, for example on actual catches, for stock assessments. It was not explicit in the discussions, but it became clear that it will only be possible to collect accurate landings data when trust and cooperation between fishers and scientists has improved. It has been emphasised that this will take time and good will on both sides. Positive action is required to take forward initiatives that bring fishers and scientists in contact and provide incentives for them to work constructively together. Common projects, and the funding to underpin these, were considered a good means of building up trust and understanding.

The Study Group recommended the development by ICES of generic documents, describing the principles of stock assessment, the methods for formulating management advice and the development of recommendations on TACs. It called for non-technical summaries of the state of each stock at the beginning of each stock assessment report and in each section of the ACFM report. It suggested that sea-going scientists should be fully briefed on the outcomes of recent stock assessments and should communicate more effectively with fishers. It recommended that national meetings should be held between fisheries assessment scientists and the fishing industry to discuss the stock assessments. Data should be collected in collaboration with the fishing industry and joint strategies should be developed for improving data quality. ICES should use the most up-to-date information on catch and landings in assessments and stock projections. Working Groups should explicitly test alternative assessment models and modify their existing models. Information on major changes in fishing patterns should be provided by fishers for discussion with scientists before the stock assessments. Regional and/or stock specific project groups should be set up jointly by scientists and fishers to assist the stock assessments and improve their transparency. Greater use should be made of commercial Catch-per-Unit-Effort (CPUE) data in the assessments, and scientists and fishers should work together to evaluate survey strategy and design, and develop working procedures for joint abundance surveys, involving both fishing vessels and research vessels. Finally, the results of the North Sea Fishers' Survey should be more fully evaluated by scientists, and the results should be compared with those from other sources.

Fishers' organisations and fisheries research institutes from every country around the North Sea have committed themselves fully to the Partnership. Both ICES and the European Commission have sent representatives to meetings and have given their support. Interest in the work of the Partnership has been expressed by the Fisheries Committee of the European Parliament. The Ministerial Declaration of the Fifth International Conference on the Protection of the North Sea draws attention to the value of the work done by the Partnership.

In looking at the progress made by the Partnership, and the difficulties it has experienced, it is possible to identify some of the factors that have ensured its success. The partners themselves have had a common picture of the purpose of the Partnership, and how its aims can best be achieved. The individual partners have worked well together, and have developed a degree of mutual trust. Such trust can be especially difficult to sustain as both scientists and fishers have to defend their particular positions in other fora, and cannot always resist the temptation to criticise other parties outside Partnership meetings. There are problems with ensuring continuity of membership, and problems over the language used. Much work is sometimes necessary outside the formal meetings by the co-coordinators and facilitators to ensure that all views are taken into account. Funding opportunities have to be seized, and a great deal of administrative work and entrepreneurial skills exercised outside the actual Partnership meetings. However, realisation by all the parties involved that it is important to make progress, and to achieve consensus views on how to manage the North Sea fisheries, has acted as a stimulus for the work of the Partnership.

4.9 Regional Advisory Councils (RACs)

From the very beginning, discussions between scientists and fishers within the NSCFP focused on the suggestion that a fisheries management council should be established for the North Sea to provide a wide group of stakeholders with the opportunity to participate in governance of the fisheries. This idea was not new, but followed a suggestion in the Green Paper on reform of the CFP (EC 2001b) that there was a need for greater stakeholder participation in fisheries management.

In its 'Roadmap' on the reform of the CFP (EC 2002a) the Commission indicated its wish to make decision-taking more effective and participative, by the establishment of Regional Advisory Councils for fisheries management (RACs). This initiative was subsequently taken forward in a new Regulation on the conservation and sustainable exploitation of fisheries resources under the CFP (EC 2002b). RACs are intended to ensure greater stakeholder involvement at the regional and local level. Interestingly, rather than the RACs consisting of appointed individuals reflecting different national and sectoral interests, membership is based upon organisations representing the fisheries sector and other interests. In this respect, the RACs will differ substantially from other regional management bodies

RACs will be able to:

- Submit suggestions, of their own accord or at the request of the Commission or a Member State, on matters relating to fisheries management to the Commission or the Member States concerned;
- Give an opinion on Commission or Member State proposals on conservation and management dealing with a fishery relevant for the region concerned;
- Comment on and recommend improvements in the implementation of the Community legislation in the region concerned;
- Conduct any other activities which are necessary to fulfil their functions.

Although the Commission and Member States will not be bound by the recommendations of the RACs, they will have to explain how they dealt with the opinions provided. It is very unlikely that the Commission would ignore a consensus opinion given by a competent RAC.

The procedures for establishing a RAC have been set out within a Decision from the Fisheries Council (EC 2003). A group of interested stakeholders may submit a proposal for a RAC to the concerned Member States, who may agree the proposal and then pass it to the Commission for implementation and funding. The North Sea Commission Fisheries Partnership has taken the lead in formulating a RAC proposal for the North Sea. The Partnership established a RAC Development Working Group, which included a wider range of stakeholders, including environmental organisations. Advice was sought from administrators from the Member States. The Partnership then produced a draft prospectus and rules of procedure for a North Sea RAC and forwarded this to the Commission and to the North Sea Member States. Progress has been rapid. Member States and the Commission have accepted the proposal prepared by the Partnership, and the North Sea RAC will shortly begin its work.

The specification for the RACs set out in the Council Decision has not entirely met with the approval of the potential participants. The role of the RACs is purely advisory, whereas fishers would like them to play a direct role in fisheries management. The design of the RACs laid down by the Commission is unduly restrictive, with the stakeholders carefully defined and even the number of participants tightly specified. Choosing the membership of the RAC is the responsibility of Member States, rather than the RAC itself. Moreover, only limited funding has been allocated, and that funding is to be steadily reduced through the life of the RAC. There is a fear amongst fishers that too many environmental and other secondary stakeholders have been given membership of the RACs. On the other hand, environmental bodies do not like the fisheries sector being awarded two-thirds of the seats on the RAC. Nevertheless, the RACs represent an important step forward in involving stakeholders in providing advice on fisheries management.

The RACs also represent a first step towards introducing a regional element to the management of the fisheries of the EU. Previously, the Commission, greatly influenced by those Member States that benefit from the principle of open access embedded in the CFP, has resisted the introduction of regionally based fisheries management. Countries like Spain have been suspicious that regional management might exclude its wide-ranging fishing fleet from attaining effective rights to fish in areas like the North Sea. There are still sensitivities on this issue, and the definition of 'concerned Member States' for participation in a RAC is carefully phrased as 'a Member State having a fishing interest in the area or fisheries covered by a RAC'. This broad definition allows virtually any littoral Member State to participate in any of the RACs, thus partially undermining the regional element.

It is proposed by the Commission that a RAC should be established for each of the following areas:

- Baltic Sea;
- Mediterranean Sea;
- North Sea;
- North Western waters;
- South Western waters;
- Pelagic stocks;
- High seas/long distance fleet.

There is little interest in establishing a Mediterranean RAC, because of the preponderance of countries within that area which are not members of the EU and therefore ineligible to be members of a RAC.

Each RAC will consist of a large General Assembly, which will meet once per year, and a smaller Executive Committee, which will produce advice on behalf of the RAC. It is not yet clear how well each RAC will function. Large numbers of interested parties will be represented within the RACs, and there is the possibility that these will continue to act independently, resulting in a diversity of opinions rather than a strong consensus emerging. It will be important for the RACs to assume some of the characteristics of a partnership, with the participants assuming joint responsibility for the advice that emerges. If such a partnership can be achieved, then the RACs may make it possible to go beyond the narrow boundaries of the existing bureaucracy to establish new management bodies, which are both more inclusive and more effective than existing bodies.

4.10 Further devolution of management powers

Inevitably, there will be opposition within the EU to any further handing down of management powers and responsibilities to stakeholders, who may be seen by some as the unelected representatives of special interest groups. Opponents will argue that members of fishing communities are already adequately represented within the EU through their democratically elected representatives, either within their Member States or through the European Parliament. Further devolution would surely require evidence of a democratic deficit. I would argue that there is such a deficit. First, elected representatives on the European Parliament play no direct role in the management of fisheries. Neither the Parliament nor its Fisheries Committee are able to initiate legislation or to introduce fisheries management measures. Their role is restricted to commenting on proposals coming from the Commission and Council. Second, the Commission, which in practice has proved to be the most influential force in fisheries management, is not elected and is responsible only to a college of unelected Commissioners. Although, through the Council of Ministers, it is possible for the Ministers of Member States to influence decisions, the process is one of haggling and trade-offs. The key meetings of the Council are accompanied by furious lobbying by fishers and other interests, a process that is both expensive and unseemly. There is indeed a democratic deficit.

It is now important to widen consultation and to take stakeholders' views into account in many areas of decision-taking by the EU. Stakeholders cannot take over the roles of politicians and civil servants, but they do have an important contribution to make within a more participatory system of governance. In the area of fisheries management, there is a strong case for allowing stakeholders to participate directly in decision-taking. At present, the Commission and Council of Ministers are required to take short and long-term management decisions, which will have immediate effects upon fishers and upon fishing communities. They are doing so informed only by the narrow and unrepresentative advice provided by ICES and STECF. Advice is lacking on the wider economic and social perspectives, and on the impact of management measures upon the operation of the fishing fleets. Those affected by the decisions are not part of the process, leading, as we have seen, to a disinclination to follow instructions. There is no sense of ownership of the decisions taken.

The Commission also lacks advice on the ecosystem implications of the fisheries management measures they adopt. Environmental scientists and ecologists are poorly represented on the scientific advisory bodies, and stakeholders with an interest in conservation and the environment are not involved in management. More widely, the onshore industries dependent upon fishing, like boat builders, net makers, fish buyers and fish processors, also need to have their say. Consumers of fish also have a valid stake in the management of the fisheries. In these circumstances, there is a clear case for involving all the stakeholders to a much greater extent, both in terms of seeking their technical help and advice, and in terms of giving them some responsibility for taking decisions.

It is not yet certain how the proposed RACs will contribute in practice to improvements in the governance of fisheries. Some stakeholders wish to see the RACs as powerful bodies with a wide mandate, able to participate directly in the regulation of the fisheries. Others believe that there will be strong constraints placed upon their role, and that they may simply provide advice on technical matters. Much will depend upon the abilities of the participants to work together.

Involving stakeholders more fully in fisheries management will bring its own problems. There may be difficulties in obtaining adequate representation of particular stakeholders. There will be inevitable difficulties in reconciling opposing views, and consideration will have to be given to ways of promoting trust and achieving consensus. New mechanisms will need to be developed for doing this: the establishment of RACs is only the first step towards achieving a more participatory form of governance.

4.11 Conclusion

The complexity of the current problems in fisheries management means that many organisations and individuals must be involved in their resolution. There is a particular role for partnerships in resolving the current difficulties. Eventually, fisheries management will have to move away from the traditional hierarchical command-and-control form of decision-taking towards a more collective approach in which tasks and information are shared openly. In particular, stakeholders must be allowed and encouraged to participate in decision-taking and to assume ownership of the system of management. This involves a significant change of working culture on the part of all those involved, including scientists and other technical experts, administrators, fishers, representatives of other industries dependent upon fishing, environmental interests and consumers.

The trend towards public questioning of management decisions should be seen as a positive development by managers and those who advise them. It will provide new opportunities, and also new forms of responsibility, which require discussion and development. Indeed, by introducing new systems for managing fisheries the Community may be developing new, more participatory forms of governance, which can be applied elsewhere in other, more important, contexts.

References

Anderson, E (ed) (2002) '100 years of science under ICES' *ICES Marine Science Symposia* 215

EC [European Community] (2001a) *European Governance*, A White Paper COM (2001) 428, Brussels, European Commission

EC [European Community] (2001b) *The Future of the Common Fisheries Policy*, A Green Paper COM (2001) 135, Brussels, European Commission

EC [European Community] (2002a) *Communication from the Commission on the Reform of the Common Fisheries Policy* ('Roadmap'), COM (2002) 181, Brussels, European Commission

EC [European Community] (2002b) Council Regulation (EC) No 2371/2002 of 20 December 2002 on 'the conservation and sustainable exploitation of fisheries resources under the Common Fisheries Policy' *Official Journal of the European Communities*, L 358/59 31.12.2002, European Commission

EC [European Community] (2003) *Proposal for a Council Decision establishing Regional Advisory Councils under the Common Fisheries Policy*, COM (2003) 607 final, Brussels, European Commission

EC [European Community] (2004) *Draft Treaty Establishing a Constitution for Europe,* Conference of the Representatives of the Governments of the Member States, CIG 86/04, Brussels, European Commission

Midgley, M (2001) *Science and Poetry,* London, Routledge

Symes, D, Steins, N and Alegret, J-L (2003) 'Experiences with fisheries co-management in Europe' in DC Wilson, JR Nielson and P Degnbol (eds) *The Fisheries Co-Management Experience,* Dordrecht, Kluwer

CHAPTER 5
REGIONALISATION OF FISHERIES GOVERNANCE: AN EMPTY VESSEL OR A CORNUCOPIA OF OPPORTUNITY?

DAVID SYMES
Department of Geography, University of Hull, HU6 7RX, UK

Abstract

In May 2004 the Council of Ministers gave its final approval to the establishment of Regional Advisory Councils (RACs) to assist the Commission in developing appropriate policies for sustainable fisheries. The extent to which RACs are successful in giving fisheries policy a stronger sense of regional direction will depend on a range of factors including their structures, terms of reference, financial support, internal relations and the extent to which the Commission is willing and able to act upon their advice. Different interpretations of the role of RACs are beginning to emerge. Are they to act as technical committees dealing specifically with detailed fisheries regulation? Or should they serve as a wider point of reference for implementing the Commission's commitment to environmental integration and an ecosystem-based approach to fisheries management under the revised Common Fisheries Policy (CFP)? The chapter concludes that RACs could prove instrumental in transforming the style of fisheries management in Europe but that the path ahead is likely to be challenging and potentially difficult.

5.1 Introduction

One of the successes of the recent reform of the Common Fisheries Policy (CFP) from a British point of view was the decision to establish Regional Advisory Councils (RACs). The proposals appear to fall somewhat short of the aspirations of the National Federation of Fishermen's Organisations (NFFO) and the Scottish Fishermen's Federation (SFF) to see RACs granted greater executive powers. Nonetheless, the hope is that RACs will be in a position to influence the Commission and Member State governments in the development of regionally sensitive management policies. Somewhat surprisingly the Commission showed little inclination to fast track their implementation. Whether the delay was the result of more pressing business, nervousness on the part of the Commission as to what kind of demon it was releasing, or a desire to see the contentious cod and hake recovery plans settled before RACs were put in place, are moot points. A more likely explanation is that the Commission had not sufficiently thought through the role and structure of RACs at the time the final reform package was being presented to the Council of Ministers towards the end of 2002.

Late in 2003 the Commission published its detailed proposals in the form of a draft Regulation (EC 2003).[1] These revealed something of the intended scale, constitution, content and funding of RACs, but little about how the regional advice would be handled within the Brussels policy apparatus. Not until these essential parameters become

[1]At the time of writing, the Regulation setting out the details of the structure of the RACs had not been published; it is understood that it will follow the draft proposals closely.

T. S. Gray (ed.), Participation in Fisheries Governance, 85–102.

clearer - through practice rather than design - will the true purpose of RACs be evident. Nor will the question be answered as to whether the fishing industry has been duped into helping to prop up a failing system of management or handed a genuine opportunity to transform the process, content and outcomes of policy making. Much will depend on the extent to which RACs see their role essentially as technical committees reacting to policy proposals generated by the Commission, or regard themselves as empowered to take the initiative through 'own accord' recommendations presented to the Commission.

In itself, regionalisation changes very little beyond adding a new and potentially awkward feature to the institutional architecture of management. It need not imply any significant shift in the direction of management. On the other hand, it can serve both as a means of fine tuning policy to suit the particular conditions of the regional seas which make up the EU's 'common pond' and as a vehicle for developing a more integrated form of management based around an ecosystem approach which uses our expanding knowledge of ecosystem functioning to achieve the shared goals of sustainable fisheries and healthy marine ecosystems.

The aim of this chapter is to explore some of the tensions, risks and opportunities implicit in developing a regional perspective in the context of a reformed CFP. It begins by examining the role of regionalisation in policy implementation and the ways in which RACs may help in the delivery of the 'new' CFP, before looking more closely at possible undisclosed agendas for RACs and finally focusing on what could become a key issue, namely the relationship between regionalisation and environmental integration.

5.2 Regionalisation and the CFP

In its simplest terms the aim of a regionalised approach to fisheries management is to escape the imposition of a 'one size fits all' straightjacket, commonly associated with centralised direction of policy, by allowing for differentiation and diversification of the ways in which the living resources of the sea are managed. The CFP is no stranger to a regionalised approach. After all, the current regulatory system involving total allowable catches (TACs) and quotas is predicated on the management of spatially defined pressure stocks. But more strikingly the recovery plans currently being developed for cod stocks in the North Sea, West of Scotland and Irish Sea, considered by some to be the precursors to RACs, are prime examples of a spatially discrete management approach though not, it should be added, of integrated regional management. It is not yet clear how far the European Commission intends to further the cause of regionalisation through the setting up of RACs.

The problems of regionalisation focus largely on interrelated questions of scale, defining characteristics and the delineation of boundaries. Where the underlying purpose of regionalisation is to assist the implementation of centrally determined policy – as in the case of the CFP – the preference will be for fewer but larger regional units in order to reduce the disparities in the application of policy. In practice, the Commission has opted for predetermined 'regions' through the adoption of ICES areas used in stock assessment and the calculation of TACs and quotas. Such geometrical designs can only offer a very crude approximation to verifiable regions based on the distribution of the major commercial fish stocks, natural ecosystems or patterns of fishing activity. But, in truth, drawing meaningful boundaries in the sea is an impossible task: the shifting distribution of fish populations and the dynamic nature of ecosystems mean that natural boundaries are both permeable and unstable. In a maritime context, therefore, regions are bound to be socially constructed rather than naturally occurring and their boundaries inevitably reflect a compromise between overlapping sets of distributions and ecosystems. But this does little or nothing to diminish their value in the implementation of policy. In fact a number of putative regional seas occurring wholly or partly within the EU's 'common pond' like the semi-enclosed Baltic, North, Mediterranean and Irish seas – despite being defined as ICES areas – do conform quite well to recognisable marine ecosystems. Elsewhere, along Europe's Atlantic coast the delineation of boundaries is likely to be more arbitrary and the results possibly less satisfactory.

Integrated regional management has been a feature of many traditional inshore areas such as those in the Pacific based on concepts of territorial use rights (Ruddle *et al* 1992) or – closer to home – the comprehensive administrative network of Sea Fisheries Committees initially established for England and Wales in the 1880s (Symes and Phillipson 1997). Much less common is the application of regional management to extensive offshore areas, made possible only by the declaration of Exclusive Economic Zones (EEZs) or Exclusive Fishing Areas (EFAs) in the second half of the 1970s. Indeed, leaving aside the short lived experiment in New Zealand abandoned in favour of rights based management in 1986, the only significant example is the United States.

In 1976 the *Magnuson Fishery Conservation and Management Act* laid the basis for the regional management of US marine fisheries requiring Fisheries Management Plans (FMPs) for the implementation and monitoring of national policy to be drawn up for each of eight newly designated regions. The rationale behind this move was that it is simply not possible to develop a single management plan covering all US fisheries given the very significant differences in fish stocks, fish behaviour and fishing activities between the east and west coasts and through several degrees of latitude from Alaska in the north to the Mexican border in the south. Subsequently, in a report to Congress outlining an agenda for the incorporation of an ecosystem-based approach into US fisheries management (NMFS 1999) it was proposed that the eight regional councils should prepare Fisheries Ecosystem Plans (FEPs) to complement the existing FMPs. According to Rosenberg (2003), the Regional Management Councils (RMCs) in the US have met with varying levels of success. While Alaska's adherence to scientific advice and a cautious approach to permissible levels of fishing effort have laid the foundations for a prosperous, relatively stable and essentially sustainable fishery, in New England the strength of vested interests on the Council has until recently prevented the issue of persistent overfishing from being tackled in an effective manner.

The Magnuson Act serves as both a guiding light and a warning beacon for the development of regional management in European waters. Just as, in the United States, FMPs are intended to ensure the implementation of national policy, so too in the EU high level management objectives, broad strategic goals and global targets set by the Commission and Council of Ministers can be more readily applied when full cognisance is taken of the particular conditions of the regional seas. On the other hand, there are features of the US situation which one would not wish to see repeated in the European context. In America, it can be argued that RMCs have become overly strong players.

The federal government finds it difficult to implement a management policy without the backing of the councils' recommendation. And because the administration has no authority to alter a FMP but can only refer it back to the council for amendment, the councils have the ability to delay the implementation of national policy if they so choose (Rosenberg 2003).

Rather more disturbing is the emergence in the United States of 'management by litigation'. A number of judgements in cases brought against the federal government for failure to manage stocks on a sustainable basis have contained rulings directing the management authorities to undertake specific courses of action in what are sometimes impossible time frames (McCay 1999). What may be good law does not necessarily make for good fisheries management.

Linked to the regionalisation of fisheries management is the recently revived concept of strategic spatial planning and management of the oceans, also viewed by some as an essential element in the implementation of an ecosystem-based approach to the management of the living resources of the sea. In this context, what strategic spatial planning can seek to achieve is a rational, well coordinated network of marine protected areas (MPAs). MPAs are defined by the IUCN (1994) as areas "dedicated to the protection or maintenance of biological diversity and natural and cultural resources and managed through legal or other effective measures". They embrace a range of situations varying from strict protection ('no take zones' or NTZs) to more permissive regimes where the emphasis is on sustainable exploitation rather than environmental protection. Closed areas are already commonly used for the protection of critical fish habitats (spawning and nursery grounds) to assist improved recruitment. But for purposes of environmental protection, the present system in Europe based largely on Special Areas of Conservation (SACs) under the EU Habitats Directive (EC 1992) is considered inadequate. The designated areas are seen by some as too small, largely confined to inshore waters and the level of protection weak. Networks of MPAs covering more extensive areas of sea are thought to offer considerable long-term benefits for both commercial fisheries and marine nature conservation (see, for example, Gell and Roberts 2003). Moreover, the development of a zoning system, deploying different levels of restriction in environmentally sensitive marine areas in order to define the level and type of use for fishing and other forms of exploitation, could help to ensure not only a more sustainable marine environment but also the peaceful coexistence of potentially conflicting uses.

Implicit in the concept of spatial planning and management is the extension of the principles of strategic environmental assessment (SEA) and environmental impact assessment (EIA) - already applied to other forms of maritime development including oil and gas extraction, wind farms and major aquaculture projects - to include fishing activities. SEAs are perhaps the more relevant to the present discussion: they refer to comparatively large geographical areas and involve the identification and assessment of environmental interactions arising from particular forms of activity (such as fishing) and the implementation of mitigation measures where the interactions are deemed to have negative impacts on the environment.

5.3 CFP reform and Regional Advisory Councils (RACs)

There can be little doubt that, on paper at least, the revised CFP marks a considerable advance on what has gone before. However, much will depend on the detailed interpretation of the new Regulations and proposed Action Plans and on the way decisions are implemented as to whether the rhetoric of policy reform is translated into new forms of management practice. The reform package (see Box 5.1) is not limited solely to the content of the 'framework' Regulation No 2371/2002 (EC 2002a) dealing with the conservation and sustainable exploitation of resources, and the amendments to arrangements for structural assistance to the fishing industry contained in Regulation No 2369/2002 (EC 2002b) which call a halt to the subsidised building of new fishing capacity from 2005. These basic Regulations are to be complemented by a raft of Action Plans, lacking the legal force of Regulations but outlining further policy developments in relation to environmental integration, sustainable aquaculture, discards and the socio-economic impacts of the reform package. In practical terms, the future development of Europe's fishing industry will also be guided by the new financial framework, currently being constructed under the Financial Instrument for Fisheries Guidance (FIFG) for 2007 to 2013.

Box 5.1: Key elements of the new CFP

A Core legislation	· **Council Regulation (EC) No 2371/2002** on the conservation and sustainable exploitation of fishery resources under the CFP · **Council Regulation (EC) No 2369/2002** amending Reg. (EC) No 2792/1999 laying down the detailed rules and arrangements regarding Community structural assistance in the fisheries sector
B Contingent legislation	· **Council Regulation (EC) 2370/2002** establishing an emergency Community measure for scrapping fishing vessels
C Action plans	· **Biodiversity Action Plan for Fisheries**: COM (2001) No 4 Vol iv · **Communication setting out a Community Action Plan to integrate environmental protection requirements into the CFP**: COM (2002) 186 final · proposed strategy for the sustainable development of European aquaculture · proposed action plans: (a) to counter the social, economic and regional consequences of restructuring the EC fishing industry (b) to reduce discards
D Other relevant considerations	· Sixth Environmental Action Programme · **Communication from the Commission to the Council and the European Parliament: Towards a strategy to protect and conserve the marine environment** COM (2002) 532 final · FIFG programme (2007-13)

Underlying the reform of the CFP is a number of fairly consistent messages which define the salient features of the 'new' policy approach. In the first place, there is the shift from a preoccupation with short term crisis management identified with the annual round of TAC fixing – which had induced considerable uncertainty and instability in the previous regime – to a system based on medium term (or multi-annual) planning.

However, a strategic long-term vision for the future of the Community's fisheries and fishing industries remains sadly lacking. Second, there is a much stronger commitment to the development of an environmentally responsible fisheries policy identified particularly in references to the incremental development of an ecosystem based-approach, emergency measures where threats to the conservation of resources and/or the marine ecosystem become evident, and to Action Plans for biological diversity, environmental integration and discards. And finally, there are signs of a tentative move towards a relaxation of the 'centralising tendency' in EU fisheries management through the creation of Regional Advisory Councils, though the move stops short of actual decentralisation or devolution of decision-making.

If the purpose of RACs is to assist the Commission in the achievement of the CFP's stated objectives – namely the sustainable exploitation of living marine resources, the application of the precautionary principle, the progressive implementation of an ecosystem-based approach and the creation of an economically viable and competitive fishing industry (EC 2002a:Article 2.1) – then there are grounds for arguing that the specific roles ascribed to RACs by the Commission are likely to fall short of their full potential. Box 5.2 suggests that the Commission sees these roles as restricted mainly to advice on matters included in Chapter II of the framework Regulation (Conservation and Sustainability). Specifically their roles are to advise on multi-annual management plans, the introduction of emergency measures and the management of fisheries in the 12nm fishing zone. Such a restrictive view of the RACs' core activities would seem to confirm a widely held view that RACs are conceived as technical rather than policy committees. There is, for example, no specific reference to RACs' involvement in relation to either the adjustment of fishing capacity (Articles 11-16) or, more remarkably, to the rules on access to fishing grounds and the allocation of fishery resources (Articles 17-20). These are precisely the kinds of policy areas where regionally based professional advice can make a significant contribution. Given the proper scope and structure, regionalised management can provide the essential framework for addressing the issues of sustainable fisheries and environmental integration. In the context of the CFP, however, the question which still needs answering is the underlying reason for the establishment of RACs.

5.4 RACs: The hidden agendas?

In seeking to explore the underlying reasons for introducing what could be an important new element into the architectural design of the CFP, three inter-related scenarios – good governance, policy process and institutional structures – are examined in detail.

5.4.1 RACs AS GOOD GOVERNANCE

According to the Commission's Green Paper (EC 2001a), one of the reasons for the failure of the CFP has been the lack of effective consultation and meaningful dialogue with the principal stakeholders. Thus the fishermen came to be seen as the object rather than the subject of fisheries management. Largely ignored in the formulation of policy, fishermen had become alienated from the policy process, engendering weak levels of commitment to its outcomes and, in some instances, a cynical disregard for the regulatory measures. The need, therefore, was to turn this situation around, to engage fishermen more fully in the formulation and implementation of policy in the hopes not only of capturing the vital practical experience of fishermen but also of regaining the industry's confidence in the policy process.

Good governance features quite prominently in the new CFP (EC 2002a: Article 2.2) where the key principles are defined as:

> (i) a clear definition of responsibilities at Community, national and local levels (ii) a decision making process based on sound scientific advice which delivers timely results (iii) broad involvement of stakeholders at all stages of the policy from conception to implementation and (iv) consistence [sic] with other Community policies…

The juxtaposition of the second and third of these principles is interesting, particularly in view of the growing lack of confidence in the science of stock assessment felt by much of the fishing industry.

According to Rhodes (1996) modern concepts of governance also involve the "hollowing out of the state" and the transfer of some responsibility for management decisions from the state to responsible stakeholder-led organisations. In the eyes of the fishing industry, the involvement of 'non-accountable bureaucrats' in the Commission and 'self-seeking politicians' in the Council of Ministers lies at the heart of the CFP's failure. The horse-trading that reportedly takes place around the time of the December Council of Ministers meetings discredits the CFP as an objective, scientifically grounded and transparent policy in the eyes of most stakeholders. Politicians are no longer to be trusted: they are culpable of selling untested solutions on the basis of a quick fix, as with the development of the first generation of cod recovery plans introduced at the time of renegotiating the CFP. The industry therefore sees the transfer of key areas of decision making from Brussels to industry-based regional management organisations as a means of 'taking the politics out of fisheries management' and reestablishing the basis of a sound, transparent and workable management system. From the industry's perspective, however, RACs do not really provide this opportunity: they are advisory in function and involve no transfer of executive powers to stakeholder groups as the industry had originally hoped (SFF and NFFO 2000). Only the sources of professional advice are being decentralised and there are no guarantees as to how much influence RACs will be able to bring to bear on the final decision-making.

5.4.2 RACs AS PART OF A REFORMED POLICY PROCESS

Despite a tendency to regard the adherence to the principles of good governance as something of a cosmetic exercise in the presentation rather than the substance of the new CFP, it is arguable that RACs do have a potential to alter the ways in which advice is recruited, managed and translated into policy. Perhaps most immediately important is the opportunity to rebalance the contributions of science and practical experience.

Recent events concerning attempts to put in place stock recovery plans have widened the gap between fishermen and scientists not only over the assessment of the current and future states of the cod stocks in the waters around the UK but also in relation to the interactions between cod and other commercial species in a mixed fishery. Though such variance can in part be accounted for by differences in the time horizons and levels of precautionarity adopted by fishermen and scientists, there is a disturbing lack of respect now shown by many fishermen for the science that must underpin policy decisions. There is a sense in which science has unwittingly obscured the wider truth. It has led us to believe in the past that we knew what was happening to the fish stocks and that scientific advice, correctly applied, could steer us away from the ultimate disaster. Since the science has become undermined by increasing uncertainty and from the time scientists started to become 'prophets of doom', the industry stopped believing in the science. It rejected both the message and the messenger and turned instead to divine providence. RACs offer the possibility of rapprochement and bridge building between the two protagonists, especially if they are able to incorporate some of the procedures for dialogue and mediation between the industry and those responsible for stock assessments already adopted by the North Sea Commission.

Hitherto, the European Commission has relied rather too heavily on a technocratic approach to fisheries management based on a largely unchallenged acceptance of scientific advice from ICES. Consultation with stakeholders has been limited to periodic meetings of the Advisory Committee on Fisheries and Aquaculture (ACFA). Otherwise the industry's main hope of influencing the outcomes of decision making has lain with its ability to put pressure on the fisheries ministers to act in the industry's interests at meetings of the Council of Ministers. The extent to which RACs can in the future open up a new and constructive channel of influence and make a real contribution to the policy process will depend on two unresolved factors: the scope which they are given (and which they also define for themselves) in terms of their remit; and the mechanisms adopted for handling regional advice within the Commission.

Although the specific tasks assigned to RACs (see Box 5.2) are rather narrowly defined, a good deal of latitude is granted in terms of 'own account' recommendations and a catch-all statement of "any other activities necessary to fulfil their functions' (EC 2002a:Article 31.5). In effect, RACs will be in a position to set their own terms of reference. Rather less certain is a guarantee that the advice from RACs will be listened to and acted upon. Apart from a requirement that RACs "shall transmit an annual report to the Commission, the Member States ... and the Advisory Committee on Fisheries and Aquaculture" (EC 2003), there is no clear indication of the system for absorbing regional advice within the central decision making machinery. Moreover, there is no stated requirement for the Commission to take account of the advice it receives from RACs when formulating proposals for fisheries management and, therefore, no obligation on the part of the Commission to explain how the advice has been viewed and, in the event, why it may have been discounted.

Box 5.2.Council Regulation (EC) 2371/2002: The framework for the new CFP

Article	*Description*	*Detail*
1	Scope	... shall cover conservation, management and exploitation of living aquatic resources, aquaculture, and the processing and marketing of fishery and aquaculture products ...
2	Objectives	... ensure exploitation of living aquatic resources that provides sustainable economic, environmental and social conditions ... apply the precautionary approach ... aim at a progressive implementation of an eco-system-based approach ... be guided by the principles of good governance.
5/6	Multi-annual plans	multi-annual recovery plans for stocks outside SBL and multi-annual management plans for other stocks, including the setting of catch targets, effort limitations and technical conservation measures.
7/8	Emergency measures	... evidence of a serious threat to conservation of living aquatic resources or to the marine ecosystem resulting from fishing activities and requiring immediate attention, the Commission ... may decide on emergency measures which shall last for not more than six months; ... in waters under the sovereignty or jurisdiction of Member State where any undue delay would result in damage that would be difficult to repair that Member State may take emergency measures ... not exceed three months.
9/10	MS measures within 12 nm zone	... non-discriminatory measures for conservation and management of fisheries resources and to minimise effect of fishing on the conservation of marine ecosystems within 12 nm of its baselines; ... where measures to be adopted are liable to affect vessels of other Member States ... be adopted only after the Commission, Member States and Regional Advisory Councils concerned have been consulted.
11/12	Adjustment of fishing capacity	Member States shall put in place measures to adjust fishing capacity of their fleets in order to achieve a stable and enduring balance between such fishing capacity and their fishing opportunities ... (involving) ... withdrawal of licence and fishing authorisations; ... Commission shall establish for each Member State reference levels ... for the total fishing capacity.
13/16	Entry and exit schemes	
17/18/19	Access rules	In the waters up to 12 nm from baselines ... member States shall be authorised ... to restrict fishing to vessels that traditionally fish in those waters from ports on the adjacent coast; ... for species of special importance in the [Shetland Box] ... fishing activity by Community vessels ... of not less than 26 m ... shall be governed by a system of prior authorisation [to be reviewed by 31 December 2003].

Box 5.2. (cont)

20	Allocation of fishing opportunities	The Council ... shall decide on catch and/or fishing effort limits and on the allocation of fishing opportunities among Member States as well as the conditions associated with those limits ... in such a way as to assure ... relative stability of fishing activities...; each Member State shall decide on the method of allocating the fishing opportunities assigned ... inform the Commission of the allocation method.
21-28	Control and enforcement	
30	Committee for fisheries and aquaculture (ACFA)	The Commission shall be assisted by a Committee for Fisheries and Aquaculture (see Decision 1999/468/EC)
31,32	Regional Advisory Councils (RACs)	... shall be established to contribute to the achievement of the objectives ... and to advise the Commission on matters of fisheries management in respect of certain sea areas or fishing zones ... composed principally of fishermen and other representatives of interests affected by the CFP, such as ... fisheries and aquaculture sector, environment and consumer interests and scientific experts from all Member States having fisheries interests in the sea areas ... may be consulted by the Commission in respect of proposals for measures ... without prejudice to the consultation of STECF and the Committee on Fisheries and Aquaculture ... may submit recommendations and suggestions of their own accord ... inform the Commission or the Member State ... of problems relating to the implementation of Community rules ... and conduct any other activities necessary to fulfil their functions ... Regional Advisory Councils shall inform the Committee for Fisheries and Aquaculture of their activities ... The Council shall decide on the establishment of a RAC [which] ... shall cover sea areas falling under the jurisdiction of at least two Member States.
33	STECF	... Scientific, Technical and Economic Committee for Fisheries shall be...consulted at regular intervals ...[and] the Commission shall take into account the advice from the STECF when presenting proposals on fisheries management.

For regional advice to make its greatest impact it must be disseminated to the widest possible audience including the Commission, relevant member states and the European Parliament, *inter alia*. For it to succeed it will need to engage the support of at least the majority of Member States directly involved.

5.4.3 RACs AS STRUCTURES

Central to the detailed design of RACs is the question of whether they are to function simply as technical advisory committees - a position favoured by the fishing industry - or to be considered as part of the essential apparatus for strategic regionalised management of marine resources. Only if they are portrayed in this broader context will they be able to exercise real influence over issues of sustainability, environmental integration and the future of fishing dependent communities. Despite some concessions to the principles of good governance in defining a relatively broad constituency of membership, the balance of evidence to date seems to point to the narrower view of their functions.

A key structural problem is how to contain the array of national and sectoral interests which seek representation on the RACs within a necessarily compact decision making body. One solution to the 'dilemma of numbers' lies in what can loosely be termed as a bi-cameral model (see Box 5.3). This involves an open annual conference, allowing all relevant interest groups to meet, debate the issues and draw up an agenda for action, and a much smaller 'executive council' meeting throughout the year to respond to requests for comment and advice from the Commission and/or the member states and to formulate specific recommendations. The work of the 'executive council' could also be assisted by expert permanent or *ad hoc* working groups able to provide more detailed information.

The Commission's proposals, adopted by the Council of Ministers, follow the bi-cameral model quite closely but restrict access to the annual assembly to delegates nominated by the Member States, with the much smaller executive body chosen by delegates to the assembly. Two thirds of the seats on both the assembly and the executive are to be occupied by representatives of the fisheries sector (vessel owners, crew members, producers' organisations, processors and women's networks), with at least one seat for each member state with fishing interests in the area. Non-fishing interests, embracing aquaculture, recreational fisheries, marine conservation and consumers, are to take up the remaining one third of the available seats. Although the Commission's draft proposal (EC 2003) originally recommended that the executive body should comprise 12-18 members, it is widely expected that the final Regulation will see the upper limit raised to 24. Otherwise, it will be difficult to believe that the Commission has got its sums right. The size of each RAC should be determined by the complexity of the regional sea in question. A council of *circa* 15 members might be adequate for an area like the Irish Sea with only two coastal states involved (and a limited number of other member states with active fishing interests) but a council of twice that number may seem scarcely sufficient to encompass the range of interests represented in the North Sea (with at least seven coastal states as well as other member states and non-member states with fishing interests in the area).

Again, the size and balance of the councils should reflect the intended scope of activity. A small council with a strong majority of fishermen would suit the limited agenda of a technical advisory committee, but it is hard to imagine a council of 18 members, with only six representing non-fishing interests, being sufficiently well equipped to engage in a wider range of tasks.

For some time there has been a lively debate as to who precisely are the relevant stakeholders, how to balance the representation and how to select the members. There is general agreement that representatives of the fishing industry should form the largest single group and possibly occupy the majority of seats, though probably not to the extent favoured by the national fishing federations at 75 per cent (Beveridge and Morrison 2003). In the event, the Commission and Council of Ministers have resisted such claims. Despite recommendations from the influential Economic and Social Committee to restrict non-fishing interests to one fifth of the seats and an even more extreme view expressed by the European Parliament that non-fishing interests should be relegated to the status of observers, the original 66:33 allocation has prevailed. It is worth alluding once again to the situation in America where commercial fishing interests make up 49 per cent of voting members on the eight Regional Fishery Management Councils and recreational fisheries a further 33 per cent. Here, Okey (2003) argues that this skewed representation of interests has helped to generate perverse incentives for management decisions that conflict with sustainability goals.

Box 5.3: Regional Advisory Councils: Comparative structures

	Commission Proposals	*Irish Sea RAC*
Functions	· to advise Commission and Member States on matters relating to fisheries management with particular reference to multi-annual fishing plans, emergency measures and inshore management · own accord recommendations to Commission and Member States · inform Commission and MS re problems relating to implementation of Community rules	· an independent 'think tank' responsible for (c) long term strategy for development of sustainable fisheries (d) integrated fisheries management plans · principal reference point for consultation by Commission and Member States in relation to proposals for regulation of fishing activities
Architecture	· an annual general assembly (two thirds of members represent fishing interests) · an executive council of 12-18 members (two thirds fishing interests)	· an annual conference open to all stakeholders · an executive council of 12-15 members, supported by · expert working groups (fisheries; environment)
Membership	· all Member States having fishing interests in the sea area have a right to participate as members or observers	· UK and Ireland as coastal states; France, Belgium and the Netherlands as other MS with active fishing interests
Composition of Executive Council	· membership should include b) representatives of the fisheries sector (66%) viz vessel owners, small scale fisheries, FPOs, aquaculture producers, processors and merchants c) 'other interests' (33%) viz environmental groups, consumers, recreational fishing and general interest (e) scientists from national/international institutions invited as experts (but not full members)	· not more than two persons from the national administrations · not more than two scientists · not fewer than 8 representatives of the fisheries sector (including aquaculture, inshore fisheries, other commercial fisheries, processors etc) · two representatives of the environmental interests (NGOs and/or country agencies) · one representative of recreational fishing interests (ie: fishing interests 53%; non-fishing interests 47%)
Funding	EU funding for the first 5 years only, reducing from _200,000 in year 1 to _110,000 in year 5	European Commission (40%); Member States (40%); stakeholders (20%).
Sources	Council Regulation No 2371/2002; Draft Proposal COM(2003) 607 final	Symes *et al* (2002)

Representation of fishing interests could prove problematic for certain member states at least. In an industry more accustomed to fission rather than fusion and where strong differences of opinion between inshore and offshore sectors or different gear groups or neighbouring districts are quite normal, experience suggests that it will be difficult for all fishing interests to coalesce around a single nomination. This may cause difficulties for the selection of representatives, the mindset of those appointed and for the development of consensus within or between the different stakeholder groups. Although there are strong grounds for preferring seats to be occupied by active fishermen so as to tap directly into their reservoirs of local knowledge, experience and good practice, it seems likely that in most cases those seats will fall by default to officials of fishermen's organisations who – though well versed in the arts of negotiation and the rough and tumble of fisheries politics – may lack the experience and practical expertise of professional fishermen.

Even more difficult is the definition of, and selection procedures for, the non-fishing interests. The range of relevant interests is itself difficult to delimit, bearing in mind the need to ensure that council members should add value to the discussions. They should certainly include environmental groups (either NGOs or official conservation agencies) and recreational fishing interests which in some areas may make a stronger contribution to the local and regional economies than the commercial fishing industry (Nautilus Consultants 2000). In both these instances, separate national representation may be unnecessary, though it will still be essential for the appointed members to have well-founded knowledge and experience of the particular regional sea. Some relief for the potential overcrowding of the councils is afforded by the decision that scientists and administrators should attend in the capacity of expert advisers rather than full members of the councils. This does, however, have the possibly unfortunate effect of distancing both the scientific community and the Member State administrations from any decisions which the council may eventually take.

Managing the work of the councils, ensuring that they do not disintegrate under the stress of internecine warfare, and framing advice in the form of strong consensus rather than weak compromise will call for inspired leadership on the part of the chairperson. To fulfil such credentials suggests that the chair would need to be appointed as an independent person with no history of allegiance to any one sectional interest within or outwith the fishing industry and, as far as possible, free from active political connections with any Member State involved. Notwithstanding, the Commission and Council of Ministers have determined that each RAC shall designate a chairperson – presumably from among its own membership – by consensus. Impartiality, therefore, may prove difficult to guarantee.

At this stage, estimating the annual operating costs of RACs can be little more than speculation. Costs will be determined by the workload and that, in turn, by the frequency with which they are called upon to give advice and by any self-imposed tasks. A high workload will probably require a dedicated secretariat and support staff or at least the means to buy in appropriate expertise. It is also worth noting that RACs may be required to make rapid responses ('within five working days') to the Commission or Member State concerning proposals for emergency measures. In 2004 the Council of Ministers agreed a five year schedule of funding for RACs from EU sources. The level of public funding will be reduced annually from a maximum of €200,000 (or 90 per cent of expected costs) in year 1 to €110,000 (50 per cent) in year 5. Thereafter RACs are intended to be self-funding, except for an additional yearly grant of €50,000 to cover costs of translation and interpretation. While there may be merit in self-funding as a means of ensuring value for money, it also serves to confirm the impression that the Commission sees RACs as low key rather than high profile additions to the institutional architecture.

However they are constructed, it is unlikely that RACs will conform to a standard pattern of behaviour. In some instances, partnerships between the key players may be easily forged, leading to a more coherent view of the region's needs and a more proactive approach, while in others the emergence of a sense of partnership may be constrained by cultural differences, contentious issues or old rivalries. It is clear from the negotiations leading up to the revision of the CFP that the proposed introduction of RACs was received very differently across Europe. Attitudes varied from the strongly supportive views of the UK, Sweden and the Netherlands, through the more cautious responses of Denmark, Belgium and France, to the deeply suspicious if not openly hostile reactions of the fishing industry in Spain. Interestingly, some of the RACs will bring together representatives from Member States with rather contrasting perceptions of RACs and their functions.

5.5 RACs: The handmaidens of environmental integration?

One distinctive feature of the new CFP is the added emphasis on environmental issues. This is evident not only in its objectives (Article 2.2) but more especially in the constellation of action plans which will have a direct bearing on matters of environmental protection. This change of emphasis was partly foreshadowed in a Commission policy document on the integration of environmental protection requirements which stated that:

> Conservation of marine ecosystems should be central to an environmentally integrated policy for fisheries. Action should be taken to conserve and, where possible, rebuild commercial fish stocks and non-commercial biota and habitats, but also with due attention to the consequences for the whole ecosystem, in order to restore their functionality and productivity when these have been damaged. Similarly…where non-commercial biota and habitats are threatened, action to remove threats should take account of its implications for fisheries and the wider productivity of the ecosystem. (EC 2001b:12)

It is reasonable to infer potentially strong links between the regionalisation of fisheries management and the means for delivering environmental integration. The 'regional seas' form a logical spatial framework, particularly in the case of the semi-enclosed Baltic, North and Irish seas which may be considered as reasonably self-contained ecosystems. Early advocacy of the ecosystem based approach to fisheries management in Europe was presented in the specific regional context of the North Sea:

> In the management of living resources [an ecosystem approach] means…that the decisions are based upon the best scientific knowledge of the functions of the ecosystem, including the interdependence of species and the interactions between species (food chains) and the abiotic environment, as well as knowledge of the temporal development of the ecosystem…The North Sea should be seen as a whole and managed as one ecosystem. Any future ecosystem approach to the North Sea management would imply that management decisions are based on the precautionary approach, taking into account all ecosystem effects of human activities as well as the impact of the environment on its resources. (Svelle *et al*, 1997:101)

This 'whole ecosystem' approach has more recently been taken up in the JNCC's Irish Sea Pilot Project, commissioned by DEFRA. It has as its goal the development of a prototype for an integrated marine nature conservation framework at the regional scale.

In the context of the CFP and the emergence of a regional dimension there is a need to explore a little more fully the ideas of environmental integration and an ecosystem-based approach to see how far RACs may be able to assist in the implementation of the environmental agenda. Although in some respects the two concepts are quite closely related, they have very different objectives. Environmental integration is concerned with the broad commitment to ensuring that robust environmental protection measures are incorporated into all areas of EU policy. In a fisheries context, it seeks to minimise the negative impacts that fishing activities may have on the marine environment. This will involve a series of 'priority actions' (i) to reduce fishing pressure to sustainable levels; (ii) to improve fishing methods so as to reduce incidental bycatches; and (iii) to raise the level of understanding of marine ecosystems through monitoring and assessment (EC 2002c). To these one needs to add a fourth 'priority action' namely the regulation of fishing practice in environmentally sensitive areas through a coordinated network of MPAs.

The new CFP is also pledged to a progressive implementation of an ecosystem-based approach. Its application in fisheries management is not concerned with the conservation of threatened habitats and species *per se*. Nor does it necessarily imply a radical agenda of intervention. Instead, it seeks to use our developing knowledge and understanding of ecosystem behaviour to create sustainable fisheries through safeguarding the essential ecosystem processes and functions on which all living resources of the sea depend. This knowledge should permit the fine tuning of regulatory measures and lead eventually to abandoning the heavy handed system of TACs and catch quotas and placing greater reliance on technical conservation measures adapted to the specific requirements of particular ecosystems and the management needs of the commercial fisheries contained therein.

The successful accomplishment of environmental integration and the development of the ecosystem-based approach quite clearly call for spatial planning and management within a defined regional framework. As marine ecosystems are highly dynamic and as our understanding of their behaviours can be expected to increase rapidly over time, the system of spatial planning and management will need to be flexible and adaptive in style. In elaborating a possible model for an Irish Sea RAC, Symes *et al* (2002) suggested that its advisory role should comprise two distinct functions: first as an independent 'think tank' capable of developing a long term strategic vision for the region's fisheries and second as a reference point for consultation on technical proposals for fisheries management. In discharging the first of these functions, the RAC should be in a position to follow the US approach and develop both FMPs and FEPs - or rather, a single integrated management plan - for regional sea activities. It is, however, difficult to envisage RACs, as outlined in the previous section, having either the resources or the competence to undertake such activities. As a result, the tasks associated with environmental integration will be much harder to achieve.

5.6 Conclusions

In May 2004, approval was given for the establishment of five Regional Advisory Councils (Baltic Sea, Mediterranean Sea, North Sea, North Western Waters and South Western Waters) covering EU waters, together with two Sectoral Councils - one for pelagics (herring, mackerel, horse mackerel and blue whiting) and the other for distant water fisheries. Despite the fact that many in the industry remain sceptical of the potential for RACs to add real value to the management of EU fisheries, the response to the green light - at least in northern Europe - has been positive and immediate. It is likely that a North Sea RAC will be in place before the end of 2004. Discussions on a North Western Waters RAC are progressing, though significantly consideration is already being given to a sub-regional structure to take account of the area's diversity (*Fishing News*, 2 July 2004). And moves to establish the pelagic council have been initiated by the Northern Pelagic Working Group of the European Association of Producers' Organisations (*Fishing News*, 18 June 2004).

Such alacrity is encouraging. But have the fishing industries of the EU been duped, in the name of political correctness, into accepting a new element of the policy process which has more to do with the presentation of the CFP than with its fundamental reform? Or does the creation of RACs signal the intention to grant the industry – and other stakeholders – a real measure of influence over the formulation and implementation of policy at the regional level? Only time will tell.

The 'empty vessel' scenario posed in the title of this chapter will only be true if RACs are used simply to perpetuate a failing system of management. The Commission's reluctance to accede to the industry's wishes to be given direct responsibility for certain areas of management is understandable. RACs will need to prove themselves capable of embracing consensual politics in providing coherent, relevant and timely advice in a trans-national, regional setting. The risks of this limited experiment in regionalisation destabilising the overall management system are minimal; the ultimate responsibility of the central institutions of the EU in matters of decision-making remains undiminished.

Indeed the new European Constitution, still to be ratified, confirms the unchallenged supremacy of the Commission in the formulation of policy. Fisheries remains one of the very few areas of 'exclusive competence' for the Commission, along with monetary policy, commercial policy and the customs union - a rather strange set of bedfellows.

Risks to the political image of the fishing industry, on the other hand, are somewhat greater. The exposure of divisions within a region's industry in defence of vested interests and a consequent failure to make its mark on fisheries policy will diminish the industry's credibility concerning its commitment to sustainability. Collaboration, partnership and consensus are the essential ingredients of success - but forging partnerships between disparate national and sectoral interest groups may prove a difficult task, and certainly not one to be hurried. It is therefore easy to understand why the fishermen's federations in the UK argue the need to build up the strength of RACs through a number of small victories rather than attempt to scale the heights of major issues in the early days of their existence (Beveridge and Morrison 2003).

The metaphor of 'cornucopia' is equally misplaced. For those who assume that RACs will immediately open wide the doors to a radically new approach to fisheries management the disappointment is likely to be much greater. Regionalisation can provide a new framework for fisheries management but it can only lead to the Holy Grail of sustainable fisheries in diverse, productive and well integrated ecosystems if it is allowed to become a vehicle for delivery of a more holistic approach embracing integrated management, precautionarity, an ecosystem-based approach, strategic long term planning and an ability to cope more effectively with scientific uncertainty. Such a radical agenda will be enormously difficult to manage politically, especially if the majority opinion on the RACs remains unconvinced of the benefits to be derived.

But the value of regionalisation is precisely that it allows things to be done differently, to jettison some of the historical ballast associated with an old style, command-and-control CFP built on a very narrow perception of management and trusting to a limited range of policy tools - dominated by TACs and catch quotas - which place unrealistic demands on the science of fish stock assessment. In time, RACs may open up a route towards the peaks of integrated management, but they will first have to negotiate the difficult terrain to be encountered in the foothills of a revised CFP.

References

Beveridge, D and Morrison, H (2003) 'The industry view', a paper presented to the *IEEP Workshop on CFP Regional Advisory Councils*, held in Aberdeen 2/3 June

EC [European Community] (1992) *Council Directive 92/43/EEC of 21 May 1992 on the conservation of natural habitats and of wild fauna and flora* (The Habitats Directive), Office for Official Publications of the European Communities, 22nd July

EC [European Community] (2001a) 'The future of the Common Fisheries Policy', Brussels, European Commission

EC [European Community] (2001b) 'Communication on elements of a strategy for the integration of environmental protection requirements into the Common Fisheries Policy' *COM (2001) 143 final*, Brussels, European Commission

EC [European Community] (2002a) *Council Regulation (EC) No 2371/2002 of 20 December 2002 on the conservation and sustainable exploitation of fisheries resources under the Common Fisheries Policy*, European Council of Ministers

EC [European Community] (2002b) *Council Regulation (EC) No 2369/2002 of 20 December 2002 amending Regulation (EC) No 2792/1999 laying down the detailed rules and arrangements regarding Community structural assistance in the fisheries sector*, European Council of Ministers

EC [European Community] (2002c) 'Communication setting out a Community Action Plan to integrate environmental protection requirements into the Common Fisheries Policy' *COM (2002) 186 final*, Brussels, European Commission

EC [European Community] (2003) 'Proposal for a Council Decision establishing Regional Advisory Councils under the Common Fisheries Policy' *COM (2003) 607 final*, Brussels, European Commission

Gell, FR and Roberts, CM (2003) *The Fishery Effects of Marine Reserves and Fishery Closures*, Washington DC, WWF-US

IUCN [International Union for the Conservation of Nature] (1994) *Guidelines for Protected Area Management Categories: Commission on National Parks and Protected Areas*, Gland, IUCN

McCay, B (1999) 'Multi-disciplinary research in fisheries from a social science and North American perspective' in D Symes (ed.) *Multi-Disciplinary Research in Fisheries Management*, Hull, ESSFIN:111-120

Nautilus Consultants Ltd (2000) *Study into Inland and Sea Fisheries in Wales*, Edinburgh, Nautilus

NMFS [National Marine Fisheries Service] (1999) *Ecosystem-Based Fishery management: A report to Congress by the Ecosystem Principles Advisory Board*, Washington, NMFS

Okey, TA (2003) 'Membership of the eight Regional Fishery Management Councils in the United States: are special interests over-represented?' *Marine Policy* 27:193-206

Rhodes, RAW (1996) 'The new governance: governing without government' *Political Studies* 44(4):652-667

Rosenberg, AA (2003) 'Experiences and observations on the use of regional fishery management councils in the United States', written evidence submitted to the House of Lords European Union Committee (Sub-Committee D)

Ruddle, K, Hviding, E and Johannes, RE (1992) 'Marine resources management in the context of customary tenure' *Marine Resource Economics* 7:249-273

Scottish Fishermen's Federation and National Federation of Fishermen's Organisation (2000) *Zonal Management: A New Vision for Europe's Fisheries*, Aberdeen and Grimsby

Svelle, M, Aarefjord, H, Heir, TH and Øverland, S (1997) *Assessment Report on Fisheries and Fisheries Related Species and Habitats Issues*, Oslo, Fifth North Sea Conference Secretariat

Symes, D and Phillipson, J (1997) 'Inshore fisheries management in the UK: Sea Fisheries Committees and the challenge of marine environmental management' *Marine Policy* 21:207-24

Symes, D, Ridgway, S and Crean, K (2002) 'Integrated Regional Management of the Irish Sea: A Feasibility Study' *CCW Contract Scienhce Report No 508*, Bangor, Countryside Council for Wales

CHAPTER 6
FISHERIES GOVERNANCE, SOCIAL JUSTICE AND PARTICIPATORY DECISION-MAKING

HANS-KRISTIAN HERNES*, SVEIN JENTOFT** AND KNUT H. MIKALSEN*
**Department of Political Science, University of Tromsø, Norway*
*** Department of Social Science and Marketing, The Norwegian College of Fishery Science, University of Tromsø, Norway*

Abstract

Controversies over distribution of access rights are a distinctive feature of fisheries management. Who should be the beneficiaries and what are the relevant criteria for awarding such benefits? We find it rather surprising that principled fisheries management debates on social justice are so rare. We are equally perplexed that so little attention is paid to issues of justice within social science fisheries research. In this article we try to remedy this, first by outlining some of the arguments in the justice literature to demonstrate their relevance for fisheries governance. Second, the establishment of a particular allocative mechanism – the so-called quota ladders - in Norwegian fisheries is used as an example of how different conceptions of justice can be applied in concrete management settings. We argue that much would be gained if a principled debate among involved stakeholders occurred prior to the actual allocation process; that is if stakeholders would agree on some general rules with regard to what constitute socially just distribution of access rights. In fact, we believe that the issue of participatory decision-making through devolvement of authority and responsibility to stakeholder groups, which is now on the agenda in many countries, would be much easier to realise if a social contract for just fisheries were established at the root.

6.1 Introduction

The politics of fisheries management is particularly zealous on the issue of the distribution of access rights. The reason for this is fairly obvious: with individual livelihoods and the survival of local communities at stake, distribution raises fundamental issues of social justice and fairness. The question, then, of what constitutes a just management regime, and what criteria are relevant for assessing its fairness, becomes pertinent. While this is basically a philosophical question, it can also be approached empirically - starting with questions such as: What principles of social justice do user-groups apply when they claim rights of access to fish resources? What justice principles do governments refer to when they defend their management policies?

Given the pertinence of such questions, we find it quite remarkable, one may say paradoxical, that the debate within political theory, spurred by John Rawls' seminal treatise 'A Theory of Justice' (1972), has attracted so little attention among students of fisheries management. With a few exceptions, such as Gray (1998), little has been written on the normative issues pertaining to fisheries governance. Rawls targeted his contribution at a philosophical audience, but his book also revitalised modern political theory and made quite an impact in the social sciences. For the last ten years or so social

T. S. Gray (ed.), Participation in Fisheries Governance, 103–118.

justice has become an important theme in environmental politics (Schlosberg 2003), rubbing off on the fisheries management discourse (Hernes and Mikalsen 2002).

We believe that fisheries management would benefit from a more principled debate on social justice standards and from what Rawls and other philosophers and social theorists have written on this issue. Fisheries management cannot be reduced to a technical exercise that should only be seen from a means-end perspective. As any other social practice, fisheries management must be subject to moral scrutiny. Good governance, in fisheries as well as in other sectors, should start from a reasoned contemplation on some fundamental principles of justice. From a decision-making perspective moreover, the key issue in determining what principle(s) of social justice that should be implemented in fisheries governance is the procedural problem of involving all stakeholders in the process. Any management regime that fails in this regard will have a justice 'deficit'. In this sense, we consider democracy a key ingredient in the lexicon of social justice.

In this paper we make an attempt at applying principles of social justice to fisheries management - using a fairly unique allocation scheme in Norwegian fisheries, the so-called 'quota ladder', as a case in point. The ladder originated in 1989 from within the Norwegian Fishers' Association as a response to the distributional difficulties brought on by the Barents Sea cod crisis. The crisis spurred a debate among fishers on how to share the Total Allowable Catch (TAC) between different sections of the fleet. This eventually led to a (albeit fragile) consensus; a 'social contract'-like agreement on a fundamental and contentious issue – thus avoiding a potential split within the association.[1] Since then, the ladder has been both refined and extended to incorporate other fish stocks than cod. As a 'social contract' - establishing long-term commitments to certain principles and 'rules' for the parties involved - the ladder has reduced the level of conflict and dispute among user-groups by bringing some interactive order into the decision-making process. In condoning this scheme, the government, for its part, has relieved itself of a highly controversial task. From a co-management perspective, this is significant because without the contract implied by the ladder, the government would not have been able to delegate decision-making power on such a socially important issue to the fishers. Agreement on fundamental principles of allocation is thus conducive to participatory decision-making. The key question, though, is whether the 'contract' holds from a justice perspective.

We start by depicting some of the main theoretical positions on social justice and how they may apply to fisheries. Then we summarise the idea, substance and effects of the Norwegian fisheries quota ladder, and the political turmoil surrounding its creation and design. Thereafter, we 'challenge' the ladder from a justice perspective. Did it come about through a democratic process? What are the normative principles underpinning it? How consistent are they? Finally, our concern is with the lessons for fisheries governance that can be drawn from this Norwegian experience. More specifically, we raise the issue of whether 'social contracts' of this type can work as a management

---

[1] The essence of this 'scheme' is that the relative share of the two basic segments of the fleet (offshore and inshore) should vary by the size of the Norwegian TAC. In its original version, the ladder – as an example – implied that with a Norwegian TAC of up to 150,000 tonnes, the shares of offshore and in-shore were 25 and 75 per cent respectively. With a Norwegian TAC of 300,000 or more, these shares were set to 35 and 65 per cent.

instrument. Do contract-like agreements negotiated by user groups and government have a potential in fisheries governance – in Norway and elsewhere?

6.2 Fisheries and theories of justice

Fisheries managers must take into consideration that there are several heterogeneous user groups and that rules and regulations may affect them differently. As Armstrong and Clark (1997:203-204) point out, "all management regimes have underlying equity implications in the shape of different distributional effects". Fisheries management thus raises important issues of social justice. Finding a management system to be unjust, a user group would tend to resist it, either through 'voice' or 'exit' (Hirschman 1970). Such a management system is likely to be ineffective. Thus, justice is not an issue managers can ignore but one that must be addressed from the very beginning. For instance, managers must decide why some users qualify for access while others do not, and who should be allowed to fish what, when and where. In order to be legitimate, such decisions must satisfy some basic criteria – or principles – of justice.

According to Campbell (2001) there is no 'true' or 'correct' meaning of justice. He also points to the danger that the notion of justice becomes too broad to have any real impact on public policies. In the absence of a precise definition, Sen believes that we at least need "a working agreement on some basic matters of identifiable intense injustice or unfairness" (Sen 1999:254). Perhaps it is not so difficult to agree on what these are when fisheries are concerned. It may, however, be easier to agree on certain principles under 'a veil of ignorance' than in the real world, where people make decisions on the basis of cards that have already been dealt. In other words, in real-world settings such as the fisheries, user-groups and other stakeholders have things at stake and, hence, something to lose. Neither can we be sure, as Rawls also points out, that justice will be served when people only look after themselves – even if they should agree on what constitutes a common good, such as sustainable fisheries. From Rawls' perspective, the maximum good is not necessarily the maximum right; right is defined independently of the good, and it is a concept that is prior to that of the good. We shall try to clarify what this means.

Common property theory regards property rights as essential. The 'Tragedy of the Commons', in Garrett Hardin's seminal exposé, is basically an outcome of a resource free-for-all. "Freedom in a commons brings ruins to all" (Hardin 1968:1244), is among his most quoted statements. Clearly, limiting this freedom by installing a rights-based management regime cannot be criticised from a justice perspective if it makes everyone better off. In reality, however, some lose while others win. Fisheries management is more of a zero-sum than a plus-sum game. However, there are many ways to restrict resource users' freedom, and not all of them are necessarily just. Even though a particular rights-based system should prevent the worst-case scenario – the tragedy of the commons – it may still be criticised from a justice perspective if some stakeholders lose relative to what they would otherwise have gained in another property rights system (Kymlicka 2002). The essence of property is the right to reserve for oneself, and exclude others from, the benefits that can be drawn from the resource. Although effective as a management device in limiting access and preventing a 'race for fish', a property rights system still divides by including some while excluding others. It is for exactly this reason that such systems are so controversial.

Property rights are a more complex issue than management regimes typically recognise. They are perceived as a technical device, but they also infringe upon basic 'natural' or moral rights. There is, for instance, a human rights aspect to fisheries management that is rarely recognised, as is clearly demonstrated in the case of indigenous peoples. The Draft Declaration on the Rights of Indigenous Peoples currently being developed within the auspices of the United Nations, article number 26 reads as follows:

> Indigenous peoples have the right to own, develop, control and use the lands and territories, including the total environment of the lands, air, waters, coastal seas, sea-ice, flora and fauna and other resources which they have traditionally owned or otherwise occupied or used. This includes the right to the full recognition of the laws, traditions and customs, land-tenure systems and institutions for the development and management of resources, and the right to effective measures by States to prevent any interference with, alienation of or encroachment upon these rights.

Rights to natural resources, such as fish and coastal territories, are here regarded as essential for the preservation of indigenous peoples' culture and material existence. These are rights that state authorities cannot change or abolish with a stroke of the pen without committing an act of legal and moral injustice. If deliberately designed to do so, property rights regimes may well support these basic rights, as, when indigenous peoples *qua* people are positively discriminated against, they are granted quotas or management authority (Jentoft *et al* 2003). Should the Declaration be adopted, we are not talking so much about 'distributive justice' as justice of 'rectification' – a compensation for previous colonisation and discrimination.

Natural rights are, of course, not unique to indigenous peoples. When Marx launched the justice principle of "to each according to his need", he was thinking along similar lines. The idea is that people have inalienable rights – individually and collectively. People have a right to exist and to what that right entails, materially as well as culturally. Also, when dependency – as in 'fisheries-dependent regions' (Symes 2000) – is thought to give priority to fishing rights, such a justice principle is alluded to. In other words, those who are most vulnerable should come first. Quota allocations are often based on the premise that those who can prove a history in fisheries should have a first right. Again dependency serves to justify certain decisions.

In fisheries management, other justice principles are also in use. A fisher, for instance, is entitled to his catch. Once the gear is in the ocean at a spot where he is allowed to be, the fish that gets tangled is his property (provided that he stays within his quota). His labour investment and the risk he takes upon himself make him the rightful owner. He has earned his catch. Here, in other words, a 'merit' principle is employed. This is justice by desert, which is to be distinguished from a principle of equity (Campbell 2001). The problem, however, is that fishing technology is never neutral, but may well be used as a tool of power. For instance, a trawl is a more powerful gear than a gill net or a long line, as can be seen when a trawl is used on fishing grounds where other gears are employed as well. Some gears are more costly and more catch effective than others and, therefore, fishers have unequal opportunities to employ them. The dilemma is well captured by Bavinck (1996:482) in his study of fisheries regulations along the

Coromandel Coast of India and the banning of certain gear from the fishing grounds:

> Two principles were seen to underlie fisher regulation along the Coromandel coast: (1) common access to inshore fishing waters; and (2) the right of every community to impose conditions on fishers using adjacent waters. The gear ban, which is an elaboration of the latter principle, has a dual purpose. First, bans apply to types or applications of fishing gear that are felt to affect important fish stocks negatively. Second, they are rooted in conceptions of social justice – gear that benefits a minority to the detriment of the majority of fishers (or the weaker sections of fishing society, particularly the aged), are not tolerated.

Scarcity, as when demand for some 'good' exceeds its supply, brings issues of social justice to the surface. Claimants to a resource find themselves having to settle for less than they want, and possibly less than they need to sustain themselves. Clearly then, economics and markets have a lot to do with social justice (Sen 1987). In fisheries, due to externalities, markets will not automatically bring social justice, as for instance between generations (Sumaila and Bawumia 2000). There is a justice principle behind market based management systems such as Individual Transferable Quotas (ITQs). In this case, you are entitled to what you have 'paid' for – a principle people may find fully acceptable in this particular sphere of life. Within an ITQ regime, rights are only loosely – if at all – coupled with dependency, which is also a reason why ITQs tend to be so controversial.

Thus, social justice cannot easily be attributed to some universal standard that can be applied in all settings. Rather, as Walzer (1983) points out, justice is derived from – and should be analysed according to – a specific context. The principles of justice backing up various distributional rules are not the same for every good or burden, for every social and political community, and for every situation or circumstance. They may also be highly 'local', as Elster (1992) demonstrates. Neither are justice principles always stable – even within the same sphere. As Armstrong and Clark (1997) show for the Norwegian codfish quota ladder, different justice principles apply depending on circumstance – as when the TAC is high or low. Walzer (1983) argues that injustice may be done when goods are converted into other goods, by a transgression of 'spheres'. For instance, there are things in life that money cannot, or should not be allowed to, buy (votes, love – some would even say fish quotas). Transgression of boundaries between spheres often triggers a moral response. Practices may be highly effective but still regarded as contrary to established social and cultural norms. Thus, some management and resource extracting practices (such as: draggers, explosives, under-reporting.) are condemned on moral rather than on 'functional' grounds. ITQs are usually not criticised because they are ineffective but because they have substantive consequences that people may find hard to accept, as when quota rights become geographically concentrated and end up as property of larger operators (Pálsson 1998).

Following Rawls, Miller (1999:14f) suggests that we should understand justice "as what people would agree to in advance of knowing their own stake in the decision to be reached." Miller argues that justice does not only relate to substantive outcomes, but also to institutions and their governing principles and procedures. He thinks that justice "must include aspects of social relations that do not fall readily under the rubric of distribution." Procedural justice has merits of its own and cannot simply be reduced to substantive fairness. This suggests that institutions for regulatory decision-making may affect the fairness of substantive outcomes and should be evaluated from a justice perspective. In this sense, democracy is prior to desert or need in the lexicon of social justice.

For example, is representation of user-groups and other stakeholders in management decision-making bodies fair? How does one justify the exclusion of groups who claim to be legitimate stakeholders? If it is true, as Schattschneider (1960, 30) points out, that "organisation is itself a mobilisation of bias", the question is how a particular bias can be justified. If there are limits to how inclusive co-management institutions can be, exactly where should the limits be drawn in order for decision-making procedures to be just and fair? This issue is relevant to the Norwegian fish quota ladder, to which we now turn.

6.3 Resource allocation

As instruments of allocation, quota ladders have – since they were first introduced in the early 1980s - been refined and made more comprehensive. As will become clear, the move towards quota ladders has not only been a major achievement for the Norwegian Fishers' Association; it has also been condoned by the Ministry of Fisheries and turned into a key instrument in the management system. Table 6.1 depicts the initial ladder agreed upon in 1989 by the members of the national board of the Fishers' Association. As can be discerned, the allocation key between the two groups, inshore and offshore, changes with TAC volume: the smaller the TAC, the larger the share of the coastal fleet.

Table 6.1. Cod Quota allocation rule

Norwegian cod, TAC in tonnes	*Coastal (%)*	*Trawlers (%)*
Under 100,000	80	20
100,000 – 150,000	75	25
150,000 – 200,000	72	28
200,000 – 300,000	69	31
Over 300,000	65	35

The quota ladder system has evolved through three phases. The *first phase* was initiated by the decision of the board of the Fishers' Association to establish the quota ladder for cod North of 62°N latitude for the five-year period 1990-94. This first allocation rule was a response to a precarious distribution conflict after the unexpected Barents Sea cod stock collapse, which led to the lowest TAC ever. Not all, however, were equally happy with the percentages set. Inshore fishers – from North Norway in particular – perceived the allocation key, shown in Table 6.1, as basically unfair. They did not, however, succeed in making changes during the five-year period. The argument that they had lost part of their historical 'entitlement' was somewhat undermined since the coastal fleet did not manage to take its designated quota as the TAC was increased. Another objection – procedural rather than substantive – was related to what some perceived as lack of internal democracy in the Association. According to some, the National Assembly, not the Board, should make decisions over issues of such great importance (Armstrong 1998; Landsmøtesak 7/2001).

In the *second phase,* criticisms with regard to the internal process of decision-making were met. In 1994, the National Assembly of the association had the final say. The new quota ladder was, like its predecessor, a fragile compromise. As with the previous ladder, inshore fishers from Northern Norway also disputed the new one – to no avail. The allocation rules agreed to earlier were renewed for seven years, putting the dispute to rest until the 2001 National Assembly meeting. Moreover, even though allocation rules had been both fragile and disputed, the ladders had met fertile ground in an organisation well known for internal struggles between gear groups and regions. In addition to renewing the quota ladder for cod, the Association established similar allocation rules for haddock, saithe, herring, mackerel and capelin.

The *third phase* was initiated in 1999 when the Board, as a follow-up of the 1994 National Assembly decision, established a special task force – 'The Resource Distribution Task Force' – with a mandate to review the allocation rule experience. The task force counted eight members: four from Northern Norway, two from Southern Norway, and two from the Boat Owners' Association.[2] This composition illustrates the intent to balance different interests, an intention further emphasised by the appointed chairman being recruited from outside the organisation: the chief executive of the sales organisation for the pelagic fisheries and a former Ministry of Fisheries state secretary (junior minister).

6.3.1 AUTOMATIC ALLOCATION

In its report, the task force recommended the continuation and expansion of the ladder agreement. In addition to the eight fisheries included in the 1994 decision, the task force considered eighteen new fisheries and recommended allocation rules for four of these. Thus, the 2001 assembly decision included twelve different fisheries, as shown in Table 2. The main reason for excluding the rest on the list, fourteen categories of fisheries, was the absence of allocation problems among Norwegian fishers and thus no need to solve management conflicts. As shown in Table 6.2, the process starting in 1989 is one of gradual inclusion of the large and, in economic terms, important fisheries, into the ladder system.

The report from the Resource Distribution Task Force is predominantly 'technical' in the sense that it basically presents and comments on statistics pertaining to the sharing of quotas (TAC) within the different fisheries for a ten-year period. It is also technical in so far as the proposed allocation rules constitute a system for 'automatic' distribution of resources among different groups of fishers. These groups are widely encompassing as implied by the distinction between coastal fisheries and offshore/trawling. That said, one should also note that the National Assembly did follow up on the Resource Distribution Task Force's proposal for a more elaborate system for the coastal fleet by dividing it into sub-groups according to boat size. The main reason for dividing this fleet into four groups was to improve the economic viability of smaller vessels by giving them a set share of the inshore quota. The aim, then, was to avoid a situation where the larger and most effective boats took the lion's share of the inshore quota.

---

[2] The Boat Owners' Association (Fiskebåtredernes Forbund) is a functional sub group within, but also partly independent from, the Fishers' Association. Inclusion of functional groups in the late 1960s was a break with the principle of territorial representation used since Association was founded in 1926. The new organisational structure has been a matter of internal controversy ever since.

Table 6.2. Development of quota allocation rules

1989	*1994*	*2001*
Cod north of 62ºN		
	Haddock north of 62ºN	
	Saithe north of 62ºN	
	Norwegian spring spawning herring	
	Herring south of 62ºN	
	Mackerel	
	Capelin in the Barents Sea	
	Capelin by Jan Mayen, Iceland, and Greenland	
		Saithe south of 62ºN
		Greenland halibut north of 62ºN
		Demersal fish by Greenland
		Demersal fish by the Faeroe Islands

The task force divided the vessels into four groups according to length. Then, for each group, quota shares were proposed for six fisheries; cod north of 62ºN, haddock, saithe, Norwegian spring spawn herring, mackerel, and herring south of 62ºN. By so doing, the task force added complexity to the ladders already in use. Table 6.3 illustrates the arrangement for cod north of 62ºN, where the Norwegian TAC is first divided between trawlers and the coastal fleet and then within the coastal group according to a key defined by the task force.

Table 6.3. Allocation rules for distribution of cod north of 62ºN set by the Fishers' Association National Assembly in 2001

Norwegian Allowable Catch (NAC) in tonnes	*Trawlers (%)*	*Coast (%)*	*Coastal group - internal distribution (%)*
< 130,000	29	71	
130,000-330,000 Linear increase/decrease in percentage of shares	Max. 33	Min. 6	0-9.99 metres: 14 10-14.99 metres: 37 15-20.99 metres: 32 21-27.99 metres: 17
> 330,000	33	67	

The allocation rules agreed upon within the Fishers' Association have had a great impact on the actual distribution of resources. As pointed out, the Ministry of Fisheries approved the 1989 quota ladder, and the same happened in 2001. As shown in Table 6.4, the actual cod quota allocation to a large extent followed the proposals of the Resource Distribution Task Force and the adjustments made by the Fishers' Association.

6.3.2 BENEFITS

The extension of allocation rules to an increasing number of fisheries is a major achievement, not least for the Fishers' Association, for several reasons. An overall goal for the formulation of long-term allocation rules has been to create stability for the fishing industry, especially for full-time fishers. The lack of predictability in the fisheries has been perceived as a problem in need of a solution – by government as well as by the Fishers' Association. The Standing Committee on Business and Industry in Parliament has repeatedly argued that a stable resource allocation between different groups of vessels and gear was necessary to stimulate long-term planning and structural adjustment of the fishing fleet (Innst. S. nr. 93 1998-1999:13). By underlining the need for structural adjustments, the committee identifies a recurrent and unsolvable political issue. The subsidy scheme established in 1963 and negotiated by the Ministry of Fisheries and the Fishers' Association, accelerated the fleet overcapacity problem. However, when the subsidy arrangement was phased out in the early 1990s, the government lost an important policy instrument (Hernes 1999). Despite later efforts to find new ways to reduce excess capacity, the means have either been politically unacceptable, like ITQs, or insufficient, especially when considering the rapid technological developments in the coastal fleet (St.meld.nr. 20 2002-2003:40ff).

Table 6.4. Actual distribution of cod north of 62°N for 2003 set by the Ministry of Fisheries (St. meld. Nr. 20 2002-2003)

Norwegian Allowable Catch (NAC) in tonnes	*Trawlers*	*Coast*	*Coastal group - internal distribution*
195,435	57,919	137,516	
			Group I: 106,836
			0-9.99 metres: 14,156 (13.3 %)
	(29.6%)	(70.4%)	10-14.99 metres: 40,865 (38.3 %)
			15-20.99 metres: 32,051 (30.0 %)
			21-27.99 metres: 19,765 (18.5 %)
			Group II: 13,064
			Boats 28 meters+ 17,616

The Fishers' Association has expressed dissatisfaction with frequent quota level changes. The scientists' problem with stock assessments, and frequent examples of rapid and unexpected alterations in quota recommendations, have been met with scepticism towards scientists and the biological models currently in use. As a better solution, the Association reiterated its request for fixing TACs for more than a year at a time, in order to improve the industry's capacity for long-term planning. As this claim was never met, defining allocation rules can be seen as a second best solution.

Although the quota ladders have clearly been beneficial from a long term planning perspective, the process of allocation rule negotiations has been cumbersome. The Association is a fragile coalition, and reaching viable compromises has always been hard. On several occasions – for instance at the National Assembly meeting in 1992 when some members, completely out of schedule, tried to renegotiate the quota ladder – the organisation has been bursting at the seams. There was always the risk that the Association would fall apart. All the same, the closing of the commons has benefited the members and thus discouraged their exit. Also, the obvious advantages of maintaining the unity and political clout of the Association, has worked in this direction.

The collapse of the cod stock in the Barents Sea mobilised coastal communities against government management practices and scientists (Jentoft 1993). The corporatist arrangement, where the Fishers' Association had been a key player, was put under pressure. A reorganisation of this arrangement, especially of the Regulatory Council,[3]

[3] The Council is a 'corporatist' body that advises the Directorate of Fisheries on management issues. The industry – the Fishers' Association in particular – holds a majority on the Council, which also includes representatives from science and the Sami parliament. The Council is generally considered to be influential, as the Directorate, and eventually the Minister, tend to follow its advice.

could have ended the Association's prominence and paved the way for the representation of other stakeholders. The Association has always rejected such claims, whether they implied a seat for environmental groups in the Council or a say for Parliament in quota management. For instance, the introduction to the 2001 National Assembly decision on allocation rules contains a policy statement expressing the need to keep quota issues a matter for fishers as a group, and the Association's willingness to shoulder management responsibilities. On this matter, the Association could refer to support by the Parliament's Standing Committee on Business and Industry, who emphasised that the industry's organisations should cooperate in the process of allocating quotas among different groups.

6.4 Justice and rights

Given the growing prominence of long-term quota allocations in Norwegian fisheries management, we would expect a principled debate on what constitutes a just distribution and participation among those affected. However, it is difficult to find examples of such a debate. The Resource Distribution Task Force, for example, made a straightforward argument in favour of 'historical rights' or entitlements. The task force obviously took for granted that extending the system of quota ladders should not cause radical changes to the existing pattern of fishing rights. The argument was twofold. First, using historical catch as a basis for initial allocations would promote the wanted stability. Second, continuing established practices would be the least controversial within in the fishing industry. Thus the task force recommended a ten-year reference period, but with the possibility of variation in allocation rules from one fish stock to the other (Ressursfordelingsutvalget 2001:13f).

Thus one may be tempted to conclude that in Norwegian fisheries management, justice is basically a 'philosophical' issue with few practical implications – and as such of little interest to empirical social science. However, such a conclusion would be hasty and superficial. Like David Miller, we think it is both necessary and important to discuss distributive arrangements from a theoretical perspective even though the question of justice is for the most part ignored by the decision-makers themselves (Miller 1999: 42ff). The question is how the idea of historical rights and the pattern of resource distribution among vessel groups can be related to various concepts of justice. We organise the discussion as follows; first we focus on the underlying principle of justice, then we discuss how the conception of what it is to be a fisher may have important implications for the distributive pattern.

6.4.1 FISHERS AND DESERT

Within the social sciences there are basically two views on fishers as social actors (Jentoft and Davis 1993). One emphasises fishers as 'rugged individuals' – embedded in their communities, adhering to local values, norms and rules, and adapting their fishing activities to fluctuating resources and subsistence needs. The other perceives fishers as 'utilitarian individualists' – profit-seeking individuals not restrained by community norms and values, fully oriented towards the market, using the most efficient equipment, and restrained only by rules imposed from 'above', by the state.

An important question is how these views tie in with different substantive principles of justice for governmental policy. The perception of fishers as rugged individuals may lead to an emphasis on need when public goods are distributed – as was the case when the Norwegian state offered financial support to fishers in dire straits during the 1930s. But after World War II, at least from the 1960s onwards, the government and the Fishers' Association underlined the importance of awarding the most productive and efficient fishers measured in terms of catch delivered. This was also the basic principle underpinning the agreement on subsidies between the state and the Fishers' Association in the early 1960s. Most of the subsidies were dispersed according to desert, as an extra income for catch delivered. In other words, the more you caught, the larger the subsidy.

The Fishers' Association perceived the subsidy scheme as essential in keeping them financially afloat and at an income level comparable to that of industrial workers. For the state, the overall goal was to improve the economic efficiency of the fishing industry. The subsidy scheme enjoyed general support within the association until the early 1980s. After that, growing criticism of the regional distribution of government subsidies gradually came to undermine the legitimacy of the scheme (Jentoft and Mikalsen 1987) In response to this, the Association changed its strategy, and argued that the state should calculate subsidies, not on the basis of the size of the catch, but according to onboard working time, a procedure that would still be based on desert as a principle for allocation. The state, however, rejected the idea (Hernes 1999).

Turning to the quota system, the desert principle is not as easily identified. Yet, several things still suggest that desert is the underlying justice principle here as well. First, guaranteed quota rights were reserved for those who could prove a minimum catch level over a period of five years, thus proving their status as bona fide fishers. Part-time fishers and those with little activity before the quota system was introduced, were not considered eligible for guaranteed quota rights. As such, a quota, or more precisely the institutionalisation of a right to fish, has in itself been a reward for the already well-established boat owners. Second, even though this was never intended, a 'grey' market for quota rights has developed – with prospects for windfall profits for those who choose to sell. This has, by and large, been officially accepted, as when Parliament, in June 2003, supported transferable quotas in the coastal fleet. A question raised in the literature on social justice is the problem of distinguishing between need and desert in real life settings. A simple objection to the argument presented above is that the actual underlying principle is need, since those boat owners benefiting are those in need of income to meet financial obligations, salaries, reinvestment and the like. In some sense it is correct that in this instance need and desert overlap, but we will still argue that desert is the most accurate label. For example, in this case need is not synonymous with poverty but rather defined in technical terms as having an economic stake in the resource. Moreover, if we take the Fishers' Association view at face value, the most persistent and capable among fishers should be rewarded.

It is also easy to see that both for individuals and an interest group, desert is a preferable criterion compared to need outside the realm of fundamental requirements, such as health care. In the context of subsidies and quotas, the reverse side of need is the risk that beneficiaries could be stigmatised as clients or free riders. Such labels would have a negative impact in the long run, and would tend to undermine the legitimacy of political action. As the opposite, desert is based on what you deserve as a 'reward' for previous efforts.

6.4.2 RIGHT TO BE RECOGNISED

From a social justice perspective, a focus on the distributional aspects of fisheries management is too narrow. According to Miller (1999), justice should be conceived as a tripartite concept, emphasising need, desert and equality. In addition, there is (historical) entitlement, and the right to be recognised, for instance as being a legitimate stakeholder. The latter points to the procedural – or democratic – elements of justice where participation in decision-making is crucial. The arguments for extending the concept have been raised, among others, by feminist scholars and students of multiculturalism, because the traditional view – what Iris Marion Young (1990) denotes as the 'distributive paradigm' – focuses too much on material goods and does not take into account group differences and the fact that some groups are oppressed and excluded. The implication of a request for "broader and more authentic participation" is a demand for empowerment through more participatory democracy (Schlosberg 2003).

It follows from this that social justice is not possible unless peoples' uniqueness is recognised and procedures are developed to allow for their participation in decision-making. Justice, thus, has both a cultural and an institutional dimension. Even though it is possible to 'split' the concept of justice, its various parts are integrated. As David Schlosberg (2003:96) contends: "one must have recognition in order to have real participation; one must have real participation in order to get real equity; further equity would make more participation possible, which would bring further recognition, and so on" (cf. Honneth 2001).

A comprehensive concept of justice may seem irrelevant for discussing quota allocations in the fisheries, since recognition and participation are terms more appropriately applied to groups that have had to struggle to be recognised. After all, the Norwegian Fishers' Association is no political novice or amateur, as the organisation has long since been recognised as the government's most important partner in fisheries management. In this context, the development of quota management schemes illustrates the fact that government decisions are largely built on proposals advanced by the Association (Hernes 1999). Interestingly, the change from corporatist arrangements to market-based governance in recent years does not seem to have changed the partnership between the association and the state. However, from a justice perspective we will argue that this solid partnership is a barrier to recognising affected interests outside the industry realm as legitimate stakeholders. Two points support this argument.

First, the established resource management regime has been criticised for not taking into account the interests of indigenous peoples – the Sami living along the coast in Northern Norway. The Sami organised a political and cultural 'uprising' in the beginning of the 1980s that paid off in the form of a new paragraph in the Norwegian constitution confirming the government's responsibilities vis-à-vis the Sami people and the establishment of a Sami parliament. Eythòrsson (2003:159) concludes that the interests of the Sami people have been largely invisible in fisheries management: "Matters pertaining to the coastal Sami have been considered not merely irrelevant, but highly inappropriate". The rather limited appreciation of the Sami presence in the fishery was clearly demonstrated when the new quota system was introduced in 1990: almost no Sami fisher qualified. The protests raised by the Sami Parliament resulted in the appointment of a Sami representative to the Regulatory Council. This analysis by Eythòrsson and others suggests that corporatism can be an obstacle to the recognition of new and legitimate stakeholders. The Fishers' Association – like other actors in Norwegian politics and administration – has rarely recognised the ethnic dimension of certain public policies. The quota ladders and quota markets, as self-governing systems, are 'blind' to questions of ethnicity and cultural diversity. To become relevant, they must be imposed on the management system from the outside, from the government or from civil society.

Second, the co-management role of an interest group such as the Fisher's Association and its close ties with the state imply that the few governs on behalf of the many and that legitimacy rests on the results achieved. From a democratic perspective, excluding other stakeholders from the decision-making process cannot be justified. After all, the Fisher's Association does not represent more than about 60 percent of all fishers. Also, not only fishers are affected by management decisions. The problems encountered by coastal communities in the aftermath of the Barents Sea cod crisis in 1989 mobilised large segments of the population, as well as regional and municipal authorities, who claimed that fisheries management is too important to be left to the 'cosy' coalition of government and Fishermen's Association (Jentoft 1993).

The failure of the Norwegian Fishers' Association – and the management system as such - to include ethnic and territorial concerns, can be explained in different ways. First, although the organisation was established on the basis of regional associations, and controversies along geographical lines have been common, it has always been important to avoid that internal strife which ends in disruption. That would only create dissatisfaction among Association members. Second, when representing the fishing industry, members of the Association identify themselves as representatives of a specific 'functional' category such as boat owner, trawlers, purse seiners or long liners, and not on the basis of their home port or region (Jentoft and Mikalsen 1994).

Also, politicians and bureaucrats have been reluctant to give preferential treatment on the basis of territorial and ethnic characteristics. Predominantly, they adhere to the principle that fish resources are a national property and hence, that management should rest with the state in cooperation with groups such as the Fishermen's Association that are directly affected, and not with regional or municipal authorities. Furthermore, resource rights should be vested in the individual and not in a collective, such as a municipality or a local community. This is a conception of justice that gives priority to desert and not to dependency or need; the management system allows quota-holders to buy and sell their rights, often to the detriment of fisheries dependent communities, municipalities or regions. How the allocation of quotas to individuals, as private property to be bought and sold, fits with the principle that fish resources are a national property is an issue which we will not go into here.

6.5 Conclusion

Walzer's argument is that, since society has no single principle of justice, we should distinguish between different spheres of justice. The Norwegian experience with quota ladders suggests that it is difficult to draw straight lines between spheres, as actors within the industry base their activity on one principle (desert) that is perceived as dubious, if not totally illegitimate, by other stakeholders who are kept outside the management 'loop'. Therefore, fisheries management seems to work according to several principles of justice, which managers somehow must attempt to reconcile. The calibration of different justice principles necessitates a management process that is inclusive; one that allows for a broader group of stakeholders to become involved in the decision-making process. In this sense, democracy may be considered a crucial precondition of social justice. If so, the Norwegian co-management system definitely needs to be reformed. The privileged role of the Norwegian Fishers' Association is questionable, as it tends to favour functional groups and suppress territorial and ethnic interests and concerns. The quota ladders are a consequence of such a democratic deficit.

Having said this, the very idea of a 'social contract', such as a quota ladder, has some obvious advantages in fisheries management. If stakeholders could arrive at some consensus among themselves on how to allocate scarce resources, the likelihood of co-management through the delegation of management authority to user-groups increases. In other words, the more fragmented and divided user-groups are, the more it is necessary for central government to interfere. The Norwegian government has largely accepted the quota ladder that the members of the Fishermen's Association have negotiated and agreed on among themselves. In 2002, for example, the Fisheries Minister proclaimed that he would not alter the arrangement but stick to the key agreed by the partners involved. He was heavily criticised in the media for declining to intervene in such an important policy issue. One may, of course, question whether this is a sensible thing to do for a minister who is ultimately responsible for all aspects of the fisheries – and to a much wider group of stakeholders than just fishers. Nevertheless, it can be interpreted as a step towards a real devolution of management authority, signalling a great level of trust in the organisation's ability to act responsibly. (There is, of course, a less flattering interpretation: the minister – and the political system – finds it politically convenient to leave controversial issues to the parties involved.)

Whether the agreement will continue to receive support among the fishers and the government in the future remains to be seen. If it does not survive, fishers may become even more divided than they are today. If conflict cannot be avoided, it is better to have the fishers fighting each other each time the allocation key is renegotiated than having them fighting each other all the time. No doubt, the quota ladder system reduces the transaction costs of fisheries management in Norway.

Today, the allocation key pertains only to quota shares among various 'segments' of the fleet. However, the 'contract' could well be extended to include other contentious issues, such as the allocation of quotas among regions and between ethnic groups. A contract should also specify who should be considered as legitimate stakeholders with a claim to representation in decision-making. In other words, we believe that 'social contracts' of this type may have great merits in fisheries management, but that the current quota ladder system is too narrow both in focus and in representation to provide for a more comprehensive discourse on social justice principles in fisheries management. A social contract for the fishery cannot be imposed from the top down. Instead, it must be built on democratic principles, where all affected stakeholders should be allowed to voice their concerns. Only through such a contract can issues of social justice inform the decision-making process. Far too often, concerns of social justice are suppressed as fisheries management is reduced to a technical fix. No wonder therefore, that fisheries management continues to be among the most contentious areas of public policy, where selective and centralised consultation is undermining the legitimacy of management policies and decisions.

Acknowledgements

The work on this chapter has been sponsored by the EU-funded project *Sharing Responsibilities in Fisheries Management* (Q5RS-2001-01998)

References

Armstrong, C (1998) 'Sharing a fish resource - bargaining theoretical analysis of an applied allocation rule' *Marine Policy* 22:119-134

Armstrong, CW and Clark, DJ (1997) 'Just fishing? Equity and efficiency in fisheries management regimes' *Marine Resource Economics* 12:203-220

Bavinck, M (1996) 'Fisher regulations along the Coromandel coast: a case of collective control of common pool resources' *Marine Policy* 20(6):475-482

Campbell, T (2001) *Justice*, New York, St. Martin's Press

Elster, J (1992) *Local Justice: How Institutions Allocate Scarce Goods and Necessary Burdens,* New York, Russell Sage Foundation

Eythòrsson, E (2003) 'The Coastal Sami: a 'Pariah Caste' of the Norwegian Fisheries? A Reflection on Ethnicity and Power in Norwegian Resource Management' in S Jentoft, H Minde and R Nilsen (eds) *Indigenous Peoples. Resource Management and Global Rights,* Delft, Eburon:149-162

Gray, T (1998) 'Fishing and fairness: the justice of the Common Fisheries Policy', in T Gray (ed.) *The Politics of Fishing*, London, Macmillan

Hardin, G (1968) 'The Tragedy of the Commons' *Science* 162:1243-48

Hernes, H-K (1999) *Forhandlingsmakt eller argumentasjonsbyrde? En analyse av Hovedavtalen for fiskerinæringen og avviklingen av fiskeristøtten,* Tromsø, Dr. polit-avhandling, Det samfunnsvitenskapelige fakultet

Hernes, H-K and Mikalsen, KH (2002) 'From protest to participation? Environmental groups and the management of marine fisheries' *Mobilization: An International Journal* 7(1):15-28

Hirschmann, AO (1970) *Exit, Voice, and Loyalty: Responses to Decline in Firms, Organizations, and States*, Cambridge Mass., Harvard University Press

Hoel, AH, Jentoft, S and Mikalsen, KH (1996) 'User-group participation in Norwegian fisheries management' in RM Meyers, C Zhang, ML Windsor, BJ McCay, LJ Hushak and RM Muth (eds) *Fisheries Resource Utilization and Policy, Proceedings of the World Fisheries Congress, (Theme 2)*, New Delhi, Oxford & IBH Publishing CO. PVT. LTD.

Honneth, A (2001) 'Recognition or redistribution: changing perspective on the moral order of society' *Theory, Culture & Society* 18(2-3):43-55

Jentoft, S (1993) *Dangling Lines: The Fisheries Crisis and the Future of Coastal Communities, The Norwegian Experience*, St. John's, Newfoundland: Memorial University of Newfoundland, Institute of Social and Economic Research

Jentoft, S and Davis, A (1993), 'Self and sacrifice: an investigation of small boat fisher individualism and its implication for producer cooperatives' *Human Organistion* 52(4)

Jentoft, S and Mikalsen, KM (1987) 'Government subsidies in Norwegian fisheries: regional development or political favouritism?' *Marine Policy* July:217-228

Jentoft, S and Mikalsen, KH (1994) 'Regulating Fjord fisheries: folk management or interest group politics?' in CL Dyer and JR McGoodwin (eds) *Folk Management in the World's Fisheries. Lessons for Modern Fisheries Management*, Piwot Colorado, Colorado University Press:287-316

Jentoft, S, Minde, H and Nilsen, R (eds) (2003) *Indigenous Peoples. Resource Management and Global Rights*, Delft, Eburon

Kymlicka, W (2002) *Contemporary Political Philosophy: An Introduction*, Oxford, Oxford University Press

Miller, D (1999) *Principles of Social* Justice, Cambridge, Harvard University Press

Pálsson, G (1998) 'The virtual aquarium: commodity fiction and cod fishing' *Ecological Economics* 24:275-288

Rawls, J (1972) *A Theory of Justice*, Oxford, Oxford University Press
Schattschneider, EE (1960) *The Semi-Sovereign People*, New York, Holt, Rinehard and Winston
Schlosberg, D (2003) 'The justice of environmental justice: reconciling equity, recognition, and participation in a political movement', in A Light and A de-Shalit (eds) *Moral and Political Reasoning in Environmental Practice*, Cambridge Mass, MIT Press
Sen, A (1987) *On Ethics and Economics*, New York, Basil Blackwell
Sen, A (1999) *Development as Freedom*, Oxford, Oxford University Press
Sumaila, UR and Bawumia, M (2000) 'Ecosystem justice and the marketplace', in H Coward, R Ommer and TJ Pitcher (eds) *Fish Ethics: Justice in the Canadian Fisheries*, St. John's Newfoundland, Institute of Social and Economic Research (ISER), Memorial University 140-153
Symes, D (ed.) (2000) *Fisheries Dependent Regions*, Oxford, Fishing News Books/Blackwell Science
Walzer, M (1983) *Spheres of Justice: A Defence of Pluralism & Equality*, Oxford, Basil Blackwell
Young, IM (1990) *Justice and the Politics of Difference*, Princeton New Jersey, Princeton University Press

Documents & public records

Innst. S. nr. 93 (1998-99) Innstilling fra næringskomiteen om perspektiver på utvikling av norsk fiskerinæring
Landsmøtesak 7/2001 Norges Fiskarlag
Ressursfordelingsutvalget (2001) Innstilling fra Ressursfordelingsutvalget, Januar 2001, Trondheim, Norges Fiskarlag
St. meld. nr 20 (2002-2003) Strukturtiltak i kystfiskeflåten, Oslo, Det kongelige fiskeridepartement.

CHAPTER 7
BETWEEN TOP-DOWN AND BOTTOM-UP GOVERNANCE: DUTCH BEAM TRAWL FISHERMEN'S ENGAGEMENT WITH FISHERIES MANAGEMENT

ROB VAN GINKEL
Department of Sociology and Anthropology, University of Amsterdam, Oudezijds Achterburgwal 185,1012 DK Amsterdam, The Netherlands

Abstract

Since 1993, the prime goal of Dutch fisheries policy has been to enhance a responsible way of fishing and a sustainable exploitation of fish stocks. That is, economic and ecological interests should be balanced in a viable way so as to achieve both economic and ecological sustainability. This policy, which encouraged new forms of cooperation, was superimposed on a system of individual transferable quotas that was officially introduced in 1985. To arrive at devolution of specific management responsibilities to fishermen, they had to organise themselves in co-management groups, the so-called 'Biesheuvel groups'. Individual fishermen bring their catching rights or quotas into these groups, and these groups are responsible for establishing fishing plans to achieve a better distribution of sea days and quota transfers within a group. The state's aim is to enhance fishermen's responsibility and social control through self-management. This chapter will address the experiences over the past ten years with this governance system, focussing especially on the conflicting views of fishermen, biologists and state representatives regarding its efficacy. Special attention will also be devoted to the perceptions of the beam trawl fishermen concerning the benefits and pitfalls of the present governance system.

7.1 Introduction

Fisheries management in the western world is usually characterised by top-down modes of policy design and implementation that involve centralised, hierarchical, command-and-control decision-making and monitoring to make up for market imperfections (Symes 1997:107; Dubbink and van Vliet 1996, 1997). The fisheries policy of the European Union (EU) is a prime example, where measures affecting the fishing industry are determined in Brussels with little or no involvement of fishers and their organisations. Such top-down policymaking often leads to a lack of transparency, high information, monitoring and enforcement costs, as well as discontent and a lack of legitimacy and compliance on the part of the fishing industry (McCay 1995:16). On the other hand, national governments have some leeway in arranging the specifics of governance structures for sea fisheries within the framework set by the EU Common Fisheries Policy (CFP).

In the Netherlands, a devolved or co-management regime was introduced in the early 1990s, delegating considerable responsibility to the fishing industry for quota management, self-regulation and self-enforcement. Co-management usually refers to "a shift in the power for decision-making away from the government agency and a

T. S. Gray (ed.), Participation in Fisheries Governance, 119–139.

scientific élite, and toward a group of resource users or a local community" (McCay 1995:14) or to forms of shared management responsibilities of state institutions and user groups (Nielsen and Vedsmand 1999:20). The idea in the Dutch situation was that the less legalistic approach of governing at this level would leave more discretion to the fishers and firms "to adapt their conduct to 'the spirit of…public policy'" (Dubbink and van Vliet 1997:183). This in turn would give a boost to the legitimacy of government and augment compliance to its rules and regulations (ibid:184). The present chapter will address the experience over the past decade with this governance system, focusing especially on the views of fishermen, biologists and state representatives regarding its efficacy. Special attention will be devoted to the perceptions of the beam trawl fishermen concerning the benefits and pitfalls of the current governance system.

The Dutch fishing industry is relatively small in terms of number of vessels and employees. In 2002, the fishing fleet comprised 393 cutters, 17 large pelagic freezer trawlers and 101 shellfish fishing boats (Taal *et al* 2003). Total employment in the fishing and shellfish-farming industry is approximately 2,650 jobs, excluding related sectors such as the processing industry, auctions, supply sector and retail trade that provide another 15,500 jobs. This only constitutes a tiny fraction of total employment in the Netherlands. However, although these numbers are relatively modest, the Dutch fish trade occupies an important position in Europe and the fleet is up to date in every respect. The value of total fish landings in the Netherlands amounted to 463 million euro in 2002; that of exports of fish and fish products was almost two billion euro. After a prolonged period of exceptionally good results in the 1990s, 2002 brought a seventeen per cent decrease of revenue compared with the previous year and a negative net result of four million euro.

The most important sector of the fishing industry is the capital intensive beam trawl fleet, which operates mostly in the North Sea to catch sole (*Solea solea*) and plaice (*Pleuronectus platessus*) on four to five-day trips usually starting on Monday. These species contribute approximately eighty per cent to the total revenues of beam trawling, a fishing technique that is applied by about half of the cutter fleet. The Netherlands holds a significant share of the EU Total Allowable Catch (TAC) for sole (roughly 75 per cent) and plaice (38 per cent). The larger vessels (exceeding 1500 horsepower engine power, with a length of 40-45 metres and crewed by six or more) land more than ninety per cent of the total market supply of flatfish. Vessels up to 300 horsepower are allowed to fish within the 12 nautical mile zone and the plaice box. In 1998, 43 of these multi-purpose Euro-cutters with a length of 20 to 25 metres and a crew of three or more were fishing for sole and plaice as the main target species. The majority of beam trawlers are family-owned and operated. Most flatfish fishing firms own one vessel, about a third own two or three, including several of the bigger beamers. Important concentrations of beam trawl fishers can be found in Urk, Goedereede, Stellendam, Arnemuiden, Vlissingen, Den Helder and on the island of Texel. Dutch beam trawl fishers operate under a share system of remuneration with crewmembers signing a partnership contract with the owner (*maatschap*), implying that all crewmembers earn a certain percentage of the net revenues.

The Netherlands was among the first countries to introduce individually transferable quotas (ITQs). This happened *de jure* in 1985, although *de facto* the practice of trading individual quotas developed in the late 1970s. Generally advocated by neoclassical fisheries economists as an effective and efficient mode of managing fisheries, the Dutch experience of ITQs shows that there was a multitude of problems of enforcement and compliance, leading to huge transaction costs. Therefore, the state took additional measures in an attempt to improve the situation. In particular, it tried to involve fishermen in fisheries governance by devolving several management, regulation and enforcement tasks to groups of fishermen in 1993. This led to a management regime that combines theories put forward by neoclassical economists on the one hand, and by institutional economists and anthropologists on the other hand, concerning a proper, participatory and effective fisheries management system. Before turning to the co-management regime, I will first outline the post-war developments in the flatfish fisheries that eventually led to the implementation of a participatory management policy.

7.2 Fishy business, flawed policy and faltering enforcement

The Dutch fishing fleet expanded considerably after World War II, both in number of vessels and in capacity. More or less starting from scratch - most vessels were either confiscated or demolished as a consequence of the hostilities - the fleet grew at a fast rate and the fisheries flourished. Between 1951 and 1971, engine capacity increased tenfold to 250,000 horsepower. It was especially the invention, in the late 1950s, of the beam trawl aimed at catching flatfish (principally sole and plaice) that enabled the expansion and innovation of the Dutch fishing fleet. This fishing technique has been hugely successful. It is based on the use of a steel tube or beam (*boom*) to which a trawl (*kor*) and chains are attached. The vessels use two such beam trawls, one on either side. The beam trawls are dragged across the sea bottom and the chains scare the flatfish, which then end up in the nets. It was soon discovered that the heavier the chains, the larger the catch. However, heavier chains required greater engine capacity and the fishers were quick to make the necessary investments.[1]

By 1966, catches diminished slightly. This tendency was believed to be a first indication of overcapacity and overexploitation of fish stocks. Several beam trawl fishers showed concern that the development of the fleet would lead to a 'horsepower race' that would ultimately result in unmitigated competition and possible bankruptcy, while the effect on flatfish stocks would be devastating. In 1971, the Dutch Fishermen's Union (*Nederlandse Vissersbond*) discussed plans to limit engine capacity to 800 horsepower per vessel. But the plans did not become effective, mainly because by this time many vessels were already equipped with more powerful engines. Proposals to allow a maximum of 1,000 horsepower were also rejected by the administration. The Fishermen's Union emphasised that the fishers were caught in a devastating competition and that the government should intervene. However, the government refrained from intervention. The fishery was mainly controlled by minimum mesh and fish size regulations. As a consequence of certain tax measures, it was wise to reinvest in vessels and equipment. This also stimulated the expansion of the fleet and aggregate engine capacity. The government favoured a stop on fishing in certain areas during certain periods, but it deemed such a policy only feasible when measures were taken at a higher level of political integration.

In 1975, the North East Atlantic Fisheries Commission established TACs for several species of fish, including sole and plaice. Each state received a share based on 'historic

[1] Though I use fishermen and fishers interchangeably, nearly all fishers are male. It is highly exceptional to find women among the crewmembers of Dutch fishing vessels.

rights'. Dutch fishers got over seventy per cent of the TAC for sole, and nearly forty per cent of the TAC for plaice. These species represented about two thirds of the fish landed by Dutch fishermen. The national quota for sole amounted to 9,445 tonnes, the one for plaice was set at 47,020 tonnes. This implied a drastic reduction in legal landings of forty-seven per cent and nine per cent respectively compared to the previous years (Davidse 1998:58). Based on the track records of their highest catches in the three previous years, the sole and plaice fishers were assigned an individual, non-transferable quota (IQ) in 1976.[2] The idea underlying the measure was that allocation of exclusive rights would offer fishers the prospect of earning resource rent and maximising their profit, increase operational certainty and lead to a responsible mode of fishing. Initially, many fishers feared that they would lose their jobs and end up on the dole. The oil crisis and economic recession contributed to their pessimistic views. Nonetheless, their attitude was to stay in business as long as possible: "If I have to hang, then on the last tree be it" (Davidse 1998:60). To them, fishing was much more than a way of earning a living; it was an existential matter, an important marker of identity and a way of life to be cherished and to be continued by successors. Therefore, fishers continued to invest in their vessels and other equipment with the help of investment premiums, tax-free fuel and the ample availability of bank loans: "Those who did not invest began to lag behind and lost the race" (Dubbink *et al* 1994:32).

The 'catching rights on paper', the introduction of which had deeply worried the fishermen, proved to represent substantial economic value. The fishers, who had received these rights free of charge, could capitalise on them. It did not, however, lead to a more responsible and efficient fishery that enabled fishers to respond flexibly to changes in the market, while the burden of management still rested with the government (van Vliet 1998a:219). Property rights were insecure, since the flatfish fishery would be closed once the national quotas for sole and plaice were exhausted. In spite of the IQ system, the race for fish was not eliminated. The expansion of the fleet and engine capacity continued after the mid-1970s, despite the express intent of the government to decrease these through the introduction of quotas. In 1977, there were 495 vessels totalling 325,000 horsepower; a decade later, the number had risen to 611 vessels totalling 580,000 horsepower and manned by 3,036 crewmembers. By then, vessels with more than 2,000 horsepower engines were no exception. The expectation was that more engine power would result in more pulling power and thus more fish[3], while at the same time reducing steaming time and enabling vessels to continue fishing in rough weather. Many vessel owners felt a need to follow suit, while newcomers (usually crewmembers who aspired to become independent and sons of owners who wanted to set up an independent firm) bought second-hand vessels from those who had ordered new vessels.

One of the driving forces behind this growth was, therefore, the fishers' desire to become independent and to set up each owner's son who showed an interest in fishing with his own vessel to skipper. Fishing was an important source of identity and an occupation that was highly valued by those in the industry, something they – skipper-owners in particular – wanted to pass on to the next generation. The social dynamic of the family firm contributed importantly to the expansion. Depending on the stage of the

---

[2] In 1977, due to widespread dissatisfaction, the allocation rule was altered to give equal weights to the basis of the historical catch record and the engine power of the vessel.

[3] However, there is no linear relation. Beyond a certain point (1500 horsepower) productivity increases only marginally, and per unit effort, catches and revenues decrease. However, it is still attractive to have powerful vessels since costs per unit of effort are lower (Davidse and de Wilde 2001:51).

family cycle, a father and son(s) or brothers usually formed the core of the crew. When the father retired and became a 'shore skipper', brothers usually continued the firm and cooperated until their sons joined them aboard ship. There would be insufficient positions to accommodate all. In addition, cousins often did not get along when it came to fishing matters and wanted to work with their father on their own vessel. As a consequence, schisms of family firms led to an enlargement of the fishing fleet. Bank loans were amply available, since catching rights could be used as collateral. In addition, from 1978, a general investment subsidy in the form of a fiscal allowance of twelve per cent or more on newly built vessels created an important incentive to invest in new vessels (this allowance was abolished ten years later).[4] When individual quotas became officially transferable in 1985, the same tax benefits applied to the purchase price of ITQs.[5]

This period of rapid expansion of the Dutch fishing industry was characterised by reports of illegal fishing, under-reporting of catches, grey and black trade circuits and inadequate policing and enforcement by the Dutch state. Auctions helped fishers logistically and administratively to get rid of illegal fish, while local authorities turned a blind eye to what was going on. Even if fishers were caught fishing or landing fish illegally, judges usually sentenced them to low monetary fines that still made it attractive to continue the practice. 'The last haul for the judge' became an oft-used expression. The flatfish fishermen consequently overshot their quotas by far, leading to early exhaustion of national quotas and subsequent premature closure of the fisheries. The fishers who had not yet fished up their individual quotas suffered. With this experience of being caught in a prisoner's dilemma, the race for fish continued, stimulated by lenient enforcement, ample opportunities to circumvent rules, and a strong demand for flatfish abroad resulting in high prices and rising revenues.

Abroad, Dutch fishers who evaded or violated the regulations faced tighter surveillance. In the early 1980s, the British and German authorities caught many Dutch beam trawl cutters fishing in their territorial waters or using illegal nets. Consequently, they were fined heavily. These events had an impact on the fishermen's public image. For a long time, they had been regarded as 'noble commoners', toiling to eke out a livelihood. But increasingly they were viewed as irresponsible and reckless egotists plundering the sea's resources. Soon there was a call for stricter policing and punishment of those who evaded the rules and regulations, especially after the introduction of the CFP in 1983. But this proved difficult in practice. The Ministry of Agriculture and Fisheries was criticised for failing to enforce the law and even turning a blind eye to illegal fishing. In 1987, this led to an enquiry of a Parliamentary sub-committee into the role of the Ministry as regards over-quota landings. Under mounting public pressure, politicians established a tighter legal and enforcement regime, including a days-at-sea regulation based on the relation between vessel capacity and catching rights, a licensing system

---

[4] Quota restrictions on the one hand and investment subsidies on the other hand gave contradictory incentives and ambivalent messages to the fishing industry. They "placed the sector in a somewhat schizophrenic position and increased the uncertainty in the sector" (van Vliet 1999:168; on these issues, see also van der Schans 2001:425ff).

[5] *De facto*, transfers of individual quotas occurred much earlier. Fishermen simply bought a vessel with associated quotas and then sold the vessel without quota rights. Fission and fusion of enterprises also enabled transfers of quota rights. Thus, the government merely put the practice on a statutory footing. However, it stipulated that only a vessel's entire quota could be sold, though POs could purchase an entire ITQ and then resell it in parts to their members (Valatin 2000b). As McCay writes, "ITQs amount to a giveaway of public resources" (1995:14). See also Symes and Crean (1995).

and compulsory registration and control of all landings (van Vliet 1998b:70). The state also created a quota reserve of five per cent to cover individual over-fishing of ITQs and to allow others to fully use their rights.

Despite these additional measures, however, reports about fishers who not only evaded and breached the law but who even harassed inspection officers - setting fire to their vans - figured prominently in news bulletins in 1988. As a consequence of a reduction of the sole TAC in the second half of the 1980s, illicit landings increased.[6] For the government, the small fisheries sector proved the worst headache. Catches continued to exceed the national quotas and as a consequence of his failure to contain this problem and misinforming Parliament, the Minister of Agriculture and Fisheries, Gerrit Braks, had to resign on 19 September 1990. This political crisis made it abundantly clear that the "'command and control' regulation and the adversarial relations between government and the fishing industry was untenable" (van Vliet 1998b:71).

Meanwhile, the fishing industry's representatives realised that something must be done to improve the situation. Moreover, since 1985, EU pressure resulted in a reduction of overcapacity through an engine-power licence-scheme, while a maximum engine capacity of 2,000 horsepower was subsequently introduced for newly built vessels.[7] A voluntary decommissioning scheme was also adopted. This led to a decline of the number of vessels and crew in the Netherlands. While in the mid-1980s there were still over 600 beam trawl cutters with almost 3,000 crewmembers, in 2004 they number less than 400 and some 1,800, respectively. Aggregate engine capacity declined from 500,000 horsepower to around 400,000 horsepower.[8] But decommissioning grants did not lead to diminishing quotas, since vessel owners only returned their engine capacity licence; they could sell their individual quotas to other vessel owners. In addition, the days-at-sea regulation sought to adjust fishing effort to the available quotas and to prevent the need to close the fishing season before the end of the year because the national quotas were exhausted prematurely (Davidse 1998:59).

Moreover, the Dutch government strengthened enforcement in a serious effort to stop illegal practices. Since January 1988, a team of 120 inspectors closely monitored fish

[6] The national sole quota was 10,160 tonnes in 1988 (slightly more than the 1975 quota), whereas the national plaice quota was 80,570 tonnes (as compared with 47,020 in 1975) (Davidse 1998:58). Grey landings are flatfish landed in excess of ITQs; they are held outside of landing statistics but do appear in the vessel owner's accounts.

[7] By then, there were already scores of vessels with more powerful engines, while quite a few new-buildings had been ordered just before the deadline. In 1998, 63 vessels had over 2000 horsepower engines. They have to be replaced or must reduce their engine capacity to 2000 horsepower before they get older than twenty years (Davidse and de Wilde 2001:10).

[8] However, since 1990, some 75 Dutch flatfish fishermen have bought vessels and purchased fishing rights or re-registered in the UK, Germany and Belgium predominantly. This was enabled by the EU freedom of establishment rule. The re-flagged Dutch firms are fishing against these countries' TAC shares (Davidse 1998:62-63; Hoefnagel 1998; Valatin 2000a:299; Hatcher *et al* 2002). Many of these 'quota hoppers' are flying flags of convenience for one of three reasons: "to provide adequate quotas for their own enterprises [given extremely high prices for flatfish quotas in the Netherlands]; to establish sons in fishing where most fishing enterprises are family owned and crewed; or, in a minority of instances, to profit by selling their Dutch quotas and buying cheaper foreign quotas" (Hoefnagel 1998:82). Thus, re-flagging has enlarged the room to manoeuvre in a situation of national limitations (Davidse and de Wilde 2001:8). "The re-flagged fleet counts for about 20% of the demersal North Sea fishery under the Dutch flag (in 1998), in terms of vessel number, engine-power and fishing effort" (Davidse 2000).

landings for which specific places, times and conditions were set, legal mesh sizes of the cod-ends were increased and the maximum beam length for the flatfish fishery was fixed at 12 metres. Only vessels under 300 horsepower (Euro-cutters) were allowed to fish within the 12 nautical mile zone and in the plaice box with beams not exceeding 4.5 metres overall.[9] In addition, fines for violations of the rules became much stiffer, and some fishers and auction directors have even been imprisoned. As a consequence of stricter enforcement, demand and prices for catch quotas increased sharply, making them an additional production factor.[10] Investment in quotas generally led to a greater balance between catching capacity and catching rights. The fishing industry's growing regulation by national and supranational governments had a strong impact on the daily lives of fisher folk and fishing communities. Those who had first hand experience with the period of 'much liberty and few rules' looked back in nostalgia and considered the tighter measures and policing a stifling of their autonomy. Small wonder, then, that relations between politicians and state officials on the one hand, and fishers on the other, became strained in the second part of the 1980s, with non-compliance being a serious offence for the former, and a perceived survival strategy for the latter.

After the resignation of the Minister of Agriculture and Fisheries and with the growing risk of heavy fines for non-compliance, many fishers feared that their situation would deteriorate further as a consequence of draconian measures that would almost certainly befall their sector. Braks' successor, Piet Bukman, would not want to run the risk of being censured by Parliament for incompetence in harnessing the fishers. In addition, the fishers realised that they had lost control over their individual businesses and the fishing industry as a whole. Moreover, prices were far from optimal due to grey and black landings and the exhaustion of the quotas before the end of the year. While conducting anthropological fieldwork on the Dutch island of Texel from late 1989 until the spring of 1991, I observed the changing mood over the regulatory regime in the occupational community of fishermen. Initially, a majority defended illegal practices by saying that management measures restrained them too much, while sole and plaice were believed to be so abundant that they practically jumped on deck. Later on, the dominant opinion was that restrictions were necessary and should be complied with (which is, of course, not to say that all fishers actually complied): relationships among fishermen became increasingly tense as a consequence of differences of opinion over compliance with the law and suspicions that some fishermen still evaded or violated certain rules.[11]

---

[9] The plaice box, an area of some 40,000 square kilometres north of the Dutch and German Frisian Islands, was designated in 1989. Today, the plaice box is closed for vessels exceeding 300 horsepower (initially this was the case only part of the year). The growing importance of the Euro-cutter section of the fishing fleet is an unintended consequence of fisheries policy. First, many vessels have been upgraded to 300 horsepower. Second, several fishermen whose vessels were more powerful, but whose quotas were small, decided to scale down their business and use their quotas for a Euro-cutter. In 1983, Euro-cutters made up thirteen per cent of the fishing fleet, in 1998 thirty-five per cent. Over the same period, mid-size vessels (301-1500 horsepower) almost disappeared; while big beamers of over 1500 horsepower constituted almost fourteen per cent of the fleet in 1983, and more than thirty-seven per cent in 1998 (Davidse 2000).

[10] Prices of quota rights continued to be extremely high from 1988 to 1992. Quotas were traded even above the net present value of future returns from fishing, suggesting "that the fishermen wanted to stay in business despite the high costs" (van Vliet 1998b:70). The option to sell ITQs at high prices facilitated exit decisions. Although from 1993 onward, harvesting rights became less expensive, investments in them partly absorbed the depreciation for new vessels. As a consequence, the fishing fleet is aging compared with the situation in the 1970s-1980s (Davidse 1999, 2000). The mean value of harvesting rights increased from DFL 150,000 per vessel in 1983 to DFL 5,000,000 in 1998 (Davidse 2000). See also Davidse (1997) for a detailed account of trade in flatfish ITQs.

[11] Such violations included, for example, illegal fishing within the 12 nautical mile zone with vessels equipped with engines of over 300 horsepower; using nets with too small mesh sizes; or keeping part of the landings out of the books by selling on the grey or black market.

At the same time, the Dutch government was looking to rid itself of the increasingly heavy burden of implementing the rules and regulations pertaining to the fishing industry (Davidse 1998:61-62). "We have reached the end of our possibilities", the new Fisheries Minister admitted. Mutual confidence had to be restored and the legitimacy of fisheries policy regained. Hence, a period of reconciliation followed in the early 1990s, when negotiations on the establishment of co-management groups that would be closely associated with Producers' Organisations (POs) were begun between the government and the fishing industry. In addition, a large increase for the sole TAC in 1990 (from 9,656 tonnes in 1989 to 18,000 in 1990) helped to calm down the flatfish fishermen's discontent with the European and national fishery measures and led to greater compliance with quota regulations.

There was also growing pressure for the integration of fisheries concerns and marine environmental management. Since 1993, the prime goal of Dutch fisheries policy has been to achieve a responsible way of fishing and a sustainable exploitation of fish stocks. That is, economic and ecological interests should be balanced in a viable way so as to achieve both economic and ecological sustainability. The state aims at furthering fisheries through regulation, consultation with the industry and quality care. Fishermen have to take into account other functions of the seascape, especially its value as a nature area. This harmonisation policy is laid down in the white paper *Vissen naar evenwicht: Structuurnota Zee- en Kustvisserij* (Balanced Fisheries: Policy Document on Sea and Coastal Fisheries), a document embodying fisheries policy until 2003 (Vissen 2003). It reflects the fact that the environment had taken pride of place on the political agenda during the 1980s. In addition to seeking a balance between economic and ecological interests, *Balanced Fisheries* aims at giving responsibility to the Dutch fishery sector through self-management and new forms of cooperation. To arrive at devolution of specific management responsibilities to fishermen, they had to organise themselves into groups – the so-called 'Biesheuvel groups', named after the chairman of the committee that advised on the new policy, former Prime Minister Barend Biesheuvel. Parliament threatened to introduce regulations to limit engine power of each vessel should the fishing industry decide not to accept organisation in groups.

The developments and measures described above have thoroughly changed the occupational praxis of beam trawl fishers – and to a considerable extent their occupational culture as well. Once they were used to deciding where to cast their nets, but since 1975 they have been increasingly restricted in their operations. State involvement limited their freedom at sea and brought about more paper work, leading to less job satisfaction; tensions mounted due to policing and enforcement; the public image of fishers deteriorated since their activities were seen as environmentally damaging and they were perceived as notorious law-breakers and reckless egotists, leading to a declining status; mutual mistrust increased, since bumper catches were believed to be caught illegally; catch kings turned into quota kings, placing entrepreneurial skills higher than fishing skills; crew loyalty declined, since investments in quotas meant lower shares in the share system of remuneration; this in turn led to declining interest in becoming a fisherman, while at the same time it became more difficult to maintain family firms (van Ginkel 1999). In general, fishers felt they had lost control over how to run their individual enterprises. The new policy adopted in 1993 sought to give back some of the decision-making authority to the fishers. In what follows, I will describe and analyse the devolved management system;

the ways in which, and the extent to which, fishermen can participate in fisheries governance; and their perceptions of the benefits and demerits of the management regime.

7.3 Grassroots involvement: Experiences with co-management groups

The 1993 white paper, *Balanced Fisheries,* took as its point of departure that the long-term continuity of the fishery sector was the industry's own responsibility and that further intensification of state enforcement and policing would be financially and politically unfeasible (Vissen 1993:5-6). It sought to implement a policy that would fit within the EU's CFP and at the same time enhance the social and political feasibility of the regulatory regime with less, rather than more, government. The state aimed to increase fishermen's responsibility and social control through self-management. The underlying idea was that fishers' involvement in policy and management would lead to greater legitimacy, and in its wake to increased compliance with the rules and regulations and cooperation with the administration. It was hoped that in turn this would lead to a more sustainable exploitation of marine living resources in an economically responsible way. Essentially, the Biesheuvel system of public-private management was a compromise between the long-term interests of nature and the short-term interests of the fishing enterprises: "It is dominated by a desire to 'keep the social peace' within the limits set by scientific research and the public debate" (Salz 1997). It reflects the predominant Dutch practice of solving economic, social and political problems through consultation and compromise.

As mentioned, fishers had to organise themselves into co-management groups. Eight such groups, each comprising between twenty and ninety vessels, have been established. They are relatively homogeneous, since membership is mostly arranged according to the type of vessel and gear used, the species sought, the region fishers hail from and membership of one of the two national fishermen's organisations (Hoefnagel and Smit 1997:163). In addition, there is considerable overlap between specific co-management groups, POs and local fishermen's organisations. The Dutch Fish Board (*Productschap Vis*) supervises and coordinates the groups and if necessary harmonises their regulations, assists in secretarial work and is a direct link between fishers and government. It organises regular meetings with group delegates to discuss problems and performance, and advises the government. Ninety-seven per cent of all beam trawl fishermen have joined a group, even though the fishers were initially reluctant to cooperate. This high percentage came about as a result of Parliament's threat to take coercive structural measures (a general horsepower reduction) should the percentage remain below seventy-five per cent, leaving the fishers little choice but to join.[12] But there were also positive incentives. Group members were, until its recent abandonment, entitled to more days-at-sea than non-members and the period in which the latter can trade ITQs is restricted. This 'sticks and carrots' approach proved successful. The aim of the management groups was twofold: first, to arrive at an effective and efficient system of quota compliance that is supported by the fishers; second, to improve economic performance within the quota restrictions.

[12] The leaders of the national fishermen's organisations wanted to start with 'a clean slate', and they visited the fishing communities in order to convince fishers that they had few alternatives than to join a group. In addition, a new head of the ministry's Fisheries Directorate made it abundantly clear that the fishing industry would suffer if it refrained from accepting the Biesheuvel system.

A board of directors that supervises group management and has certain rights and obligations, heads each group, considerably reducing state involvement and enforcement costs. An independent chairperson who is not involved in the fishing industry heads the board; the other board members are producers who are elected by group members. Group members must sign an agreement committing them to abide by the group's rules, remain in the group for the entire year, and provide logbook and auction data to the group and the General Inspection Service. The board controls and manages the quotas of individual members at the group level, controls quota transfers within a group and warns individual fishers when their quota take-up has reached eighty per cent. It also sees to it that fishermen do not speculate with ITQ shares and that redistribution takes place according to need rather than to the highest bidder (van der Schans 2001:364-367). Hence, to some extent, the board is governing a socially embedded moral economy type of exchange. Individual fishermen bring their quotas and days-at-sea entitlements into these groups but remain owners, and they are responsible for establishing annual fishing plans to achieve a better distribution of days-at-sea and quota take-up over the year. The Fish Board must approve of the fishing plans and determine how the group quota will be allocated and how and when it will be fished. Members may buy, sell, lease, rent or exchange individual quotas throughout the year and several extant restrictions on catching rights transfers were lifted. For example, quota rights are now fully divisible. Group members can also buy, sell or lease days-at-sea entitlements. These measures provide for greater flexibility to smooth out surpluses or shortages and to respond to unexpected events and also contribute to higher price levels. It is mandatory for group members to sell landings through the auction markets so as to ensure that the quantity and price of fish can be effectively controlled and that adequate information is available (van Vliet 1998b:71; van der Burg 2000:48).

The Biesheuvel co-management regime to a large extent hinges on the idea and practice of social control and peer pressure. The group board can prosecute members who exceed their quotas under private law, while also ensuring that group members who are unable to fully take their quotas as a consequence are compensated.[13] Fines are heavy, outweighing any gains an offender might have from non-compliance. Members who fail to comply with group rules can be expelled, or their fishing opportunity may be limited, or their days-at-sea reduced. The entire group could suffer if its quotas were exhausted prematurely due to greedy behaviour, because the state could close the fishing opportunity for the group. Moreover, groups have to apply for government recognition each year, and this may be withheld if the group quota has been exceeded. That is why fishermen keep a keen eye on the doings of their fellow members; the underlying idea is that fishers would indeed report on offenders from their own group. However, this is not always done because it is regarded as 'tale-telling' and even 'treason' (Hoefnagel and Smit 1997:164; van der Schans 2001:345,371), so there is a social code *not* to report on infractions, and fishers 'merely' resort to gossip and ostracism. For example, they complain about fishers using illegal net provisions when they think they will fail to exhaust their individual quota before the end of the year, exceeding horsepower limitations or fishing within the 12 nautical mile zone, but they do not consider it their duty to report specific cases to the group board or the General Inspection Service.[14] In addition, directors of the group boards can be exposed to social pressure not to mete out fines

[13] In addition, fishers are held accountable under public law. This may create the problem of 'double jeopardy' (the accumulation of penalties) (Berg 1999b:166).
[14] In a statement of intent issued in April 2004, the Fisheries Minister and the fishing industry agree that co-management responsibilities will be broadened to include group enforcement concerning these offences.

because they are acquainted with, or are even relatives of, offenders (Dubbink and van Vliet 1997:198; van der Schans 2001:351-352). Some have criticised the system as being inadequate in its self-enforcement aspect, requiring a statutory system of penalties and procedures (Berg 1999a; Berg 1999b).[15] But peer monitoring and self-regulation are not always a problem. In 1999, a co-management group expelled three of its members and held one vessel under arrest for ITQ over-fishing. Given the fact that the entire group would suffer from transgressions by individual firms, the incentive to report on offenders is high:

> Empowering groups of vessel operators to decide for themselves on what rules to operate and penalties to apply, the raison d'être for regulation and the impact of non-compliance on other group members became more transparent, and fishermen's perceptions of the legitimacy of the management system increased. (Valatin 2000a:300)

Beam trawl fishers appreciate the co-governance system because it gives them a say in the management of the group and their own firm; it increases their flexibility because they can transfer quotas and days-at-sea; it provides them with the certainty to take their quota share at the time they deem economically most rewarding; and the likelihood that others will dodge the rules and regulations has decreased. They value the current stability in the sector and the regained control over their day-to-day operations. Transparency has increased by mandatory auctioning and by making the effects of over-fishing of a single vessel operator visibly felt on all the other group members (Davidse 1997:270; Valatin 2000b). One of the bonuses that attracted fishermen into the Biesheuvel management system was that, on average, prices and, therefore, economic results would be better. Mandatory auctioning has indeed led to higher prices, since price-undermining illegal landings belong to the past (van der Schans 2001:344). In general, the economic performance of the beam trawl fleet has been rather good between 1991 and 1998 due to high fish prices and low fuel prices, although higher fuel prices in 1999 caused some problems for profitability (Davidse and de Wilde 2001), but 2000 and 2001 were good years again. Of course, it is much easier to accept a new governance regime when the outcomes are beneficial. In the Dutch case, it has led to a willingness to accept and work within the quota rules. Having experienced several benefits of the co-management system, the fishers are generally satisfied with the way it functions and feel relieved that they have left behind the 1970s and 1980s 'wild west' period of quota busting. As Dick Langstraat, then chairman of the Fish Board, states: "Transparency, greater flexibility and improved profitability of the fishing enterprises have convinced the fishermen that the system is an attractive one" (1999:77). Other factors that contributed to the success of the Biesheuvel system are undoubtedly that the flatfish sector is relatively homogeneous and that it is entitled to a large share of the European TACs for sole and plaice. The fish processing industry and fish traders are also happy with the devolved management system, since they can be sure of a steady supply of fish throughout the year. The Dutch beam trawl fishers even take pride in 'their' self-governance mode of operation. They boast about their 'good behaviour' (forgetting why the system was

---

[15] Symes also points to these problems in writing that "the devolving of specific management functions to the industry begs questions over both the commitment of voluntary user group organisations to undertaking additional responsibilities and their robustness in disciplining their own members in the event of non-compliance" (1997:111). Yet, he regards the Biesheuvel system as coming "close to providing a template for group management" (ibid.:113).

introduced in the first place) and point to the fishing industries of other EU member states where, in their opinion, illicit practices and lack of enforcement are still the order of the day.

However, although the Biesheuvel regime has delegated considerable responsibility to fishers for quota management, government control with regard to the fishing industry as a whole is still large and, in effect, the Fisheries Directorate "determines the conditions under which the groups are allowed to manage their own affairs" (van Vliet 1998b:72), while national regulations still have to live up to the requirements of the CFP. In addition, the government is still strongly involved in enforcement through its General Inspection Service (Berg 1999b:152), though the Biesheuvel regime has led to a substantial reduction of control and enforcement costs (van der Schans 2001:358,371ff). The fishermen are of the opinion that rules and regulations should apply to each and all equally. Therefore, they deem state coercion necessary, although they are ambivalent about specific rules they deem 'unworkable', 'bureaucratic' or simply 'silly'. After reviewing the co-management system in 1996, the government decided to make the cooperative governance structure permanent, mainly because the problems with exceeding the national quotas had been brought to an end and the number of violations of the rules had decreased spectacularly. A recent review reiterates the successes of the Biesheuvel regime (van Geffen *et al* 2002).

Both vessel owners and state institutions thus seem to be satisfied with the outcome of the co-management process (cf. Hoefnagel and Smit 1997:175). There has been "...a progressive development from non-compliance with the regulators towards a greater willingness to accept and work within the rules, especially where the rules are being set by, and for the benefit of, the fishermen rather than imposed by the administration"(Davidse 1998:66). Whereas, for a long time, fishermen wanted to escape from the restrictions by evading or infringing them, there has been a development "towards rights they want to be protected" (Davidse and de Wilde 2001:33). Managing the group's share of the national quotas does indeed enhance a feeling of responsibility, peer pressure and social control, and enables a better spread of flatfish landings over the year with positive effects for market prices, while quota owners can be certain that they can catch what they are entitled to. It has reportedly led to complete compliance with quota regulations, a drastic reduction of offences and therefore to "administrative and political stability in and around the sector" (van Vliet 1998b:72). Consequently, enforcement officers meet with much less resistance than previously and can do their job under less inimical circumstances. From having a reputation within the EU as being completely unable to enforce quota regulations, the Netherlands is now "being widely regarded as a model of good landings enforcement, even if individual violations still occur" (Valatin 2000a:300). When visiting the Netherlands in 1995, EU Commissioner Emma Bonino even suggested that the Biesheuvel system should be adopted by other member states. Fishermen feel satisfied because the co-management system has reduced uncertainty and created stability, provided for flexibility and put an end to the race for fish that characterised the 1970s and 1980s fishing industry and that frequently led to illegal practices and early fishing closures because national quotas were exhausted. Today's mantra is *beheerst vissen*, that is, fishing in a disciplined manner. Generally, then, the Biesheuvel system of group management has been heralded as a successful example of fisheries co-management (Hoefnagel and Smit 1997:175; Berg 1999b:159; Langstraat 1999; van der Burg 2000:48; Symes *et al* 2003:124-126; Hoefnagel *et al* 2004).

Yet, at a more general level, fishers feel that they are not taken serious enough by state institutions and that their involvement and participation in fisheries matters leaves much to be desired. Through their national organisations, the Fish Board, POs and local voluntary associations, fishers can attempt to exert some influence on the policy and management process, but the rank-and-file members of the national organisations are not very pleased with what their leaders have been able to achieve. The fact that there are two such organisations is generally perceived to be an obstacle in gaining political clout, especially now that the fishing industry is becoming smaller and smaller. The organisations' leaders maintain that the distance between them and the Ministry is considerable, and fishers state that they do not have a say at all and are not listened to (Hoefnagel *et al* 2004:36-37,41,44,46). In general, fishers feel that groups and POs should be given more influence on the policy-making process (Hoefnagel and Smit 1997:172). They are in favour of less, more workable and uniform regulations within the CFP framework that should be enforced in equal measure in all member states. As it stands, they have the impression of being the 'most virtuous pupil in the class'. The government agrees and has promised to make a strong effort in EU negotiations to arrive at harmonisation of rules and enforcement. Fishermen are less than satisfied with some aspects of the CFP and its demerits for the Dutch co-management regime. As an Urk fisherman summarised to a newspaper reporter: "Brussels decides about us without us, but especially against us" (*NRC Handelsblad*, 10 December, 1999:13).

7.4 Heading for troubled waters?

> March 1, 2001. Early in the morning, angry Dutch flatfish fishermen in a concerted action block a number of important waterways leading to the ports of Amsterdam and Rotterdam with more than a hundred large beam trawlers. They are extremely dissatisfied with the sudden cod recovery measures of the European Union, that consist of the closure for the duration of ten weeks of fishing areas in the southern and north-eastern parts of the North Sea, important sole and plaice fishing grounds. Not only fishers but also Members of Parliament and biologists contest the measure. The fishers fear for their livelihoods, the politicians think that the measure will hit the Dutch fishing industry disproportionately hard, whilst the biologists doubt the efficacy of the closure that comes after the cod spawning-season, is not targeted at the main cod fishing grounds and in their view will merely lead to concentration of fishing efforts elsewhere in the North Sea. Earlier on, the fishers had proposed to voluntarily refrain from sailing for four weeks, but this proposal had been rejected. With the closure pending and the Deputy Minister for Fisheries, Geke Faber, refusing to compensate the fishers, they have 'spontaneously' decided to take the action that is subsequently backed by the national fishermen's organisations. Later that day, the fishers dissolve the blockade. Under threat of being held accountable for the damages and having to pay recognisance, they capitulate and return to sea. They achieve something later that night, when negotiations between the leaders of the fishermen and the Deputy Minister lead to some concessions regarding compensations.[16]

---

[16] See Quist (2001) on the blockade. What aggravated the fishers was the fact that on top of the closure, sole and plaice quotas had been reduced considerably, while fuel prices were extremely high. However, by the end of the year, economic returns proved less disastrous than the fishers had feared, although the crews of larger beamers faced a considerable drop in income (de Wilde 2003b).

Despite vehement contestations of EU measures they deem useless, unfair or unacceptable, Dutch fishers have become increasingly aware of ecological problems. In 2002, they voluntarily agreed to take turns to fish plaice during the first quarter of the year, while in the same year they held a symposium and discussed among themselves how to achieve more sustainable fisheries. But generally, fishers are not satisfied with stock management measures. The lengthy annual political ritual of setting TACs by the European Council of Fisheries Ministers – usually a payoff between biological and national fisheries interests – is often accompanied by anxiety on the part of the Dutch flatfish fishers. Substantial fluctuations make it extremely difficult to develop long-term investment plans and lead to a sceptical view about the future of the fisheries and the role of biologists in assessing fish stocks.[17] The flatfish fishers are aggravated by the 'guesstimates' of biologists and the short notice on which quota measures for the new season are publicised (Smit 1996:39), while they are often disappointed in what the Dutch Fisheries Minister has achieved for them. From one year to another, the quotas for sole and plaice can vary hugely (see Figure 7.1).

The dramatic cuts of the plaice quotas for 1995 and 1996 caused much turmoil in the flatfish fishery sector. It brought the moment of truth for the resilience of the co-management system, for chances of non-compliance would be considerable. As Hanna hypothesises, "even against a background of ongoing industry participation, participatory management processes increase in difficulty as resource scarcity increases" (1995:42). However, the Dutch flatfish fishers adapted to the circumstance: "Evidently, fishermen are more willing to comply with rules laid down by their own management group than with those previously set by the administration", fisheries economist, Wim Davidse, concludes (1998:64). Some of the pressure was alleviated because plaice prices were on average seventeen per cent higher due to the scarcity of supply.

[17] Lower quotas are, however, partly compensated for by the usually higher market prices for sole and plaice due to scarcity.

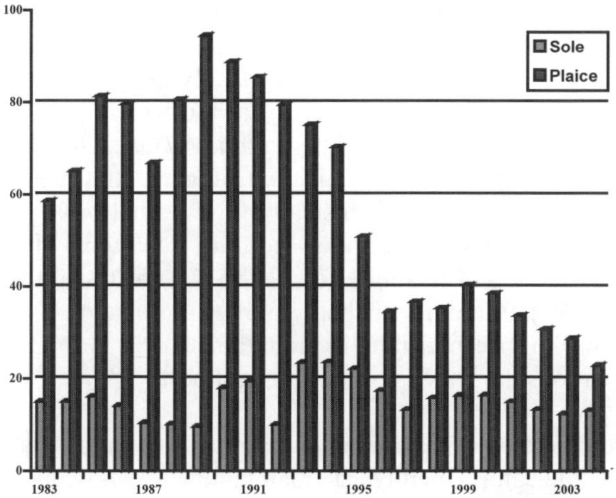

Fig. 7.1 Dutch Sole and Plaice Quotas (x 1000 tonnes), 1983-2004.[18]

Nonetheless, there is a communication problem between managers, scientists and the fishermen:

> There is still a large gap between the managers' world of models, stock sizes and TAC recommendations and the daily reality of fishermen. Often, conclusions of fisheries biologists on the status of the fish stocks and the resulting management measures are in contradiction with the perceptions of fishermen on their own catches and the economic status of the fishery. This contradiction damages the confidence by fishermen in the managers and scientists, which may cause problems in the co-management process. (van Oostenbrugge and van Hoof 2003:2)

For example, I have often heard fishermen say that there were plenty of plaice and sole in the sea, and that biologists just used the wrong methods to assess these flatfish stocks. What also disturbs them is the fact that TACs are sometimes adjusted *ad interim*, making it hard to conform to their initial fishing plans. They would prefer multi-species, multi-annual and more stable TACs that would enable them to provide for long-term planning instead of being continually yo-yoed. To circumvent the problem of receiving the same percentage of the TAC for sole and plaice – a consequence of the EU CFP's principle of relative stability – Dutch flatfish fishers and their organisations would also like to see freely transferable quota rights across member states. This would give them the opportunity to buy or lease sole and plaice quota entitlements abroad.[19] Lastly, they are in favour of having some role in fixing TAC levels.

---

[18] Based on Davidse and de Wilde (2001) annex III; van Wijk (2000); Taal *et* al (2003).

[19] However, this is still not legally allowed, although, in fact, the quota hopping and re-flagging practices amount to much the same.

Biologists, for their part, distrust the Dutch flatfish fisheries co-management system. They maintain that its efficacy is mainly due to the fact that fishers simply cannot fish up the national quotas for sole and plaice; the exhaustion percentage has been below a hundred per cent since the early 1990s. Some "cynically suggest that compliance and the political-administrative rest is bought off with far too lenient TACs" (Dubbink and van Vliet 1997:192). In their reports to the International Council for the Exploration of the Sea (ICES), biologists adhere to the objective of what they perceive as rational exploitation: maximum sustainable yields instead of a minimal biologically acceptable level of exploitation that is currently the state's point of departure. This would in their view require a drastic reduction of fishing effort. They feel frustrated by national states' attempts to obtain the maximum for fishers in negotiations over TAC and quota allocations, which lead to irresponsible exploitation levels (Corten 1996). Corten (1996:5) expressly mentions the Dutch co-management policy:

> …in which the objective of rational exploitation was explicitly abandoned... The new policy would aim…merely at maintaining stocks above a minimum biologically acceptable level…Ministers and administrators increasingly consider quotas as amounts of "paper fish", which should be large enough to avoid any inconvenience to the national industry.

In addition, the ITQ system leads to the wasteful practice of high grading and discarding of low-value fish. In recent years, biologists and fishers have strongly contested each other's views in the weekly *Visserijnieuws* (Fishery News) and more and more fishermen have stopped cooperating with the Netherlands' Institute of Fisheries Investigation's biologists.

In addition to the uncertainties that go along with the present manner of determining TAC levels that continue to make it difficult for the fishers to understand and accept broader policy and management frameworks, they experience contradictions between this layer of the CFP and its other backbone, the five-year Multi-Annual Guidance Programme (MAGP). In the Netherlands, the preference has been for days-at-sea regulations and voluntary decommissioning schemes. The administration leaves it to the fishers whether they would like to fish with powerful vessels or not, as long as they stay within their quota restrictions. Fully implementing MAGP targets, for example by lowering the number of days fishers are allowed to be at sea, would jeopardise the take-up of individual fishing rights and national quotas and this may undermine the Biesheuvel system and enforcement of legal landings (Valatin 2000a:301). The centralised EU targets of reducing capacity become increasingly at odds with the decentralised quota management responsibilities (Davidse 2000). As chairman Dick Langstraat of the Fish Board relates, "the MAGP straitjacket threatens to undermine the fishermen's support for our co-management system" (Langstraat 1998:12). The fishing industry argues that it was agreed under the co-management arrangement that fishers should be allowed to take their share of entitlements, and that this right should prevail over the days-at-sea regulations. The fishermen feel that they are being punished for good behaviour. However, for the time being at least, the problem of mandatory decommissioning has been averted, since in 2003 it was agreed that the Dutch fishing fleet would be allowed to continue its days-at-sea arrangement to comply with MAGP targets.

In the issues mentioned above, fishers feel that their voice is insufficiently taken into account. As a matter of fact, participatory management is extremely limited and is first and foremost directed at quota management. Moreover, both the ITQ system and the Biesheuvel regime have had considerable social consequences. The ITQ system has led to a gradual concentration of fishing rights in the hands of fewer fishermen who have been in a position to acquire additional quotas.[20] The owners of large beam trawlers of 1500 horsepower and over own some eighty per cent of the flatfish ITQs. Those who could not buy catching rights to match vessel capacity have been ousted from the business. From a neoclassical economic viewpoint, this would be explained as a desirable outcome; secure property rights are supposed to increase efficiency, while (self)-enforcement would also become easier and less expensive. Sociologists and anthropologists, however, point to equity problems and social dramas (McCay 1995). Marginalised fishermen maintain that those who have obtained a large share of the catching rights have done so through initial illegal fishing, and investing the perks of their practices in quotas. Those who bought or leased quotas say that they had the right kind of entrepreneurial and managerial spirit and that it was those who did not have sufficient catching rights who indulged in illegal fishing. In general, the system tends to be to the advantage of vested interests with subsequently little room for change due to the owners' "campaigning skills and ability to mount legal challenges" (Valatin 2000a:306).

Although related to the ITQ system, and not specifically to the Biesheuvel regime, the co-management groups tend to look after the interests of their **present** members who are likely to resist any change that would diminish the value of their property rights. They do not hesitate to go to court to seek compensation if new measures undermine the value of their assets (Valatin 2000b). The state automatically renews their ITQ for the new fishing year, and as long as the entitlements retain their value, the owners will remain content. To become a group member, a fisher must already have an ITQ, a fishing licence and a vessel. Starting a firm from scratch is impossible, since outsiders cannot obtain a licence and quota rights unless taking over another firm. The value of catch rights makes the costs prohibitively high. Therefore, aspiring newcomers are effectively barred from entry into the fishing industry, while the initial introduction of IQs has unintentionally created a 'millionaire's club'. Even continuing a family firm is extremely difficult because high prices for quota have to be paid (Dubbink *et al* 1994:33). Presently, the value of quota rights usually exceeds the value of the vessel. Because tax inspectors took into account the value of quota rights, succession-duties and other taxes rose phenomenally. Many owners have changed the juridical form of their firms into limited liability companies, among other things to better allow for succession of ownership (Davidse and de Wilde 2001:33). Although special tax arrangements have made succession from father to son easier, if tangible and intangible assets are passed on from an uncle to a nephew (when, for example, a vessel and ITQs are owned by brothers who fish together but one of whom has no successor) there is still a heavy tax burden. In several instances, it has forced agnates to continue fishing together much longer than used to be the case. Heirs who inherit quotas face the need of borrowing to buy out those heirs who do not fish. Thus, a new generation of skipper-owners has to produce at higher costs (Davidse 1997:107,217).

---

[20] The concentration trend slackened after 1994. There was less trade in ITQs since the number of right-holders remained fairly constant and prices continued to be rather high (Davidse 2000). The concentration trend was much weaker than in the Icelandic cod fishery (Pálsson and Helgason 1995).

The organisational character of the family firm has also changed to a considerable extent. Whereas management decisions used to be relatively simple and were made with short time horizons, with the introduction of fishing rights the fishermen increasingly need the knowledge and skills of external specialists who can advise them on juridical and fiscal matters related with quota rights. Several so-called 'couch' fishermen can live comfortably by leasing and thus cashing in on their quota entitlements without going to sea (young fishermen have to buy or lease quotas at disproportionate prices – sometimes pitting the younger against the older generation of fishers).[21] Many fishers argue that holders of quota entitlements who are not actively fishing should hold on to these rights for a period of no more than five years. But there are other loopholes that enable the couch fishers to continue their practice. Non-propertied crewmembers also face the consequences of the new regime in that owners have altered the division between the vessel share and the crew share, lowering the percentages given to the former. Officially, the partnership contract gives joint responsibility of skipper-owners and crewmembers when it comes to fishing, but in practice, crewmembers have little influence in the running of the business; owners usually decide on issues such as where and how to fish, investments, and quota transactions. Owners argue that the net incomes of crewmembers will still be high since more harvesting rights imply greater revenues. All in all, the introduction of transferable fishing rights has been socially divisive.

Of late, there are signs of reluctance to join fishing crews, and even owners are said to be encouraging their sons 'to learn a trade' instead of following in their footsteps. Pessimism about the fishing industry's future appears to be on the increase. Whereas for a long time it has not been a problem to recruit crewmembers for the family firms – either within or without the circle of kin – in recent years, fishery schools face declining numbers of enrolling students and several vessel owners have had difficulties finding successors and replacements.[22] The amount of time spent away from home, the relatively poor remuneration given the long work hours, the bad public image of fishing as an occupation, and especially the uncertainties inherent in the annual delimitation of national and thus individual quotas with all the financial insecurities that entails, are some of the main reasons for poor recruitment, while the booming economy of the 1990s created lots of alternative job opportunities. The EU and Dutch state policies to reduce the number of fishermen finally appear to be 'successful'. Fishers find it difficult to cope with the precariousness that these institutions have created for the fishing industry's future. In 2002, due to low quotas and high fuel prices, the owners of eight big beamers applied for the decommissioning programme. This was the first time in all the years of fleet reductions that vessels of this size were decommissioned (de Wilde 2003a).

7.5 Conclusions

Modern-day owners of beam trawlers have to be entrepreneurs and managers who invest in quotas, draft fishing plans, and co-operate with colleagues in the Biesheuvel group.

[21] ITQs could be held separately from ownership of a vessel for a maximum period of two years. However, many owners circumvented this rule by keeping their vessels without going to sea.
[22] This happened despite the fact that the number of fishermen working on the fleet declined considerably in the 1990s. Some vessel owners even recruited Polish crewmembers to cope with a shortage of hands.

These new realities of the trade have affected some of the key values in the fishermen's occupational culture. For example, thoroughly enjoyed competition for recognition as a top skipper has made way for a less exciting managing and harvesting of quotas. To be a catch king today, you need to be a 'calculating quota-manager' (Davidse *et al* 1999:543) (although many a fisherman would add that 'you still have to catch the fish'). And to be a quota king means complying with the state's regulatory regime and with the rules of the group, whose social control is tight. Unlike two decades ago, today Dutch sole and plaice fishers generally seem to acquiesce to EU and national quota rules and regulations, which is in large measure due to co-management of ITQs.

However, the Biesheuvel regime is not a panacea.[23] The economic and political advantages sought with the co-management regime have largely been fulfilled, but not without considerable ecological and social costs. The quota-busting problem is largely solved through a mix of input measures, output measures and technical measures and devolved management tasks. This brought an end to political turmoil and fisher resistance to enforcement; quota holders can be relatively sure of catching what they are legally entitled to, while fishing plans have led to more continuity in landings and higher prices. National quotas for sole and plaice are even under-utilised. But there is still industry pressure to set TACs higher than maximum sustainable yield levels, while the rights-based beam trawl fishery leads to discarding and high grading. The public-private governance system is not about managing fish stocks, since this is done at the supranational level. Moreover, as an unforeseen and unintended consequence, fishing within the 12nm zone with Euro-cutters has increased considerably, leading to pressure on inshore sole and plaice stocks.

Despite these weaknesses, the co-management regime meant an important improvement in comparison with the tragedy of the commons, the prisoner's dilemmas and free riding that characterised the period before 1993, in that compliance with quota regulations has improved tremendously. However, it still remains to be seen whether the Biesheuvel system can withstand major shocks that may hit the flatfish fishing industry. As McCay maintains, "[r]esilience of management systems, including their flexibility and adaptability in the face of uncertain and changing social, economic and ecological conditions, is critical" (1995:18). So far, the conditions for the economic and political success of the Dutch co-governance regime have been extraordinarily favourable, in spite of occasional problems and setbacks. But this could change if persistent storms undermine its legitimacy among the fishers. These storms may come in the form of further quota reductions, days-at-sea limitations, area or seasonal closures, or mandatory decommissioning. Another Achilles' heel is unequal enforcement across member states. Yet the real problem is that fishers feel that their involvement in governance matters is extremely restricted. Despite all the rhetoric, the responsibilities devolved to fishers are actually few and limited, so that participatory management is a grand designation for what remains, in essence, a command-and-control type of regulatory regime.

Acknowledgements

I owe thanks to Tim Gray for superb editing and to Derek Johnson for useful comments.

References

Berg, A (1999a) *Implementing and Enforcing European Fisheries Law. The Implementation and Enforcement*

[23] Nor are other co-management regimes. For an interesting account of some of the problems, see Singleton (2000).

of the Common Fisheries Policy in the Netherlands and in the United Kingdom, The Hague, Kluwer Law International

Berg, A (1999b) 'Shifting boundaries between public and private law enforcement: the case of fisheries self-enforcement' in JAE Vervaele (ed.) *Compliance and Enforcement of European Community Law*, The Hague, Kluwer Law International

Corten, A (1996) 'The widening gap between fisheries biology and fisheries management in the European Union' *Fisheries Research* 27(1):1-15

Davidse, W (ed.) (1997) *Property Rights in Fishing: Effects on the Industry and Effectiveness for Fishery Management Policy*, The Hague, Agricultural Economics Research Institute

Davidse, W (1998) 'Property rights and regulatory systems: the strategic response of Dutch fishermen' in D Symes (ed.) *Property Rights and Regulatory Systems in Fisheries*, Oxford: Fishing News Books/Blackwell Science

Davidse, W (1999) 'Fisheries management by property rights' in P Salz (ed.) *Proceedings of the 10th Annual Conference of the European Association of Fisheries Economists*, The Hague, LEI

Davidse, W (2000) 'The effects of transferable property rights on the fleet capacity and ownership of harvesting rights in the Dutch demersal North Sea fisheries' in R Shotton (ed.) *Use of Property Rights in Fisheries Management*, FAO Fisheries Technical Paper 404/2, Rome, FAO (available from FAO website)

Davidse, W and de Wilde, JW (2001) *The Dutch Beam Trawl Fleet for Flatfish*, The Hague, LEI

Davidse, W, McEwan, LV and Vestergaard, N (1999) 'Property rights in fishing: from state property towards private property? A case study of three EU countries' *Marine Policy* 23(6):537-547

de Wilde, JW (2003a) '"Government, hold us, 'cause we're heading for trouble!" Observations on thirty years of change in Dutch sea fisheries' presented at the MARE Conference *People and the Sea II*, Amsterdam, 4-6 September

de Wilde, JW (2003b) 'The 2001 cod recovery measures: economic consequences for the Dutch fishing fleet' *Proceedings of the 15th EAFE Conference*, session 6, Brest, 15-16 May

Dubbink, W and van Vliet, M (1996) 'Market regulation versus co-management? Two perspectives on regulating fisheries compared' *Marine Policy* 20(6):499-516

Dubbink, W and van Vliet, M (1997) 'The Netherlands: From ITQ to Co-management. Comparing the usefulness of markets and co-management illustrated by the Dutch flatfish sector' in *Towards Sustainable Fisheries: Issue Papers*, Paris, OECD:177-202

Dubbink, W, van der Schans, JW and van Vliet, M (1994) *"Vissen is veel meer een berekening geworden, maar nog niet al mijn collega's zijn zich hiervan bewust"*, Rotterdam, Faculteit Bedrijfskunde, Erasmus Universiteit Rotterdam

Hanna, S (1995) 'User participation and fishery management performance within the Pacific Fishery Management Council' *Ocean and Coastal Management* 28(1-3):23-44

Hatcher, A, Frere, J, Pascoe, S and Robinson, K (2002) '"Quota-hopping" and the foreign ownership of UK fishing vessels' *Marine Policy* 26(1):1-11

Hoefnagel, E (1998) 'Legal but controversial: quota hopping, the CFP and the Treaty of Rome' in D Symes (ed.) *Property Rights and Regulatory Systems in Fisheries*, Oxford, Fishing News Books/Blackwell Science

Hoefnagel, E and Smit, W (1997) 'The Netherlands: Experiences in Dutch co-management of marine fish resources' in *Towards Sustainable Fisheries: Issue Papers*, Paris, OECD

Hoefnagel, E, Visser, L and Vos, B (2004) *Drijfveren van vissers en duurzaam visserijbeheer. Een verkenning*, Den Haag, LEI

Langstraat, D (1998) 'The economic contribution in policy process', P Salz (ed.) *Proceedings of the 10th Annual Conference of the European Association of Fisheries Economists*, The Hague, LEI

Langstraat, D (1999) 'The Dutch co-management system for sea fisheries', in D Symes (ed.) *Alternative Management Systems for Fisheries* Oxford, Blackwell Science

McCay, BJ (1995) 'Social and ecological implications of ITQs: an overview' *Ocean and Coastal Management* 28(1-3):3-22

Nielsen, JR and Vedsmand, T (1999) 'User participation and institutional change in fisheries management: a viable alternative to the failures of 'top-down' driven control?' *Ocean and Coastal Management* 42(1):19-37

Pálsson, G and Helgason, A (1995) 'Figuring fish and measuring men: the individual transferable quota system in the Icelandic cod fishery' *Ocean and Coastal Management* 28(1-3):117-146

Quist, C (2001), 'Enough is enough!' *Samudra* April 2001:24-28

Salz, P (ed.) (1997) *Fisheries Management Policies Towards 2010 – Scenarios and Science and Technology Needs*, The Hague, NRLO Report 97/29

Singleton, S (2000) 'Co-operation or capture? The paradox of co-management and community participation in natural resource management and environmental policy-making' *Environmental Politics* 9(2):1-21

Smit, J (1996) 'De vis wordt duur gehaald' *Intermediair* 32(36):37-39

Symes, D (1997) 'Fisheries management: in search of good governance' *Fisheries Research* 32(2):107-114

Symes, D and Crean, K (1995) 'Privatisation of the Commons: the introduction of individual transferable quotas in developed fisheries' *Geoforum* 26(2):175-185

Symes, D, Steins, N and Alegret, JL (2003) 'Experiences with fisheries co-management in Europe' in DC Wilson, JR Nielsen and P Degnbol (eds) *The Fisheries Co-management Experience*, Dordrecht, Kluwer Academic Publishers

Taal, C, van Wijk, MO, Klok, A, de Wilde, JW and Smit, MH (2003) *Visserij in cijfers 2002*, Den Haag, LEI

Valatin, G (2000a) 'Quota trading systems in EU fisheries' *Reciel* 9(3):296-306

Valatin, G (2000b) 'Fisheries management in the Netherlands: an overview' presented at the *12th Annual Conference of the European Association of Fisheries Economists*, Esbjerg, 13-15 April

van der Burg, T (2000) 'Neo-classical economics, institutional economics and improved fisheries management' *Marine Policy* 24(1):45-51

van der Schans, JW (2001) *Governance of Marine Resources. Conceptual Clarifications and Two Case Studies*, Delft, Eburon

van Geffen, J, Hagendoorn, A, Water, J and Boon, A (2002) *Op weg naar evenwicht. Evaluatie van de Structuurnota Zee- en Kustvisserij (1993) "Vissen naar Evenwicht"*, Ede/Wageningen, Expertisecentrum LNV

van Ginkel, R (1999) '"A nice shipload of fish" and other issues: on the work world of Dutch beamtrawl fishermen' *Europaea* 5(2):121-142

van Oostenbrugge, JAE and van Hoof, L (2003) 'CPUE as an integrated bio-economic indicator for the Dutch beam trawl fishery', *Proceedings of the 15th EAFE Conference*, session 18, Brest, 15-16 May

van Vliet, M (1998a) 'From economic fish stock management to integrated management: communicative miracle or a Tower of Babel?' in D Symes (ed.) *Northern Waters: Management Issues and Practice.* Oxford, Fishing News Books/Blackwell Science

van Vliet, M (1998b) 'Fishing as a collective enterprise – ITQs and the organisation of fishermen' in D Symes (ed.) *Property Rights and Regulatory Systems in Fisheries*, Oxford, Fishing News Books/Blackwell Science

van Vliet, M (1999) 'The tragedy of the commons model and the revisionist criticism: an assessment on the basis of an analysis of flat fish sector regulation in the Netherlands' in P Salz, P (ed.) *Proceedings of the 10th Annual Conference of the European Association of Fisheries Economists*, The Hague, LEI: 157-188

van Wijk, MO (2000) *Visserij in cijfers 1999*, Den Haag, LEI

Vissen (1993) *Vissen naar evenwicht. Structuurnota Zee- en kustvisserij*, Den Haag, Ministerie van Landbouw, Natuurbeheer en Visserij

CHAPTER 8
CO-MANAGEMENT AT THE ELEVENTH HOUR? PARTICIPATION IN THE GOVERNANCE OF THE NEW ENGLAND GROUNDFISH FISHERY

MADELEINE HALL-ARBER
MIT Sea Grant College Program, 3, Cambridge Center, NE 20-368, Cambridge MA, 02139

Abstract

The process that led to the adoption of the latest amendment to the New England Fishery Management Council's (NEFMC) Multispecies (groundfish) Fishery Management Plan (FMP) provides an excellent case study of the movement from primarily top-down management to a variation on adaptive co-management. The contributions of a policy entrepreneur and institutional leader to this process are noted as critical. Factors constraining the participation of fishing industry members in the development of groundfish regulations, a brief history of groundfish regulations, and the various combinations of rules offered as options by the Council are reviewed. In response to the harsh criticism and controversy over the degree to which those options would restrict fishing and be likely to devastate communities, the Council offered fishing industry members a last chance to recommend a different combination of management tools – as long as they adhered to the tools that had been discussed at public hearings. Three fishing organisations offered plans that were considered. The Council ultimately selected a plan from the Northeast Seafood Coalition, a broad-based industry group, which emphasises flexible or adaptive mechanisms and optimism for the future. This case suggests that the negotiation of power and authority is important in the context of management in a complex setting with a diverse constituency, and, equally important, communication and outreach are essential elements for change.

8.1 Introduction

A look at the development of New England Fishery Management Council's Amendment 13 to the Multispecies (groundfish) Fishery Management Plan (FMP) reveals an intriguing interplay between the hierarchical and participatory forms of fisheries governance, shaped in part by a consideration of economics. While the largely top-down management process dictated the range of management tools that could be selected, ultimately, a participatory process developed that allowed an innovative approach to the selection or combination of tools that were agreed upon. The choice made may be considered a form of adaptive co-management (Olsson *et al* 2004).

A number of factors contributed to the movement towards the more inclusive form of management. As Olsson *et al* (2004) described in the case of a change to adaptive co-management of wetlands in southern Sweden, the change in New England was triggered by perceived crisis. The fishing industry, writ large, realised that proposed management

T. S. Gray (ed.), Participation in Fisheries Governance, 141–162.

changes were likely to financially ruin a large percentage of the existing harvesting and processing sectors and could decimate the infrastructure so that any effort to rebuild the industry in the future would be compromised.

This chapter will discuss the role of a key individual, a **policy entrepreneur**, who led the industry effort to redesign the management package with the cooperation of an **institutional leader** who was willing to foster change.[1] Neither individual would have been able to achieve what they did without the sense of crisis permeating the whole industry. Pinkerton (1989:4) observed similar reactions: "Co-management agreements between government and fishing interests have arisen out of crises caused by rumoured or real stocks depletion or from political pressure resulting from claims that the government's ability to manage is insufficient to handle specific problems". Jentoft and McCay (1995) cite a number of cases in which co-management is adopted in response to crises. Furthermore, as Berkes *et al* (2003:19) point out, crises have a constructive role in resource management, in that they can lead to renewal. This is an example of the broader claim that the social, political and economic context in which fisheries operated has a significant bearing on the form that co-management arrangements may take (ENRC 2001:21).

The significance of successfully developing adaptive co-management in a crisis situation should not be underestimated. Sustainability requires adaptive capacity, or resilience, for societies to deal with change (Holling & Meffe 1996). A primary goal of the industry groups who contributed proposals for Amendment 13 was to assure the sustainability of the industry and communities that are supported by groundfish and the associated ecological system. Whether or not the groundfish fishery and the communities will be sufficiently resilient to survive remains a serious question. What is explored here is the attempt to design a system with the flexibility to respond to changing ecological conditions.

8.2 The origin of the New England Council

The Magnuson Fishery Conservation and Management Act of 1976 established eight regional Fishery Management Councils in the United States.[2] Some portion of each Council consists of obligatory members (such as state marine resources department heads and the National Marine Fisheries Service (NMFS) regional director), and another portion is appointed by the Secretary of Commerce from nominees provided by the governors of each of the states in the region. The Act requires that the nominees be knowledgeable about conservation and management or the harvest of fisheries resources in the region. Furthermore, the secretary must "To the extent practicable, ensure a fair and balanced apportionment, on a rotating or other basis, of the active participants (or their representatives) in the commercial and recreational fisheries under the jurisdiction of the Council" (USC 1996:Sec. 302 (b)(2)(B)).

[1] The concept of the *policy entrepreneur* is discussed in Olsson *et al* (2004). The concept of the *institutional leader* was characterised by Stein (1997).

[2] This Act later became known as the Magnuson-Stevens Fishery Conservation and Management Act or just the Magnuson-Stevens Act, which was amended in 1996 by the Sustainable Fisheries Act.

Thus, the Councils were devised as a nascent co-management system with representatives from state and federal managers as agents for the public good and representatives of the fishing industry, bringing their expertise and local knowledge to bear on the deliberations. In practice, rarely in almost three decades has the New England Council been successful in tapping the potential for co-management in this system. The latest amendment to the New England Council's Multispecies FMP (groundfish), however, does reflect some movement in that direction.

The New England Council consists of seventeen voting members, including eleven appointed by the Secretary, from Maine, New Hampshire, Massachusetts, Rhode Island, and Connecticut. This Council has the responsibility of managing the multispecies (groundfish) complex. Included in the complex are cod, haddock, pollock, yellowtail flounder, winter flounder, Windowpane flounder, white hake, redfish, American plaice (dab), and witch flounder (gray sole). To complicate management, different geographic stocks of the species are recognised and assessed separately: for example, Georges Bank cod is differentiated from Gulf of Maine cod. A further complication is that some stocks in the groundfish complex are at record highs, while others are at near record lows.

Fishing for groundfish off New England's coast was one of the earliest occupations in the region; indeed, it was the primary reason this area was settled. In addition to the "Yankee" fishermen who settled along the coast from Maine to Connecticut, several ethnic groups have congregated in particular ports, often specialising in certain gear and/or fishing styles. Commercial vessels typically range in size from about 30 feet to 120 feet. Smaller vessels tend to be used for day fisheries or short trips, while the larger vessels may choose day or trip fishing. Accessible grounds include the inshore Gulf of Maine and offshore Georges Bank, while the habitat runs the gamut from sand and mud to complex rocky ledges, deepwater and shoal. Groundfish gear includes otter trawls, midwater trawls, various configurations of hooks (such as hand, long-line, and jigs), traps, pots, and gillnets. Communities with fishing ports include some wholly dependent on the fishing industry; others that gain significant financial benefit from the fishery; and still others that host only a small fleet, but benefit from the tourism draw of a working waterfront.

8.3 The concept of fisheries co-management

The benefit of cooperative management of such common property resources as fisheries has been thoroughly discussed in the literature (Jentoft and Kristoffersen 1989, Pinkerton 1989, Ostrom 1990, Dyer and McGoodwin 1994, Jentoft and McCay 1995, Acheson 2003). In one of the earliest and best known analyses of co-management in fisheries, Pinkerton (1989) pointed out that data gathering, harvesting decisions, allocation decisions, protection of habitat, compliance, enhancement and planning and broad policy-making could all be improved or enhanced by more participation. Jentoft and McCay (1995) argue that greater participation is essential if hard decisions are to be taken:

> Make the decision-making process more open, less hierarchical and more decentralised than would otherwise be the case, provide a two-way channel for communication of information and knowledge between industry and government [and] a means of producing support and of sharing responsibility for hard decisions that…pose a challenge to every management system (quoted in ENRC 2001:24)

It is now almost a cliché to note that those who participate in the development of regulations are more apt to abide by them.[3] Given the usual conditions of fishing (including independent-minded fishers working far from view), it is not surprising that managers are willing to try any method that promises compliance. McCay and Finlayson (1995:12) also make a strong argument for opening up the policy process and scientific debate as well as de-centering science "so that the information, knowledge, and concerns of fishermen and community members can play more direct roles." Collaboration in the development of stock assessments as well as the policy responses may help lead to more successful fisheries management.

Many researchers have found that the kind of collaboration necessary for effective co-management develops most easily in a homogeneous sector of a local community where, for example, there are ties of kinship and ethnicity. In the case of fisheries, a similar homogeneity may be found among users of a single gear type in one port. The reasons are basic to community development: crosscutting ties among community members make both peer pressure and educational outreach more practical and effectual. Furthermore, a sense of unity, a belief that others of the group can understand and identify with the issues, as well as the knowledge that all will be affected similarly by any regulations (equity) pervades such groups.

A place-based definition of community was reinforced in U.S. fisheries management law when the Magnuson-Stevens Act was amended by the Sustainable Fisheries Act (SFA) that established specific national standards to be achieved by management. Among other things, National Standard 8 required managers to analyse the impacts of regulations on place-based communities. Researchers have found, however, that not all communities are place-based. There are communities of interest (such as communities based on fishing site or gear types) and virtual communities. Odell (2004) suggests that a social movement or particular issue can galvanise a group sufficiently to form a community for purposes of co-management. Citing McCay's (1989) and Dale's (1989) chapters in her edited volume, Pinkerton (1989:29) also noted that "co-management is more likely to develop if there is an energy centre: a dedicated person or core group who applies consistent pressure to advance the process." A review of the process that led to an industry-selected package of regulations for Amendment 13 reveals how this can work. At the core of the "energy centre" was a **policy entrepreneur** who was able to work within the constraints of a legal framework to design an adaptive strategy and, equally important, who successfully drew on his social capital to promote the idea.

[3] For a description of a test of this hypothesis see Honneland (1999).

In his case study of the Maine lobster industry, Acheson (2003) points out that other variables can play crucial roles. Distribution fights, a low discount rate (long-term perspective of those involved in the industry), a sense of mutual vulnerability, as well as political entrepreneurship, were keys to the development of rules that eventually led to co-management in Maine's lobster fishery via an area management system referred to as the "zone system".

8.4 Evolution of New England groundfish regulation[4]

Council management of groundfish began in 1977 when quotas on landings were set, as were minimum sizes of fish and meshes in nets. Incremental changes were made for several years. Access to the haddock-rich grounds of the Northeast Peak of Georges Bank was lost to Canada in 1984 when the Hague Line was drawn. The Multispecies FMP was implemented in 1986. Minimum sizes of fish and mesh size increases were the principal techniques used to control fishing effort on the groundfish. Three years later, the Council acknowledged that several of the groundfish species were overfished. In 1991, the Conservation Law Foundation sued the Department of Commerce to force compliance with regulations that required an end to overfishing. In 1994, Amendment 5 to the Multispecies Plan was supposed to phase in a set of much stricter regulations over the course of five years, but by the end of the year, emergency regulations had been imposed due to scientific warnings of the imminent collapse of Georges Bank cod. By 1996, Amendment 7 had been developed, codifying many of the emergency regulations. Also in 1996, the SFA set higher standards for management. Stricter regulations with total allowable catch limits, limits on the days-at-sea, increased closed areas, increases in minimum size, elimination of exemptions and other stipulations were imposed. A pilot project to buyback fishing vessels and permits was established to reduce capacity, and the next year, this program was substantially expanded. Various framework adjustments were made to cut days-at-sea, impose trip limits and other restrictions for the next three years.

In 1999, Amendment 9 to the plan established overfishing definitions and set Optimum Yield (OY) for twelve species in an effort to comply with the SFA. Also, a Federal Register notice in February 1999 announced that the Council was beginning work on Amendment 13.[5] In 2000, American Oceans Campaign and other conservation groups sued the Council [Civil Action No. 99-982 (GK)], arguing that essential fish habitat and impacts of fishing gear were not being adequately addressed by the FMPs. The court ruled that though the Essential Fish Habitat (EFH) amendments met Magnuson Stevens Act's requirements, the environmental assessments were inadequate and in violation of the National Environmental Policy Act (NEPA).[6]

While stock assessments showed that Amendment 7 targets for controlling effort on

[4] Based on a timeline created by Eric Brazer (2003) and NEFMC (2003a:1-5).

[5] Amendments 11 and 12 addressed EFH and management of certain species with a separate small-mesh Multispecies FMP implemented in 2000.

[6] Northeast Multispecies Amendment 13 SEIS:. 3

Georges Bank cod, haddock and yellowtail flounder had been met, in December 2001, the Conservation Law Foundation and other organisations successfully sued again, arguing that the stocks rebuilding plans implemented by NMFS were not consistent with Amendment 9's overfishing definitions. Furthermore, they argued that the management plans failed to establish bycatch-reporting, efforts to reduce bycatch and minimise bycatch mortality. A negotiated settlement was agreed to – referred to as the Interim Rule – pending implementation of Amendment 13. Amendment 13 is intended to "address stock rebuilding issues, greatly reduce fishing effort and capacity in the multispecies fishery and implement additional measures to specifically address habitat protection." (NEFMC 2003a:1-5). In addition to the eleven amendments before Amendment 13, the Multispecies FMP has been altered 30 times since 1994 through framework adjustments.

8.5 Participation in management before Amendment 13

While the design of the regional Councils implicitly recognised the value of stakeholders' participation in fisheries management, typically only a few individuals in New England regard the Council process as truly participatory, and even fewer regard it as a form of co-management. Three aspects stand out as constraints on participation and co-management:

1. The Council designs management plans, but recommends them to NMFS and the Secretary of Commerce. In other words, the Secretary has a veto power and makes the final decision about whether or not a plan fulfils the legal requirements and can be implemented. That places the Secretary and NMFS in a superior position to other stakeholders rather than 'sharing authority', and leads industry members to question whether or not their choices are truly considered. It is true, as Pomeroy and Guieb (2004) note, that co-management systems around the world differ in the degree of responsibility and/or authority vested in the state versus the community, but the key is "negotiated power where the interaction of the state and non-state actors would be an important factor in defining a common and acceptable balance in sharing power and allocating responsibilities." There has been a perception among industry members that they have little power in comparison to NMFS. This perception is heightened by NMFS' control over the science that is utilised as the basis for management, even when it is obviously or apparently flawed.[7]
2. The proliferation of committee, subcommittee and whole Council meetings requires a tremendous investment of time, making it difficult to sustain a small business while participating in management. Furthermore, industry advisory panels are consulted only at the whim of the committees' chairs, so even for those willing and able to devote time to the process, the opportunity is not always offered in a setting where options are often designed and/or selected.
3. The room set-up and the style of interaction during the public meetings can be

[7] In September 2002, the Northeast Science Center revealed that assessments had been carried out with misaligned warps on their vessel resulting in a scandal referred to as "Trawlgate." NMFS has also been criticized for selecting uncritical reviewers for peer reviews of their research/assessments.

> intimidating for stakeholders. Moreover, as Smith (1990:1) points out, there are very basic differences in the world views or cognitive models of the two major parties to management negotiation, that is, "the public sector personnel (for example, administrators, scientists, technicians) and…the user groups – particularly members of the commercial fishing industry…", making each side's view virtually incomprehensible to the other.

When interviewed about the fisheries management process during research for *New England's Fishing Communities* (Hall-Arber *et al* 2001), fishermen and fishing community members often commented that they did not feel a part of the process, nor did they feel well represented. This was particularly true after the resolution of the lawsuits resulted in much stricter regulations, despite assessments that indicated that the twelve major groundfish stocks in the multispecies complex had tripled in biomass since 1994 (NEFMC 2003b), suggesting that existing restrictions were effective. They did not feel that the Council system provided a true opportunity for co-management, but instead was, in practice, a hierarchical, top-down approach to management.

Some fishermen and community members do note that changes in Council staff and policy since 1999 have contributed to increased opportunity for input. Staff members, for example, held community meetings explicitly seeking comment on social and economic impacts of regulatory change, and sent letters and met with industry groups encouraging participation in and/or proposals for Amendment 13.

8.6 Amendment 13

In its ruling on Conservation Law Foundation *et al v.* Evans *et al,* the U.S. District Court ordered NMFS to submit an FMP to "comply with the law." Essential criteria for Amendment 13 included a halt to overfishing, commencement of stock rebuilding within a specified period and a decrease in bycatch.

Representatives of fishing industry organisations and some individual fishermen followed the lengthy process of amendment development. One early conflict between NMFS and the Council, on the one hand, and the fishing industry, on the other, developed after a change in biomass targets was announced. The Northeast Fisheries Science Center revised the groundfish biomass targets in 2002 after re-evaluating the stocks using different models. The groundfish industry cried foul, accusing the Science Center of "moving the goalposts". Because these targets had not been peer reviewed before the Reference Point Working Group's recommendation that they be instituted, some industry members believed that they would be overturned, and the Council would reinstate the previous biomass targets. The new targets stood, however, despite some question about whether or not an early 2003 independent peer review actually supported them, and they were regarded by NMFS as an essential element in Amendment 13.

Eventually, the Council went to public hearing with a host of options, in four alternative packages. The draft management plan document and environmental impact statement was approximately 1500 pages long. It seemed to some observers, however, that the way the alternatives were packaged would result in a consolidation of the industry and a loss of diversity no matter what choices were made.

8.6.1 THE FOUR ALTERNATIVES PROPOSED BY THE COUNCIL

The proposed series of alternatives was very complicated, with potential actions or restrictions much qualified by 'if so' scenarios. The proposals all build on the management measures that resulted from the settlement of the Conservation Law Foundation *et al v.* Evans *et al* law suit. At the risk of oversimplifying, the following is a summary of the choices presented in the public hearing document:

> *Alternative 1: Up to 65 per cent Reduction in Used Days-at-sea*
> Option 1: 55 per cent reduction in used days-at-sea in conjunction with constant fishing mortality or adaptive rebuilding strategy.
>
> Option 2: Phase-in of 65 per cent reduction in used days-at-sea
>
> In order to achieve the reduction, in regulated mesh areas days-at-sea were to be counted at a differential rate of 1.5:1 from December to April. Year-round and seasonal closed areas were to be increased in size and number; possession limits, trip limits, and certain gear restrictions were proposed (raised footrope trawl and changes in the numbers of gillnets allowed); and increases in the minimum fish sizes were designated.
>
> *Alternative 2: Reduction in Allocated Days-at-sea/Gear Modifications*
> Option 1: Allocated days-at-sea are baseline determined from the maximum days-at-sea used over the period 1996-2000, reduced by 20 per cent. Vessels have to sign into the Gulf of Maine at the beginning of the fishing year and have their days-at-sea allocation reduced 30 per cent from their baseline and can not use more than 25 per cent of their allocated days between May and July.
>
> Option 2: Allocated days-at-sea baseline same as Option 1, but vessels can only use 70 per cent of their baselines in the Gulf of Maine regulated mesh area. Vessels have to declare their intention to fish in the Gulf of Maine for a minimum of thirty days (all their days-at-sea used in the thirty days would be counted against their Gulf of Maine limit) and as in Option 1, could not use more than 25 per cent of their allocated days between May and July.
>
> Year around closed areas are the same as Alternative 1; seasonal closed areas are the same as Alternative 1, except for the additional areas of 148-155 in October and November. Minimum fish sizes are also the same as for Alternative 1. Possession limits and gear restrictions are significantly different. Possession limits for Georges Bank cod, for example, are set at a maximum of 500 pounds/day or 4,000 pounds per

trip, whereas Alternative 1 permits either 1,000 or 2,000 pounds/day and 10,000 or 20,000 pounds per trip, with a few more restrictive stipulations for certain periods. A quota or total allowable catch (TAC) will be implemented as a "backstop." Vessel Monitoring (VMS) is required for all vessels.

Alternative 3: Area Management
Six areas were defined: inshore Gulf of Maine, offshore Gulf of Maine, western Georges Bank, eastern Georges Bank and southern New England/Mid-Atlantic. Species-specific TACs would be defined for each area; consequences for exceeding TAC defined; and options for moving between areas suggested. Reporting requirements, area management meetings, adjustment to measures, advisory panel changes, and potential for delegation of authority from the Council were all discussed in the public hearing document.

Alternative 4: "Hard" TAC
TACs will be applied to all the stocks in the Multispecies complex and commercial activity monitored. On stocks with a significant recreational catch, TAC would be specified for the recreational sector as well.

In addition to effort controls, the Public Hearing document included information about fishery program administrative measures and choices of alternatives to control capacity (permit absorption, permit transfer, days-at-sea transfer, freeze on unused days-at-sea, days-at-sea reserve, and mandatory latent effort categorisation with voluntary flexibility options); alternatives to minimise the adverse effects of fishing on EFH; and a few other management issues. The Public Hearing document describing the proposed alternatives was eighty pages long; eight additional pages summarised potential impacts and five pages were devoted to a glossary and list of acronyms (NEFMC 2003b).

8.6.2 REACTION BY FISHERIES STAKEHOLDERS

Most industry members were appalled by what they considered extreme measures required to meet the Court's interpretation of the Magnuson-Stevens Act. Some reacted with anger and urged all to 'just say no'. Protests, including circling vessels and a parade of seafood processor trucks, were held in Gloucester and New Bedford. Almost intuitively, some members of the industry focused on the diminishment of diversity as particularly worrying since, they feared, the downsizing of the fleet and its resulting homogeneity would negatively impact their communities and the industry infrastructure. Interestingly, a parallel argument has been voiced in discussing the drawbacks of centralised management with linear models and a corresponding goal of reducing natural variation "to control nature," since by reducing the "range of natural variation in a system…the system loses resilience" (Holling and Meffe 1996:330). Similarly, when industry members are asked about their vision for the future of the industry, the author discovered that the majority would like to see a continuation of the same diversity currently extant in the northeast fishing industry (Hall-Arber *et al* 2002).

Nevertheless, the majority of the industry recognised the inevitability of the implementation of Amendment 13, and attention was turned toward mitigating the socio-economic impacts. One lobbyist for the fishing industry arranged for industry representatives from the whole region to meet with Bill Hogarth, the director of NMFS, in New Bedford on September 18th, 2003. Hogarth agreed to ask the Council to hold a special public hearing to give the fishing industry the opportunity to present what they considered the best options among those reviewed in the public hearings. Thus, Hogarth played the role of **institutional leader** willing to foster change – or at least to hear the arguments.

The rules that constrain the imposition of regulations in the United States specify required analyses, lengthy comment periods, and public hearings, all of which had already been conducted based on the four alternatives agreed upon by the Council. Consequently, the industry was not free to develop completely new ideas. The opportunity being offered was to work within the strict parameters of the already identified options and the set of requirements outlined by the agreement that had resolved the Court case.

Industry members had been participants in the development of the various options and many commented extensively both orally and in written form. Industry representatives had also signed the negotiated settlement to the Court case. However, once the Council packaged the alternatives and industry members began to analyse the likely outcomes, it became obvious that the negative impacts would be so extreme that the fishing culture of the northeast would be severely diminished. Participation to this point, therefore, had not led to the development of a mutually agreeable set of regulations.

After the meeting with Hogarth, the fishing industry had just three weeks to develop a plan on the options proposed for Amendment 13, that they could agree should be presented to the Council. Various representatives of the industry attempted to organise all the stakeholders to devise a repackaging of the options in a way that would allow the majority of participants to survive until the stocks recovered. Despite efforts to create a unified approach, the diversity in the northeast industry compelled the various organisations representing different sectors to write their own proposed alternatives.

As will be discussed below, subsequent results raise a question about the nature of co-management. If one argues that the Council system is not true co-management, in spite of participation by the commercial fishing industry (in some cases as Council members, in other cases as advisors or as commentators on issues), because of NMFS's veto power, the *ad hoc* use of advisors, and the other constraints on participation by a broad spectrum of the industry, is it possible for co-management to occur when the industry is divided into separate groups whose interests conflict or are not accommodated in the proposed management regulations? This is the heart of the argument that co-management works best in a homogeneous setting.

The next section will briefly introduce the industry's organisations and their proposed alternatives to Amendment 13 as packaged by the Council.

8.6.3 ORGANISED STAKEHOLDERS AND PROPOSED ALTERNATIVES

At least eight fisheries organisations seized the opportunity to submit written proposals to the Council by October 15th 2003. A handful of the submissions were selected for review and comment by the Plan Development Team (PDT) via a conference call on October 17th. Only two of the proposals were thought to have followed "the spirit of 'mixing and matching' elements" found in the draft amendment, and therefore could be seriously considered for adoption (Neis 2004). These and one other proposal were discussed at the special Council meeting held on October 21st. The three proposals that were discussed were submitted by the Northwest Atlantic Marine Alliance, with the Stonington Fisheries Alliance's proposal considered as a subset; the Associated Fisheries of Maine (Groundfish Group – ASF) together with the Trawler Survival Fund (TSF); and the Northeast Seafood Coalition. A proposal submitted by the Cape Cod Hook Fishermen's Association requested a sector allocation that was to be considered regardless of whichever of the packaged alternatives was chosen. I will consider these three proposals in turn, summarising the two that were not ultimately selected and offering a more detailed review of the third.

8.6.3.1 *The Northwest Atlantic Marine Alliance (NAMA)'s Proposal*

Since 1995, NAMA has been working to "generate a new voice and institutional presence that is centered on ecological and economic stability, personal responsibility and accountability, resource protection and distributed power and authority." Although not strictly a membership organisation, NAMA's supporters and participants include commercial and recreational fishermen, conservation advocates, educators, members of the New England Fishery Management Council, and six community alliances – Stonington Fisheries Alliance, Community Alliances of Interdependent AgriCulture, Saco Bay Alliance, Independent Fishermen Involved in Sustainable Harvesting, New Hampshire Marine Coalition, and Provincetown Fishermen's Association. NAMA has a fifteen-member Board of Trustees comprising representatives from each of the community alliances and a number of members-at-large. In addition, there are 'Advisory Trustees' who are former Board of Trustees members, former staff, and some government employees who could not accept an appointment to the Board. Strategic planning and agenda- setting for the organisation takes place at an annual meeting (Deese 2004).

NAMA's stated purpose is "To restore and enhance an enduring Northwest Atlantic marine system, which supports a healthy diversity and abundance of marine life and human uses, through a community-based, self-organising and self-governing institution." NAMA and its alliances are thus committed to ecosystem management, and one of their first steps in this direction is the development of the Gulf of Maine Inshore Fisheries Conservation and Stewardship Plan, an area management system. For this, NAMA staff convened meetings throughout New England, seeking industry and community input on positive alternatives to the options being developed for Amendment 13 within the Council process. NAMA also participates in a number of projects including a wild scallop stock enhancement project and FleetLink, a program using fishing vessels as platforms for the collection of environmental data.

One of their community alliances, Stonington Fisheries Alliance worked with NAMA on an alternative Amendment 13 proposal. Stonington Fisheries Alliance, Stonington, Maine, has a diverse membership that includes among its members the former commissioner of Maine's Department of Marine Resources, a minister, a former gillnet fisherman and a scientist. Though both NAMA and Stonington Fisheries Alliance submitted a proposal, the PDT chose to consider the two proposals as though they were a single proposal.

NAMA's proposal selected the options of reduced days-at-sea (modified Alternative 1) with Phased Area Management. The phase-in of area management, without area-specific hard TACs, proposed retaining the status quo for offshore Gulf of Maine, based on the Interim Agreement, while the inshore Gulf of Maine would be governed by the Gulf of Maine Inshore Conservation and Stewardship Plan. One of the critical features of this plan was that fishermen would have to "opt into the area exclusively" for one fishing year. This was an attempt to "limit vessel mobility in accordance with fish ecology in order to prevent 'pulse fishing' and encourage a stronger sense of stewardship" (NAMA 2003). Like the Northeast Seafood Coalition's, NAMA's proposal was the culmination of perspectives from a diverse group of individuals including fishermen, conservationists and academics from Gloucester to Downeast Maine. Also like the Coalition's proposal, adaptive management to better reflect current conditions was emphasised. Perhaps the greatest drawback to the proposal, given the need for immediate approval and implementation, was the lack of detail on the required development of an area management governance structure, presumably community-based.

Though considered a subset of NAMA's proposal, Stonington Fisheries Alliance's proposal, also area-based, may have been perceived as having less legitimacy, since no groundfish days-at-sea have been allocated to fishermen in the Downeast region of Maine for several years. Although traditionally active in the groundfish fishery, most fishermen moved to the lobster fishery as groundfish stocks fell and lobster stocks (and value) rose. An argument has been made that for now, while groundfish stocks are rebuilding, those dependent on groundfish should have access to more days-at-sea than those who rely on other species for their 'year's pay'. (An opposing argument states that those who moved off groundfish contributed to the rebuilding and therefore should be rewarded.)

8.6.3.2 *The Associated Fisheries of Maine (AFM) and Trawler Survival Fund (TSF)'s Joint Proposal*

AFM has worked within the Council process for many years. The Groundfish Group represents seven vessels, primarily offshore draggers. Their groundfish representative attends every groundfish committee meeting, habitat committee meeting and full Council meetings, commenting and advising whenever appropriate, trying to work proactively. The New Bedford based TSF currently represents about 100 draggers from Boston to Southern New England. The proposal that AFM and TSF jointly submitted was designed as a phased reduction plan. The two associations had jointly hired a scientist who worked closely with the NMFS Science Center to develop and vet their plan, coming up with something that all agreed was 'scientifically valid', even though industry members believed that the proposed reduction was more than what is strictly necessary (Raymond 2004). The scientist-consultant had used a different model to assess the biomass of the stocks and came up with very different numbers from those used as the basis for the federal goals for rebuilding stocks. Consequently, the industry groups rejected the revised biomass targets and based their plan on the previous targets.

AFM and TSF argued for Alternative 1, option 2 - a 'Phase-in of 65 per cent days-at-sea reduction'. This was not truly a new alternative, having been incorporated at least a year before the public hearings. It would have phased in the reductions over four years, but because NMFS required that whatever alternative chosen must include rebuilding to the revised targets, the PDT and Council did not consider that the submitted plan was sufficiently in line with that requirement and assumed that NMFS would not approve it. However, the AFM successfully lobbied for the inclusion of an alternative, allowing leasing of days-at-sea. Despite fears that only the 'wealthy' owners of large vessels would be able to lease days-at-sea, in fact, days have been leased by vessels of every size under 500 tonnes (Plante 2004).

8.6.3.3 *The Northeast Seafood Coalition's Proposal*

Cut backs in days-at-sea and extensive area closures have affected all vessels and gear types in the groundfish industry in New England. In turn, the fewer days spent at sea has translated into lower demand for many shore-side businesses. In Gloucester, Massachusetts, founded over 350 years ago as a fishing community, concerned fishermen were worried that further cuts would lead to the tipping point, driving so many out of the industry that Gloucester would no longer be able to provide the basic services required to sustain a fleet.[8] In addition, they believed that the small and medium-sized vessels were under-represented in the existing management process. Leaders in the Gloucester fleet joined with neighbouring communities' leaders and began discussing ways to become more effectively involved in management. They drew on community legal expertise and called for a community meeting. Ultimately, it was decided that all of the groundfish industry needed to be unified in one body comprising shore-side business owners, community leaders, and fishermen to have a stronger voice in management, particularly in facing Amendment 13. Without such unity, they feared that the majority of the fleet would be ruined, and groundfish ports across the northeast would be forever changed. Thus, the Northeast Seafood Coalition was founded in the winter of 2002.

---

[8] In *The Tipping Point*, Gladwell (2000) offers a fascinating look at the phenomenon of sudden societal change explained by the concept of the tipping point.

Initially, having been started by a group of Gloucester fishermen and other local business owners, the Coalition was Gloucester-centred. However, the group felt very strongly that they would be more efficacious if they were broader-based and consequently, it has made an effort to diversify. In addition, they emphasise a team approach, incorporating the experience and knowledge of fishermen, and expertise of a political consultant and lawyers. A significant amount of effort is expended on communication and educational outreach among industry members, managers and politicians. The Coalition now has about 120 vessels with approximately 300 fishermen members and 60 shore-side business members. Several municipalities have also contributed to the Coalition. Membership fees for vessels are based on landings (1 or 2 cents per pound or 2 per cent of the catch). Shore-side businesses pay an annual fee of $1000. Maine, Massachusetts, New Hampshire and Rhode Island have the most members; Connecticut and New York each have a few. Eighteen months after its formation, the Coalition hired an executive director and administrative assistant.

Decisions are made by a twenty nine-person board of fishermen and shore-side business owners that represent interests from across the northeast and approved by a vote of the membership. The focus of the Coalition is groundfish policy. However, the organisation is also actively involved in the reauthorisation of the Magnuson Act, and regulations surrounding bycatch, offshore energy development, EFH, and issues arising out of the development of marine reserves as integral concerns. The Coalition considers itself a part of the fishing community in the northeast; supports family-owned businesses, and is working toward the common good, by promoting policy that is equitable and based on sound science and legislation. It attempts to enable fishermen to become involved in the management process without devoting an inordinate amount of time attending meetings. Moreover, members of the Northeast Seafood Coalition agreed that their goal is to help achieve sustainable harvests and communities. In other words, they wanted to see a balance struck between conservation and community needs. The retention of fleet diversity was also considered a high priority. They recognised that for their Amendment 13 proposal to be viable, it had to achieve the rebuilding strategies demanded by the Court, yet they also wanted to provide opportunities that would help ensure that the family-owned fishing businesses in the Northeast could survive. The Council ultimately accepted the Northeast Seafood Coalition's proposal as the basis for Amendment 13. The specifics are delineated below in section 6.6, Final Rule.

8.6.4 THE POLICY ENTREPRENEUR

One of the individual fishermen, who had been carefully following the development of Amendment 13, became a key leader helping "direct change and transform governance". In steps that parallel those identified by Olsson *et al*'s (2004:6) local **policy entrepreneur**, this fisherman "initiated trust-building dialogue, mobilised social networks with actors across scales, and started processes for coordinating people, information flows and ongoing activities, and for compiling and generating knowledge."

This important leader began ground-fishing out of Gloucester with his father in 1977. With his Sicilian-American heritage, born and raised in Gloucester, and with fishermen on both sides of his family, he has strong links with the predominant ethnic group among both the groundfish fishermen and the shore-side industry in the community. Working with an industry lawyer and a lobbyist, the fisherman began to re-work the package of alternatives in an attempt to devise a strategy that would resolve the immediate conservation demands as required by the court case, yet offer opportunities for active groundfish fishing vessels to fish for alternative species and also preserve future access for those historically active in ground-fishing.

At a meeting of the Northeast Seafood Coalition, held at the Gloucester Display Auction, the fisherman and lobbyist introduced the new package of alternatives. As he put it, they "used the same ingredients (or tools) [as the Council], just made a different recipe." Initially, fishermen were sceptical, but after much discussion agreed that the repackaging might actually allow more of their family-owned businesses to survive.

8.6.5 SPECIAL NORTHEAST FISHERY MANAGEMENT COUNCIL MEETING

The Council and PDT made it clear that only one industry proposal could be selected at the Special Council meeting since it would have to be analysed in order to be brought forward to the full Council meeting in November. The Council had to vote on the final measures at this time in order to comply with the Court's ruling that the new regulations be in place by May 1st 2004.

The two proposals that were considered to be serious contenders were those of NAMA, together with the Stonington Fisheries Alliance; and the Northeast Seafood Coalition. While the AFM and TSF proposal was also discussed, the PDT and Council viewed it as an assertion of position (for the status quo with some add-ons) rather than a full proposal. The AFM/TSF argument that the Council should reject the revised biomass targets was the primary reason the proposal was not considered as a true alternative package. With regard to the Northeast Seafood Coalition's proposal, the Council directed the staff and PDT to do an analysis of the proposal in time for the next regularly scheduled Council meeting. Submissions from other organisations were regarded as comments on specific proposed measures, or recommendations for specific fisheries, rather than full industry proposals. Some of the specific recommendations would have demanded reliance on qualitative analysis, as there was not sufficient time for a full or quantitative analysis.

8.6.6 FINAL RULE, AMENDMENT 13

At the Council meeting on November 4-6th 2003, Council members voted in favour of an Amendment 13 that incorporated a mix of adaptive and phased reduction rebuilding strategies (variations on Alternatives 1 and 2) and added opportunities to target healthy fish stocks. "The proposed action was developed in response to comments received from the public on the Amendment 13, developed through the efforts of the Northeast Seafood Coalition" (NEFMC 2003a). So the Council meeting largely endorsed the Northeast Seafood Coalition's proposal.

Days-at-Sea are broken into three categories and the allocation is reduced accordingly to 60 per cent, 40 per cent (subdivided into 2 parts), and Fishing Year (FY) 2001 allocation less the effective effort. For example, the holder of a fleet days-at-sea permit (88 days-at-sea allocated in FY2001) with 88 days effective effort, would be allocated 52.7 Category A days (60 per cent of 88); 17.6 days (20 per cent of 88) Category B (regular) days and 17.6 days (20 per cent of 88) Category B (reserve) days and 0 Category C days (88-88). However, a fleet days-at-sea permit holder with an effective effort of 50 days would have 30 Category A days; 10 each in the two B day categories and 38 days in the C day category.

Category A days-at-sea can be used as usual, subject to the various management measures. While Category B days-at-sea are divided into two categories, B regular days and B reserve days, currently, only B reserve days have been approved. Category B days are used to target stocks that do not need a reduction in fishing mortality, subject to various restrictions including the requirement for VMS. For now, B reserve days can only be used in an approved Special Access Program.[9] Though not yet approved, B regular days are intended to be more flexible, allowing fishermen to fish without the strict controls of time, area and gear but controlled by VMS and very strict bycatch restrictions (hard TAC). B regular days will only be used if the permit holder has sufficient Category A days remaining to cover the trip in case the vessel exceeds the limit for a stock of concern. (So the permit holder can 'flip' to a Category A day/trip if necessary.)

As stocks recover (or if they diminish) the number of days in each category can be reclassified. This provides an adaptive mechanism that allows effort to be redirected depending on the condition of the stocks. Category C days will not be fished until stocks recover, but this category provides a way for those fishermen who were active participants in the groundfish fishery in the past to regain access to groundfish stocks before other fishermen are allowed in.

Other management restrictions include year round closed areas, including some specifically designated as EFH. Seasonal closed areas are specified. Possession limits vary with the days-at-sea category, stock, and gear. A number of restrictions on gillnets continue, as does a minimum hook size and circle hook requirement. Minimum fish size for cod is increased. Though not addressed in the special Council meeting, proposals for two other changes were ultimately voted for by the full Council. The Cape Cod Hook Fishermen's Association was given a sector allocation and the AFM and TSF's proposal to allow leasing of days-at-sea was agreed to.

---

[9] The healthy stocks that can be targeted with "B" days are pollock, redfish, Gulf of Maine haddock and winter flounder, and Georges Bank haddock, yellowtail flounder and winter flounder.

8.6.7 WHAT MADE THE NORTHEAST SEAFOOD COALITION'S PROPOSAL SUCCEED?

In addition to the proposed measures and packages submitted for the special meeting, it should be noted that Amendment 13 has been in the process of development for more than four years. For several of those years, a proposal submitted by the Gulf of Maine Fisheries Alliance that eventually became known as the City of Gloucester proposal was among the 'packages' of alternatives being considered. NAMA had also influenced the development of several alternatives, as did Associated Fisheries of Maine. Over the years, however, it seems these concepts just 'dried up and blew away'. What made the Northeast Seafood Coalition's proposal succeed, when these other proposals failed?

One reason for the success of the Coalition's proposal was a critical flaw in the proposal submitted by its rival, NAMA. NAMA had been attempting to move the Council towards area management of the inshore Gulf of Maine for several years. Criticisms of their specific approach in their Amendment 13 proposal focused on NAMA's apparent inability or unwillingness to move beyond the designation of boundaries to design rules or management measures for the areas. One observer commented that "they just wanted us to draw lines on a map, but I can't do that without knowing what the objective is, is it to protect spawning aggregations, juveniles, what?" What some of the fishermen did not like about the proposed area management is that it would confine them by forcing them to declare an area in which they plan to fish for the year, precluding opportunities to follow the fish or new opportunities. This is an example of strong disagreement within the industry about the costs and benefits of different fishing styles.

Another reason for the Coalition proposal's success was its silence on the subject of the controversial science. As mentioned above, the new biomass targets developed by the Northeast Science Center were criticised and rejected by most of the groundfish fleet. However, environmental groups are poised to seize the opportunity for new lawsuits if NMFS does not regulate according to the new targets. Hogarth sent a letter to the Council for their November meeting warning that Amendment 13, however designed through the Council process, must meet the new targets or it would not be approved, and, instead, NMFS would be forced to write a new version. In their proposal, the Northeast Seafood Coalition did not discuss the targets, but focused on adapting the existing alternatives to achieve more flexibility and more hope for the future.

In addition, the Coalition may have been perceived as representing a larger range of fishing industry participants (according to both geography and gear sector) and communities (the mayors of both Gloucester and New Bedford submitted comments in support) with a proposal that could be implemented much more quickly within the existing system of management than some of the other proposals. Moreover, the policy entrepreneur and the others who led the Coalition's effort, devoted a great deal of time to communication and educational outreach, explaining what is a rather complicated system to the stakeholders until they could comprehend the adaptive aspects and the proposal's underlying optimism.

Furthermore, several elements of the proposals by other fisheries organisations were either incorporated into Amendment 13 (for example, leasing and the hook fishermen's sector allocation) or left as an option to be pursued in the future through "framework adjustment."[10] The area management proposed by NAMA, for example, was designated as a "frameworkable item."

8.6.8 RESULTS SO FAR

The NEFMC staff and PDT are very optimistic that Amendment 13's provisions mean that the biological (that is, mortality) objectives for groundfish will be met. With the possible exception of Georges Bank cod, rebuilding will be achieved well within the required timeframe.

However, there remain four concerns among managers and participants:

1. The push for approval of new Special Access Permits (SAPs) so that Category B days can be used, may be too hasty. If the Council fails to hold to a strict standard of proof with supporting data, it is possible that SAPs could be granted inappropriately. Furthermore, "the burden of managing all the SAPs could quickly become overwhelming" (Neis, 2004). On the other hand, if an insufficient number of SAPs are granted, the use of Category B days may not provide enough benefit to save the fleet from economic failure.
2. The recreational catch, particularly of Gulf of Maine cod, may be too high. Trip limits and minimum size regulations are often ignored, but recreational interests tend to fight the more effective and more easily enforced closed areas and seasons (NEFMC 2003a).
3. As stocks rebuild, bycatch issues and minimum size restraints may cause the fishery to be shut down prematurely. The industry will have to be constantly adapting and adjusting gear for increased selectivity.
4. The management of groundfish, while an exception to the single species management characteristic of U.S. fisheries management, does not solve the problem of reduced flexibility due to the requirement of permits based on fishing history within relatively narrow timeframes.

8.6.8.1 *Leasing*

While it is too early to comment on the impacts of leasing days-at-sea, so far NMFS has approved 33 one-year leases of a total of 1,135.91 days-at-sea, involving 27 vessels leasing days to 28 vessels (Plante 2004:7A). Most of the leased days so far have moved to vessels home-ported in Maine and Massachusetts. 1152 vessels had groundfish permits in 2002. In June 2004, 42,989 Category A days and 28,660 Category B days were allocated to 915 vessels. Another 339 vessels received Category C days only.

Some of the leasing agreements have been for days leased between boats owned by the same company or set of owners. Contrary to expectations, the leases have so far been "equally distributed among four tonnage classes" (below 500 tons) (Plante 2004:7A). No

[10] Framework adjustment is a technique that the Council has adopted that allows some changes to be made in a management plan without going to a full amendment process that takes years to complete.

requests for permanent transfers have been received. Such a transfer has a 40 per cent conservation tax while leasing has none, so this is not surprising. Also, both leasing and transfer have tonnage upgrade restrictions. Proposed changes in the tax and the tonnage restrictions will be discussed and may be incorporated into the next groundfish framework adjustment. Opponents of leasing fear that it will increase the usage rate of days-at-sea with a potential consequence of cut backs in days-at-sea allocations across the fishery in the future. The leasing option is restricted to Category A days only and it sunsets in 2006.

8.6.8.2 *Category B Days*

The large vessels in the northeast traditionally use a larger proportion of their allocated days-at-sea than do small and medium-sized vessels. Consequently, any reduction in allocated days means that they have lost days they would most likely have used. Because the larger vessels can move offshore to fish, Category B days, that are most commonly available offshore, are 'saving' some of the boats (APC 2004). Eight million pounds of yellowtail that would not otherwise have been caught has been landed by vessels using their Category B days. In addition, vessels are allowed steaming time to reach the grounds that is not counted against their days-at-sea.

Unfortunately, there has been some misunderstanding about when or how Category B days can be used. When, for example, what is known as Closed Area 2 opened in June for fishing on yellowtail flounder, many of the boat operators thought incorrectly that they had to fish in the first week of the opening or they would not be able to fish there at all during the open season.[11] Consequently, the markets were glutted, processors could not handle the quantity of product, and prices dropped accordingly (AMPC 2004). (In fact, for the last several years, before the implementation of Category B days, too many yellowtail of moderate quality, due to their recent spawning, have been landed in June with negative impacts on prices). However, some vessel owners point out that Category B days provide a limited opportunity. Because a very small bycatch of the species of concern is allowed, some anticipate that the cod bycatch TAC will soon force a closure of the haddock SAP.

8.7. Conclusion

This chapter uses the Amendment 13 to the Northeast Multispecies Fishery Management Plan as a case study for exploring the potential for co-management in a large, diverse area with a long fishing tradition and well-organised groups with strong leadership but differing philosophies. We found that the common assumption that only place-based communities can effectively co-manage resources, could lead to missed opportunities for non-traditional, but organised groups of stakeholders to have a determining voice in resource management. Australia's Environment and Natural Resources Committee (ENRC 2001) emphasises the importance of including a broad array of stakeholders, not just certain sectors, in the

[11] In contrast to the beneficial contributions of the policy entrepreneur, another character that has arisen in fisheries management in the northeast may be appropriately referred to as 'the trickster'. In this case, a leader of a fisheries organisation warned fishermen that they had to 'use it or lose it' in reference to the 'B' days. Whether this was an intentional act to benefit his own business, a result of a cynical view towards the managers, or simply an honest misunderstanding, is not clear.

[co-management] process, including 'whole-of-community partnerships', that can create a vision and viewpoint to help the fisheries to respond to future challenges and opportunities.

The choice that the New England Fishery Management Council made in selecting the Northeast Seafood Coalition's package of management alternatives revealed a movement towards an acceptance of the benefits of co-management. While the constraints on co-management within the Council system that were detailed earlier remain, the acceptance of the Coalition's proposal, as well as elements of other industry proposals suggest that some negotiation of power and authority has taken place.

Pomeroy and Guieb (2004) who have facilitated the implementation of co-management systems in multiple, primarily artisanal fisheries in developing countries, argue that co-management is a process, rather than a "single strategy to solve all problems of fisheries management…maturing, adjusting and adapting to changing conditions over time". The Northeast Seafood Coalition's proposal was an attempt to develop an adaptive strategy for the management of groundfish within the framework of the legal demands articulated by the Court and certain environmental groups. Furthermore, in the interest of equity, the Coalition's proposal worked to ensure that those who have traditionally participated in the groundfish fishery will be the first to benefit from rebuilt stocks.

In the management of New England's fishing industry, efforts to institutionalise co-management face critical problems in determining how to assure that the diverse array of stakeholders are fully represented and that the decisions made are equitable. The Amendment 13 process reflects some of the benefits of co-management identified by researchers such as an opening of communication and knowledge transfer between managers and industry. The use of VMS (for "B" days) is likely to aid data gathering and lead to improved habitat protection. The sharing of responsibility for harvesting and allocation decisions has only begun, but the selection of one of the industry's proposals does suggest potential for enhancement of participative policy-making. The adaptive mechanisms of the amendment lend themselves to the sustainability of both stocks and the industry. Whether the most often cited benefit of co-management –compliance – is also effected is yet to be determined, but the New England Fishery Management Council has taken the first step towards adaptive co-management that reflects an awareness of these issues and an attempt to resolve them.

Acknowledgements

Jackie Odell and Vito Giacalone of the Northeast Seafood Coalition served as key informants for this report, explaining the details and rationale for the proposal the Coalition developed. Tom Neis of the New England Fishery Management Council tried to straighten out my descriptions of the sequence of events and supplied some important statistics to provide context for the results to date. Conversations with Mike Crocker and Heather Deese helped clarify information on NAMA obtained from their website. A very brief conversation with Maggie Raymond, Associated Fisheries of Maine, was also helpful. Written proposals and comments submitted to the Council by various fishing organisations on Amendment 13 were reviewed. Mistakes are, of course, my own.

References

Acheson, J M (2003) *Capturing the Commons,* Hanover, University Press of New England

AMPC (2004) Auction manager's personal communication

APC (2004) Anonymous personal communication

Berkes, F, Colding, J and Folke, C (2003) *Navigating Social-Ecological Systems*, Cambridge, Cambridge University Press

Brazer, E (2003) *Does the Atlantic Cod (Gadus morhua) Have a Future In The Gulf of Maine?*, Brown University Center for Environmental Studies Thesis

Dale, N (1989) 'Getting to co-management: social learning in the redesign of fisheries management' in E Pinkerton (ed) *Co-operative Management of Local Fisheries: New Directions for Improved Management and Community Development*, Vancouver, University of British Columbia Press: 49-72

Deese, H (2004) Personal communication

Dyer, CL and McGoodwin, JR (eds) (1994) *Folk Management in the World's Fisheries: Lessons for Modern Fisheries Management*, Niwott, CO, University Press of Colorado

ENRC [Environment and Natural Resources Committee] (2001) *Inquiry into Fisheries Management First Report: Co-Management*, Australia, Government Printer for the State of Victoria, No. 124: Session 1999-2001

Gladwell, M (2000) *The Tipping Point*, Boston, MA, Little, Brown and Company

Hall-Arber, M, Dyer, C, Poggie, J, McNally, J, and Gagne, R (2001) *New England's Fishing Communities*, Cambridge, MA, MIT Sea Grant College Program

Hall-Arber, M, Robinson, S, and Bergeron, D (2002) *Fishing Industry Economic Needs Assessment*, Massachusetts Fishermen's Partnership Report, Gloucester MA

Holling, CS and Meffe, GK, (1996) 'Command and control and the pathology of natural resource management' *Conservation Biology* 10:328-37

Honneland, G (1999) 'Co-management and communities in the Barents Sea fisheries' *Human Organization* 58(4):397-404

Jentoft, S and Kristoffersen, T (1989) 'Fishermen's co-management: the case of the Lofoten fishery' *Human Organization* 48(4):355-365

Jentoft, S and McCay, B (1995) 'User participation in fisheries management: lessons drawn from international experiences' *Marine Policy* 19(3):227-246

McCay, B (1989) 'Co-management of a clam revitalization project: the New Jersey "Spawner Sanctuary" Program' in E Pinkerton (ed) *Co-operative Management of Local Fisheries: New Directions for Improved Management and Community Development*, Vancouver, University of British Columbia Press: 103-124

McCay, BJ and Finlayson, AC (1995) 'The political ecology of crisis and institutional change: the case of the northern cod' presented to the Annual Meeting of the American Anthropological Association, Washington, DC, November 15-19

NAMA [Northwest Atlantic Marine Alliance] (2003) 'Amendment 13 – a composite alternative to address rebuilding requirements: reduced days-at-sea with phased area management', Comments submitted to NEFMC

NEFMC [Northeast Fishery Management Council] (2003a) 'Northeast Multispecies Amendment' 13 *SEIS*, 18 December 2003

NEFMC [Northeast Fishery Management Council] (2003b) 'Public Hearing Document for Draft Amendment 13 to the Northeast Multispecies Fishery Management Plan', New England Fisheries Management Council

Neis, T (2004) Personal communication

Odell, J (2004) *Perception of Legitimacy in Fisheries Management: A case study of the Atlantic States Marine Fisheries Commission's Horseshoe Crab Fishery Management Plan*, Kingston, RI, University of Rhode Island, Master's Thesis

Olsson, P, Folke, C, and Hahn, T (2004) 'Social-ecological transformation for ecosystem management: the development of adaptive co-management of a wetland landscape in southern Sweden' *Ecology and Society* 9(4):2

Ostrom, E (1990) *Governing the Commons*, New York, Cambridge University Press

Pinkerton, E (ed) (1989) *Co-operative Management of Local Fisheries: New Directions for Improved Management and Community Development,* Vancouver, University of British Columbia Press

Plante, J (2004) 'Groundfish fleet tries DAS leasing, shuns transfers' *Commercial Fisheries News* 31:12 (August)

Pomeroy, R and Guieb (2004) (in preparation) *A Handbook on Community-based Fisheries Co-management*, International Development Research Centre, Ottawa, Canada

Raymond, M (2004) Personal communication
Smith, ME (1990) 'Chaos in Fisheries Management' *MAST* 3(2):1-13
Stein, J (1997) 'How institutions learn: a socio-cognitive perspective' *Journal of Economic Issues* 31(3):729-740
USC [United States Code] (1996) 16 USC 1801-1882, April 13, 1976, as amended 1978-1980, 1982-1984, 1986-1990, 1992-1994 and 1996. Magnuson-Stevens Fishery Conservation and Management Act: Public Law 94-265

CHAPTER 9
PARTICIPATORY GOVERNANCE IN INSHORE FISHERIES CO-MANAGEMENT IN ENGLAND AND WALES

PAUL KNAPMAN
ORCA - EU, 230 Freshwater Road, St John's, Newfoundland, Canada, A1B 1B2

Abstract

First established in 1888, the Sea Fisheries Committees (SFCs) of England and Wales predate the concept of modern inshore fisheries management. However, their organisational structure, function and working practices are closely aligned with principles that are now commonly advocated and associated with good governance, not least because of the extent of participation that they provide to the fishing industry in a largely co-management system. In this chapter, I provide an outline of the institutional framework within which SFCs operate; explain their structures and functions; evaluate their governance credentials and, using my first hand experience of working at a senior level within a SFC, offer a critique of the effectiveness of SFCs as an inshore fisheries co-management model.

9.1 Introduction

For more than one hundred years the majority of inshore[1] fisheries within England and Wales have been managed by Sea Fisheries Committees (SFCs). The first SFCs were established soon after the confirmation of the *Sea Fisheries Regulation Act* in 1888. The Act allowed for the establishment of sea fishery districts and the appointment of a committee, responsible for regulating and developing the fisheries within their district. County councils were established in the same year and it was, by and large, through application by these local authorities to the government that SFCs were created and financially maintained. Today, there are twelve SFCs districts that cover the majority of the inshore fisheries of England and Wales (see Figure 9.1.) Their empowering legislation was consolidated in 1966 and further legislation has provided for additional fisheries and environmental responsibility within their districts.

While all SFCs are established and empowered by the same legislation there are distinct differences between them making it difficult to provide a standard model that accurately describes all of the SFCs (Symes 2002). For instance, there are differences in the geographical scale of the fisheries districts; in the varying complexity of the local authority structure within the districts, which in turn affects the size of the Committee; in the size and structure of the workforce; and in the number and content of the regulatory instruments used by SFCs.

In sections two to five of this chapter, I explain the structure and functions of SFCs, their links with two national fisheries management bodies, and their relationship with the fishing industry. In sections 6 and 7, drawing upon my personal experience of working

[1] Inshore fisheries in the UK are generally considered to be those fisheries within the six-mile fishery limits.

T. S. Gray (ed.), Participation in Fisheries Governance, 163–178.

as a Chief Fishery Officer for a major SFC, I provide a critical appraisal of the work of SFCs by examining, first, the extent to which they embody principles of good governance, and, second, their effectiveness in managing inshore fisheries. In the conclusion, I argue that if SFCs are to continue to play a valuable role, they must improve the calibre of local authority members and the impartiality of industry members, and they must be provided with adequate funding and regulatory flexibility.

9.2 Structure of Sea Fisheries Committees (SFCs)

9.2.1 SFC DISTRICTS

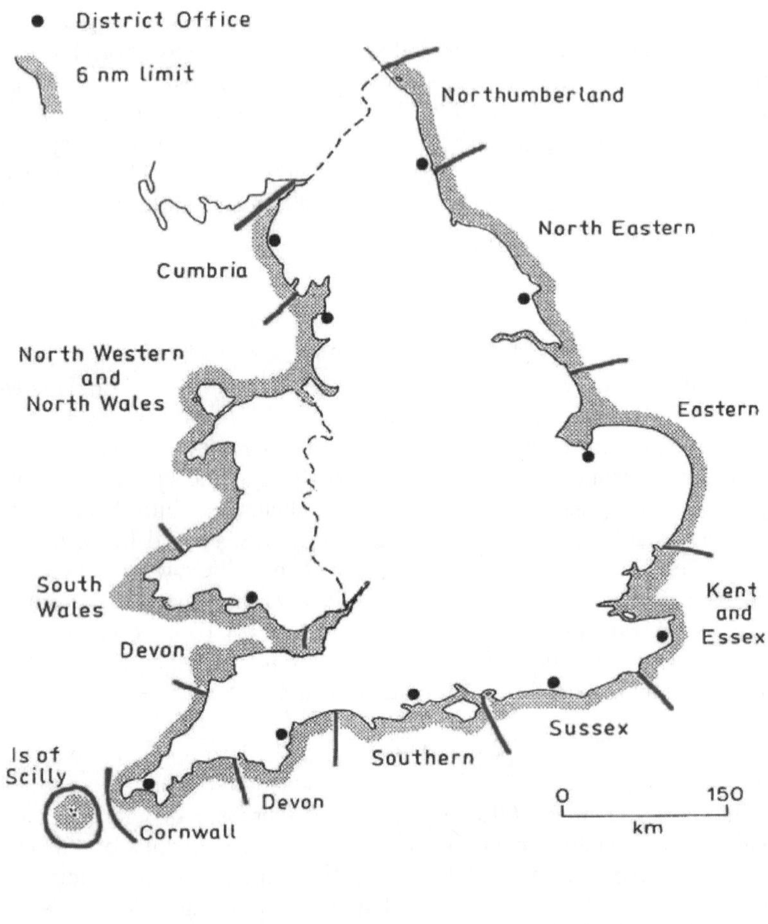

Fig. 9.1. Sea Fisheries Districts (adapted from Symes, 2002)

The SFCs districts vary in size from the smallest (Northumberland) with a coastline of 111 kilometres and sea area of approximately 1,372 square kilometres, to the largest (North Western and North Wales) with a coastline of approximately 1,713 kilometres and sea area of 6,860 square kilometres. The districts generally coincide with county council boundaries, although changes caused by local government restructuring mean there are instances where this is no longer the case. Districts extend seaward six miles from baselines (the line from which seaward limits are drawn, such as the mean low water or lines drawn between headlands across the mouth of a bay), and inland to the high water mark or, in the case of an estuary, usually to the lowest bridging point or the tidal limit.

9.2.2 THE COMMITTEE

SFCs are local government committees and, as such, are solely funded by local authorities. All of the Committees are composed of a 50/50 split between representatives from those local authorities that contribute funds for the SFCs and representatives appointed by the Fisheries Minister who are considered to be "acquainted with the needs and opinions of the fishing interests of that district or as being persons having knowledge of, or expertise in, marine environmental matters" (MAFF 2001). Ministers' appointees must include a representative from the Environment Agency (EA) and at least one marine environment expert. All of the Ministers' appointees are appointed for four years and can be re-appointed. A small number of Committees have also appointed a representative from the recreational fishing sector: such appointments reflect more a legacy, than a statutory requirement, resulting from encouragement by a former Fisheries Minister who was an enthusiastic sea angler (Symes 2002). While the latter appointment may be considered an anomaly, it makes very good sense to include a stakeholder group that has a significant economic interest in inshore fisheries: members of over a million households participate in sea angling each year and spend over £500 million (DEFRA 2004), a significant part of which finds its way into the local economies of coastal communities.

In the case of fishing industry representations, nominations are invited through public notice: for example, via the fishing industry press, and by direct invitation to fishing organisations regularly consulted on fisheries-related issues by the Department for Environment, Food and Rural Affairs (DEFRA). While named as 'Ministers' appointees', in reality, it is the DEFRA District Inspector of Fisheries, who will be acquainted with the fisheries nominees within the relevant SFC districts, who, in consultation with DEFRA administrators, confirms the appointments. The EA nominate their own representative and, in the case of the marine environmental specialist, DEFRA consult with the statutory nature conservation advisors – English Nature (EN) and/or the Countryside Council for Wales (CCW) – before making invitations and appointments.

The Committee, when considering an environmental issue, may also co-opt anybody it considers to be suitably qualified to provide advice. Some Committees have taken to inviting representatives from EN and CCW to attend meetings and contribute advice as and when appropriate. It is also regular practice for DEFRA District Inspectors of Fisheries to be invited to attend.

Local authority representatives are democratically elected Councillors. When elected to serve on a local authority, Councillors are expected or required to sit on a number of local committees. In some instances, they will be chosen by senior officials to sit on particular committees; this might be as a result of an individual's expertise or experience in the subject or issue for which the committee is responsible, or, where a committee is dealing with a highly politicised issue, members may be appointed for political reasons. In other instances, Councillors will volunteer for committees. In the majority of cases, Councillors volunteer to sit on SFCs, and, if re-elected, often continue to do so for many years. Although SFCs are apolitical, they can occasionally become politicised, particularly when Councillors who have been elected on a fisheries-related manifesto are appointed to the Committee.

The size of the Committee is dictated by the number of Councillors. Two SFCs have as many as 18 Councillors and so, with the required corresponding number of Minister appointees, their Committees have a full complement of 36 members. In contrast, the smallest Committee has only 8 members. The average size of a Committee, however, is 20. These differences are attributable to the length of a Committee's coastline and/or the complexity of the local government structures within it. Three Committees (Isles of Scilly, Cornwall and Cumbria) are made up of a single contributing local authority with the rest ranging between 2 and 11. The distribution of seats between the local authorities approximates to their relative contributions to the funding of the SFC (Symes 2002).

The Committees have their own standing orders (Committee rules), which have evolved over time, often being influenced by their constituent local authorities. The Committees are, by law, obliged to meet quarterly, give two weeks notice of their meetings, one week's notice of the agenda and keep a detailed record of the minutes. Committee meetings and papers are open to the public, unless registered as being of a private or personal nature. Most Committees have chosen to have at least one sub-committee to deal with more technical or financial aspects of the Committee's work. These sub-committees may meet on a regular basis, or, as needs require.

There is no overarching rule about how the Chairman of a Committee is appointed or their length of tenure. The Eastern SFC rotate chairs between the 3 local authorities every two years; South Wales and Cumbria alternate between local authorities and DEFRA appointees; while the other Committees leave it to their members to decide on an annual basis. The Committee's Chairman occupies an influential position, requiring a heavy time commitment and close working relationship with the Clerk and Chief Fishery Officer (CFO).

As a minimum, meetings of the full Committee involve the reporting of sub-committee meetings; a report and review of the work of the Committees Officers; and a report of the fishing and other activities (such as offshore developments and nature conservation issues) within the district. If necessary, consideration of existing and future management is undertaken, as well as discussion and confirmation of appropriate action on any relevant national or European issues.

9.2.3 THE OFFICERS OF THE COMMITTEE

Committee-appointed Officers are responsible for the fisheries management duties and for informing and providing advice to the Committee. Symes (2002) describes three different internal staffing models for SFCs (Figure 9.2). There are minor variations of these models but they serve to show the difference in size and complexity. Six of the Committees have chosen to combine the role of Clerk and CFO, while two have also chosen to break with convention and use more modern titles (Director, Chief Executive) for the same post. The role of the Clerk (a dated term which conveys the history of SFCs) is primarily to oversee the administration of the Committee and the presentation of byelaws, and serve as a point of contact for the public. The CFO is responsible for the day-to-day management of the staff and their activities.

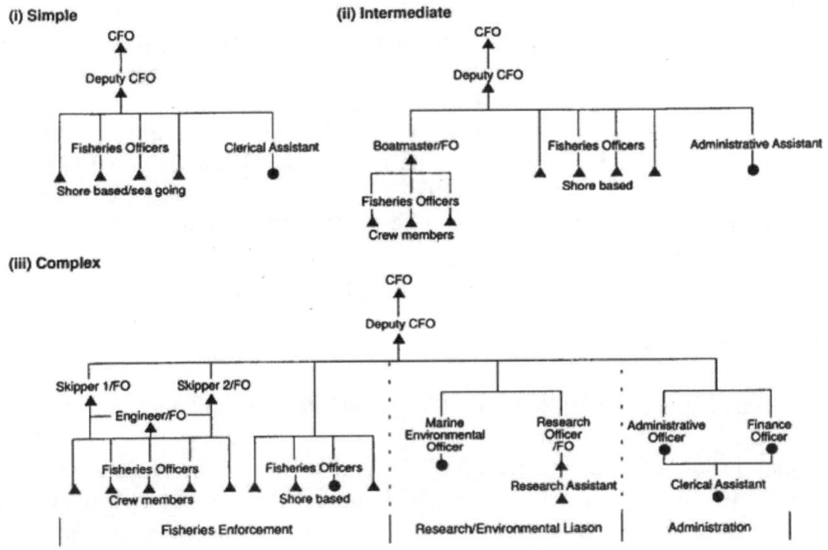

Fig. 9.2. The various internal staffing structures of SFCs (Symes 2002)

A Deputy CFO and Fisheries Officers undertake the primary SFC role of enforcing local, national and European Union (EU) regulations. This can be in either a shore-based or a sea-going capacity. All of the SFCs, except the Isles of Scilly, have a patrol vessel and at least one Rigid Inflatable Boat (RIB) used for interception and boarding of fishing vessels. This requires that the Fisheries Officers have a dual role using their seamanship skills as skipper, mate, engineer or crew-member. To be suitably qualified to undertake the duties of a Fisheries Officer, familiarity with the fishing industry, sea-going experience/qualifications and, more commonly these days, a degree are considered to be prerequisites. This often means Fisheries Officers are drawn from the fishing industry, the armed forces and the merchant navy, and from those who have gained further education qualifications. All SFCs provide training on enforcement and some also provide additional training opportunities, in particular seamanship, to ensure that officers meet the ever-increasing requirements associated with conditions imposed by insurance companies.

Some SFCs with extensive molluscan and crustacean fisheries in their districts have been able to invest in scientifically qualified research staff, thereby providing a capability to undertake stock assessments, detailed monitoring and stock enhancement. One Committee, the Eastern SFC, has been able to invest in a new vessel designed and dedicated to fisheries and environmental research.

Changes brought in during the 1990s requiring SFCs to manage fisheries with a regard for the marine environment, have also resulted in the addition of an Environment Officer for those Committees fortunate enough to be able to resource such a post. The designation of areas to protect nature conservation interests; the development of offshore industries which have to meet environmental standards; and a general move toward a more integrated approach to the management of the coastal zone, have meant a considerable increase in the time needed for SFCs to carry out this environmental aspect of their work. Where SFCs are unable to finance an Environment Officer post, the Chief and Deputy Officers often take on these duties.

A Finance Officer may be appointed to the Committee or, alternatively, some Committees may have access to a local authority Finance Officer who works on behalf of the Committee. The Administrative Officer is an essential member of staff; among their many duties, they are often the initial point of contact with the fishing industry, the public and Committee members.

9.3 Functions of SFCs

9.3.1 MANAGEMENT TOOLS – BYELAWS AND REGULATIONS

The original 1888 Act that established SFCs, and which was consolidated in 1966, enables SFCs to make byelaws to help manage the fisheries within their district. This, and subsequent Fisheries Acts, empowers SFCs to, among others things, specify particular times or seasons for fishing; restrict the size of vessels; describe types of fishing gear that are restricted or prohibited; set minimum landing size limits for fish and shellfish; and restrict fishing activities for "marine environmental purposes", which includes the conservation of marine flora and fauna (Phillipson and Symes 2001).

The creation of a byelaw often stems from a request by the industry, through a Committee member or from correspondence to the Chairman or CFO. The Chairman will call upon the Committee and use the working knowledge of its Officers to confirm, or otherwise, the necessity for action and whether a byelaw could be used to resolve the problem. A byelaw is intended to benefit the fishery as a whole, not to discriminate against any groups or individual, and cannot be less restrictive than those imposed at a national or European level.

The byelaw-making process requires SFCs to draft and advertise the byelaw for two consecutive weeks; to allow 28 days for any objections to be lodged; and to take account of any objections before giving 14 days notice to DEFRA of their intention to submit the byelaw for approval. In the case of a byelaw made on environmental grounds, the SFC also has to have consulted with EN or CCW before giving notice to DEFRA. Once the application has been received by DEFRA, they give one month's notice to the European Commission to ensure that they are satisfied that it does not conflict with any Community Regulation, and, at the same time, DEFRA assesses the byelaw to ensure it fits with the conditions set out in the SFCs' empowering Acts. Only after this does DEFRA finally confirm, by way of Ministerial approval, the byelaw.

When set out sequentially, this appears to be a straight-forward and relatively quick process, but, in reality, it can take a considerable amount of time. With increasing litigation and legal challenge by the fishing industry, DEFRA are particularly keen to ensure that any byelaw they confirm is not successfully challenged. This often requires SFCs to invest considerable time in gathering and providing substantive information to confirm the need for a byelaw, and it may require a number of attempts at drafting the byelaw before DEFRA are willing to accept it. To complicate matters, there have sometimes been inconsistencies in the legal opinion of DEFRA on similar byelaws from different SFCs. These have been attributed to differences in legal opinion of new staff within DEFRA's legal department and changes in opinion following new case law.

As well as byelaws, SFCs can use Regulating Orders to manage molluscan and crustacean fisheries. The main advantage of a Regulating Order is that it allows SFCs to licence fishing activity for shellfish within a designated area and, in so doing, set licence conditions, such as the use of a prescribed fishing method, daily quotas, and the time and areas that can be fished. A licence fee can also be levied, the proceeds of which must be re-invested in the fishery. Regulating Orders can also be combined with Several Orders – Orders that 'sever' the public right to fish – allowing fishermen to lease an area of seabed on which they can cultivate their own shellfish (such as mussels and oysters), and on which no other fisherman can legally fish. This combination of Orders is sometimes referred to as a Hybrid Order.

There are, however, two disadvantages in issuing these Orders, and this has meant that not every SFC with a shellfishery has chosen to use them, preferring to use byelaws instead. The first disadvantage is that the establishment of an Order requires consensus from the fishing industry, which is always a challenging task. If this cannot be achieved it may be resolved by a public enquiry, at the expense of the Committee. Given the financial constraints within which many SFCs must operate, negotiation is the favoured approach, which can lead to a significant dilution in the management potential of the Order. The second disadvantage is that once consensus has been agreed and the Order is in place, the number of licences can only be reduced by fishermen leaving the industry. And so, matching the number of licences to the available resource is constrained. This inflexibility is compounded by the fact that adapting management measures in the future must be re-negotiated with the fishing industry.

9.3.2 ENFORCEMENT OF BYELAWS AND REGULATIONS

The preferred option for enforcement by SFCs is one of prevention of infringements: prosecution is a last resort. This policy is reflected in the relatively low total number of annual prosecutions that SFCs chose to take by contrast to the relatively high number of Home Office written warnings (DEFRA 2004). These warnings are valid for two years; if a case is compiled against the individual during this period, and a prosecution is actioned, the Home Office warning can be used against the offender. However, it is often the case that the punishment meted out by the court is not considered by the SFC to act as a sufficient deterrent. The maximum fine for a breach of a byelaw is £5,000, and fishing gear can be forfeited, but, so far, neither of these penalties has been applied to their full extent. High profile policing at times of year and in areas where offences are more likely to occur, is common practice and results in fewer breaches of local regulations. Also, there is generally a good relationship between SFC Fisheries Officers and the fishing industry, and this helps to ensure awareness of local byelaws. Moreover, owing to changes in legal opinion caused by case law and new European human rights legislation, some byelaws and regulations associated with Regulating Orders may be *ultra vires,* and so SFCs are not inclined to prosecute, preferring to issue a Home Office warning.

9.4 Relationship between SFCs and other organisations

9.4.1 NATIONAL COORDINATION

The Association of Sea Fisheries Committees (ASFC) provides a national representative and coordinating function, as well as a central source for disseminating information, for all of the SFCs. The ASFC is constituted by the SFCs, all of which contribute funds toward its operation. The ASFC is made up of a Chairman and Vice Chairman, elected from the Chairmen of the twelve SFCs and a Chief Executive who is employed on a part-time basis. Given that resources are insufficient to provide a full time post, the Chief Executive is able to call upon CFOs for specialist support when dealing with coordinated responses to national consultations and attending national meetings.

The ASFC meets at least four times a year and is generally attended by the Chairman, Clerk and/or the CFO of each SFC. The meetings discuss and coordinate action on national and European issues that affect the SFCs, and provide an opportunity to share information and experiences between SFCs. The Fisheries Minister will address one of the ASFC quarterly meetings and discuss an agenda of issues set by the ASFC. The CFOs also meet independently at least three times a year to discuss in more detail issues that cut across all of the SFCs. Their collective knowledge and experience is particularly important in informing the ASFC on technical aspects associated with fisheries management.

9.4.2 OTHER INSHORE MANAGEMENT ORGANISATIONS

While SFCs are the primary inshore fisheries managers in England and Wales, there are two other important national organisations that have fisheries management responsibilities within the six-mile fishery limit. These are the Sea Fisheries Inspectorate (SFI) and the Environment Agency (EA). Some of their jurisdictions and responsibilities overlap with those of SFCs, which can make for a complex system of management and enforcement. Indeed, this complexity and apparent double or even triple accounting of management and enforcement has been one of the main reasons for a major review, commissioned by DEFRA in 2004, that aimed to look at the most effective organisation of enforcement in relation to the long-term needs of the fishing industry. Although it is not my intention to make comparisons between the three organisations, it is important to explain the roles of the SFI and the EA in order to understand their working relationship with the SFCs.

9.4.3 THE SEA FISHERIES INSPECTORATE (SFI)

The SFI is the fisheries enforcement arm of DEFRA. It is responsible for enforcing

European and national fisheries regulations throughout the English and Welsh territorial waters and beyond, to the limit of the Exclusive Economic Zone (EEZ) or the median line. British Sea Fisheries Officers (BSFO) – their full title reflecting their national role – are strategically based around the coast to ensure compliance with the regulations by monitoring fish landings, national quotas, fishing vessels, fishing gear and by administering fishing vessel licences. They also enforce legislation concerning fish marketing and the protection of the marine environment with respect to dumping and removal of substances at sea. Statistical data gathering, involving the collection and collation of logbooks and landing declarations, as well as biological sampling, is also an important role of the SFI. This information is used by scientists at the Centre for Environment, Fisheries and Aquaculture Science (CEFAS) and by policy-makers involved in stock assessments, quota management and national and European conservation measures.

The SFI's sea-going capability is contracted to the Royal Navy's Fisheries Protection Squadron, and chartered civilian aircraft are used for aerial surveillance of fishing activity, along with satellite monitoring of vessels greater than 15 metres in length. The SFI headquarters are in London, where an operations room coordinates offshore enforcement activity and liaises with its staff around the coast. Senior SFI staff are in close proximity to support the Fisheries Minister at short notice (for what can on occasions be a 'hot potato' of a portfolio) and to react to the needs and maintenance of the other civil service divisions responsible for developing and implementing national and European fisheries policies and legislation.

There are no formal national agreements on working practices between the SFI and SFCs. Instead their working relationships are dependent on the type of fisheries within their districts, which may or may not bring SFI and SFC Officers into regular contact with each other. Where they do, the demarcation of roles, responsibilities and working practices are usually clearly established and are often designed to reduce any inconvenience to the fishing industry. The sharing of surveillance and monitoring information is also common. Most SFCs have at least one officer who is warranted as a full BSFO or has partial BSFO powers to undertake pre-agreed enforcement work with, or on behalf of, the SFI. Another important factor is the personal relationship between the SFI's District Inspector and the SFC's Chief Fisheries Officer. If they get on well, it makes for improved liaison. District Inspectors are usually invited to attend SFC Committee meetings as a matter of courtesy.

9.4.4 THE ENVIRONMENT AGENCY (EA)

The EA was formally established in 1995, as a non-departmental public body, by the *Environment Act*, sponsored largely by DEFRA and the Welsh Assembly Government (WAG). The regionally-structured EA has a very broad remit which, among other things, includes fisheries management. As well as inland fisheries, the EA are responsible for the management of migratory species (mainly salmon, trout and eels) within the six-mile fishery limit, where it has the power to limit the number of licences for salmon and eel and to use byelaws to regulate fishing activity. The EA also has a responsibility to manage some sea fisheries. As a result of historical events, there are a number of estuaries where the EA act as a Sea Fisheries Committee, (the Dee, Severn, Taw-Torridge, all of the estuaries in Cornwall and the tidal reaches of the Thames). The EA is dependent on grant-in-aid from DEFRA to undertake its sea fisheries role, though seen in a positive light.

in some cases it uses other resources, such as income it gains from its national rod licensing scheme.

Fisheries management advice is discussed and agreed through EA-appointed Regional Fisheries, Ecology and Recreation Advisory Committees (RFERACs), though, as their title suggests, fisheries is just one of a host of issues these committees consider. SFCs may be represented on the RFERACs, but, in contrast to the constitution of SFCs, which require an EA representative, this is not statutorily prescribed. The EA's regional fishery officers employ a combination of shore-based and sea-borne inspection and enforcement using small, rapidly deployed inshore craft. As SFCs operate in many of the same inshore areas as the EA, a number of their officers are usually cross-warranted so they can enforce EA byelaws. As with the SFC/SFI working relationship, there are no formal national guidelines on working practice. The EA and SFC approach to collaboration has evolved and is influenced by the type of fisheries, the coincidence of enforcement activity and the relationship between senior staff of both organisations. Liaison between the EA and SFCs can generally be regarded as good.

9.5 Relationship between SFCs and the fishing industry

It is difficult to measure how SFCs are regarded by the fishing industry. As with any organisation charged with an enforcement and management role, they can be viewed with suspicion and treated with distain by those who are averse to authority or are willing to push the limits of lawfulness. However, drawing upon my personal experience of having worked for a SFC, my informed, if subjective, view is that the relationship between that of the SFCs and the industry is, for the most part, constructive. The appointment of industry representatives to the Committees provides an opportunity for fishermen to have a say in the management of the local fishing industry. Such participation can endorse the personal standing of individuals and the organisations; allow for a closer relationship to be established between fishers and senior SFC officers; and lead to an improved understanding of the way SFCs function and an appreciation of the constraints they work under.

For their part, SFC officers can set out to establish close relations with the fishing industry. For instance, regular shore- or sea-based inspections help to develop familiarity and allow for the sharing of information and assistance in understanding management measures and fishing activity. Polite and efficient inspections which result in minimum inconvenience are more likely to be tolerated, while responding readily to an opportunity to assist a fishing vessel in difficulty also goes a very long way in improving relations, not least because the demonstration of good seafaring skills is recognised and respected.

It is often the case that CFOs, while attending local or national meetings, find themselves in a position where they, in effect, represent the fishing industry. While it is not their role to do so it is, when inshore fishermen are not present at these meetings, the CFO who is called upon, or feels obliged, to express the views of the fishing industry and offer their technical knowledge. This is fed back to the Committee when the CFO reports to a quarterly meeting and, in turn, it may filter back to the local industry and be seen in a positive light.

The attitude or mindset of some sectors of the inshore fishing industry is more amenable to regulation and respectful of management than that of other sectors. For example, those fishermen who restrict their activities to local areas may have a greater appreciation of what the SFCs are trying to achieve. Crab and lobster fishermen, who represent a significant proportion of the inshore fishing fleet, and so fall under the jurisdiction of SFCs, appear to be more conservation-minded. This might be partly because regulations for crustacean shellfish are easier to understand and abide by (primarily based on minimum size), combined with the fact that their method of fishing allows for the return of undersize fish which have a high likelihood of surviving and contributing to their fishery in the future. By contrast, nomadic fishermen, who work on a much larger fishing area, with mobile gear that are subject to complex regulations, are less selective in their fishing methods, and have a greater tendency to fish irresponsibly, risk breaching local regulations, and move on, in the belief that any adverse effect from overfishing or damage to fish and other habitat that they might have done will have recovered by the time they return.

9.6 Discussion

9.6.1 SFCs AND PRINCIPLES OF GOOD GOVERNANCE

Turning now to a critical appraisal of the work of SFCs, I focus on two questions: first, to what extent do SFCs exemplify 'good governance'?; and, second, how effective are they at their job of inshore fisheries management? Beginning with the question of SFCs' credentials as exemplifiers of good governance, we should note that 'governance' is not a term readily used by those involved with the management of inshore fisheries in England and Wales, and probably not elsewhere. It appears to have crept into the vocabulary as a result of research conducted by academics and specialists from the economic, social, environmental and legal sectors with an interest in the issues associated with fisheries management. The use of 'new' terminology and the involvement of 'new' people does not sit easily with many who are involved with the fishing industry. Any industry steeped in tradition is likely to be suspicious of what it might perceive as outsiders with new ideas getting involved with their business. SFCs are no different. However, the reality is that the term 'governance' – meaning the sum of the legal, social, economic and political arrangements with respect to fisheries management – is what SFCs are primarily about, and the way that they function and operate follows many of the principles associated with good governance (as articulated by the FAO and DEFRA). The analysis below highlights eleven such principles; six principles relate to SFC governing **processes**; five principles relate to SFC governing **policies**. The six **process** principles are as follows:

- The SFCs are **devolved and decentralised** management bodies. While DEFRA has a role to play in appointing half of the Committees' membership and in confirming their byelaws, SFCs remain able to operate in a largely autonomous way
- SFCs provide **stakeholder involvement**. The fishing industry and environmental interests are represented and a number of Committees also have a DEFRA appointee representing sea angling interests

- The opportunity of these stakeholders to participate in SFC discussions and vote on issues that affect the management of their fisheries, realises another principle of good governance, that of **subsidiarity**
- The appointment of **democratically elected** Councillors to SFCs gives the Committees strong political **accountability**
- The appointment of EA representatives and marine environmental specialists, from academia or wildlife NGOs, to SFCs, and the invited participation of SFI District Inspectors and statutory nature conservation agencies, help to improve liaison and **institutional integration**. This is further enhanced by regular contact with these and numerous other organisations with an interest or role in the inshore region. The increased interest in integrated coastal zone management (ICZM) and various statutory and voluntary marine nature conservation initiatives, combined with the key role that SFCs play in managing fishing, has brought and, in some cases, forced SFCs, to integrate more. While this has at times caused uneasiness between organisations owing, in part, to a lack of understanding of their respective roles, this generally reflects the growing pains that new working relationships often experience
- The open meetings and the administrative requirements associated with SFCs, provide for **transparency** in the way they operate, though the selection and appointment by DEFRA of its fishing industry appointees is not transparent.

The five **policy** principles are as follows:

- The requirement for modern fisheries management to take account of the marine environment, and of the potential effects of fishing on habitats and species other than commercial fish, has recently become a facet of the work of SFCs. The DEFRA appointment of a marine environmental specialist, and, in some instances, the employment of an Environmental/Conservation Officer has provided Committees with a broader knowledge base and capability. A number of SFCs have introduced byelaws with a strong environmental component, and one SFC has introduced a byelaw specifically for environmental purposes. This represents the first tentative steps at what might be termed the **ecosystem-based approach**.
- SFCs have been more willing and able to act in accordance with the **precautionary approach**, in the past – as have DEFRA or, more precisely, their previous incarnation the Ministry of Agriculture, Fisheries and Food (MAFF) – in their role in confirming SFC byelaws. For example, some of the older byelaws that restrict access to the inshore fishing grounds were made on a precautionary basis, inasmuch as they restricted larger fishing vessels from entering the fisheries owing to their fishing potential. However, these measures were taken at a time when the precautionary approach was not formally enshrined as a principle of good governance. Today the formal application of the precautionary approach presents more of a challenge to SFCs owing to the risk of imposing contentious restrictions on the fishing industry without being able to demonstrate reasonable or measurable benefits. Given the litigious nature of some fishermen, on the one hand, and the call for application of a precautionary approach by influential environmental organisations on the other, SFCs (and DEFRA) are placed in a difficult position. Their reaction has generally been to favour the need for hard science. This reduces the risk of challenge from the fishing industry and allays

fears that once a byelaw is made on strong precautionary grounds the floodgates will open with demands of more of the same from environmental groups. But, it also negates the ability to take action before a serious problem rises.

- The management approach used by SFCs is based on restrictive access to inshore fisheries, created through byelaws, and Regulating and Several Orders. In so doing, preference for smaller vessels and the zoning of some inshore areas for particular forms of fishing have been deliberately, or, in some instances, inadvertently created. While this approach has been endorsed by the fishing industry, it challenges some elements of **equity** that are espoused as representative of good governance. For example, inter- and intra-generational equity may be compromised by the 'closed shop' effect that Regulating Orders entail by restricting new entrants to the fishery; while cross boundary equity is not always secured, because some shared stocks may be administered in different ways by adjoining districts (such as imposing different minimum landing sizes). However, since the majority of the inshore fisheries are considered to be at or close to the maximum acceptable levels of fishing, and there is a desire to manage fisheries to match the local circumstance, it is difficult to see how any other approach, given the constraints of the management tools they have at hand, can be used by SFCs.
- The diversity of fisheries and conservation issues within the inshore area, and the limited resources that SFCs command (some more than others), means that SFCs have to target enforcement action and, in so doing, ensure that action is **proportional** to the possible or likely infringements.
- The devolved management approach offered by SFCs allows for management that matches the local fisheries and the conditions within which they operate. As a result, the regulations between Fisheries Districts may be very different and, therefore a **consistency of approach** may be difficult to achieve. Where SFCs use different regulations to manage similar activities, cross-border cooperation, 'learning by doing' and the sharing of experience all contribute to improved consistency. Transparency – ensuring that the enforcement system is widely communicated, and that decisions are clearly explained – also helps to promote consistency.

From the above list of process and policy principles associated with good governance, which the SFC 'model' achieves in greater and lesser degrees, there can clearly be seen a participative mode of governance. Breaking it down further, within the participative mode, the SFC model has a strong co-management element and, to a lesser degree, an environmental stewardship element which, with time, is certain to become more prominent.

9.6.2 THE EFFECTIVENESS OF SFCs

The second question that I focus on, in my critical appraisal of SFCs, is how effective are they at managing inshore fisheries? There is a broad range of answers to this question depending on who we ask, where they are coming from, and their experience of SFCs. The following analysis is based on six key criteria, which, from my own experience of working within a SFC, I consider to be important in assessing the overall effectiveness of a SFC.

First, the calibre of the Committee members is a vital component in the overall effectiveness of SFCs as inshore fisheries managers. An ideal would be to have a

Committee made up of (1) local authority representatives with a good background knowledge and appreciation of the reality of modern inshore fisheries; and (2) industry representatives who were truly representative of the fishing industry and did not try to influence decisions to favour themselves. Alas, this is an unlikely combination: while there are some Committee members who fit these descriptions, many others do not, and, indeed, some members may unwittingly or deliberately hinder the effectiveness of the Committee.

With respect to (1) the local authority representatives, members are often appointed with limited or no background knowledge or understanding of technical matters associated with fishing, and an apparent inability to grasp many of the issues, even after a considerable length of time serving on the Committee. There may be others with an unrealistic, romantic image of small-scale, low-intensity fishing, who are unaware of the fishing and earning potential of modern fishing vessels and the 'cut-throat' nature of the business. Others may have a preoccupation with cost efficiency and may be ignorant of the reality of maintaining a sea-going capability. In the worse case, a member may fit all three descriptions.

There can also be difficulties with (2) the DEFRA fishing industry representation. It is often hard to find a working fisherman willing and able to contribute to SFCs. As a result, DEFRA may receive a limited number of suitable nominations. This might be due to the lack of financial incentive for active fishermen to forfeit a day's fishing. The reimbursement of travel expenses and loss of earnings at local government rates are unlikely to adequately compensate a fishermen and/or his crew. It may also be indicative of the fact that the majority of fishermen are not particularly comfortable debating issues in a formal setting. This difficulty is compounded by the pressure of trying to represent the views of an industry that is notoriously bad at achieving a consensus; there's a saying that, 'if you get two fishermen together you'll get three points of view'. While a confident and articulate fisherman can be very influential, he is likely to have many requests made on his time by people looking for an industry point of view and, if he is a working fisherman, he will be constrained by how much he can afford to do. A lack of suitable nominations may also reflect fishermen's perception that their input will have negligible benefits – a perception that the fishing press regularly perpetuates with regard to local, national and European fisheries management policy.

The second key issue follows on from the first: that the industry nominees who are chosen to sit on a SFC may not fairly represent the views of the industry. In the worse case scenario they may prefer to influence opinion for their own benefit, and deliberately attempt to undermine a management approach which will restrict their fishing activities. While pecuniary interests should be declared at Committee meetings, members who declare their interests are still allowed to debate issues, and, given that some local authority members may fit the profiles highlighted above, they can be misled by deliberate deceptions.

Third, there may be differences of opinion or uncertainty among Committee members about whether their management of inshore fisheries ought to be for the benefit of the local fishing industry. The Fisheries Act that establishes SFCs does not help, only referring to their regulatory role as one of preventing damage to inshore fisheries from inappropriate fishing activity. If the majority of members take the view that management is for the benefit of the local fishing industry, the general bias of management decisions is likely to be in favour of the industry rather than the resource. Given the reality of the varying calibre and motives of some Committee members, this can mean that fishing industry representatives misinform the Committee, resulting in the dilution or rejection of management proposals designed to protect the resource. For example, suppose the officers of the Committee undertake a comprehensive shellfish stock assessment and propose a total allowable catch (TAC), but the industry rejects the assessment, saying that there is significantly more shellfish available. The Committee then takes account of the industry's 'guesstimates', and agrees a revised and increased TAC. The local authority representatives feel that by negotiating and achieving a consensus they have reached a good resolution, while the industry representatives are satisfied because they are able to catch more fish. However, despite a large investment of public money in a stock assessment, considerable effort on behalf of the officers and advice given in good faith, the Committee has chosen to put the industry first rather than the resource. This is a familiar story, but one that is more often associated with the level of governance undertaken in Brussels than in inshore fisheries.

Fourth, the relationship between the CFO and Committee Chairman is a critical factor, because, together, it is their responsibility to manage this complex mix of individuals by use of strong interpersonal skills, patience, second-guessing and good preparation. With a successful chairman supported by a pro-active CFO, the Committee can be focused into achieving positive results. A good relationship between the CFO and his officers is also vitally important – CFOs have to be strong in their convictions and skilful at maintaining team spirit, because it can be easy for officers to become disenchanted or demoralised as a result of some Committee decisions. However, given the wide-ranging demands put on CFOs, it is difficult for them to constantly maintain such qualities. Likewise, SFC Chairmen do not always match up to these exacting specifications.

Fifth, the principal legislation which provides the SFCs with their regulatory powers are remnants from Victorian times, and they are no longer appropriate for the management of a modern inshore fishing sector. Some fishermen have invested in consultants and sought legal advice in order to learn how to exploit the loopholes in local regulations, and, in some circumstances, this has led to a serious undermining of the effectiveness and credibility of SFCs. The protracted byelaw-making process, and the SFCs' restricted legal scope to be proactive in dealing with a highly inventive and adaptive industry, creates a real potential for damage to fish stocks and the marine environment as fishermen develop and use new methods of fishing. This can result in a 'fire-fighting approach' in the way SFCs operate, requiring them to redirect resources from normal enforcement duties to deal with problems that could have been avoided by proactive measures. Working within this type of constraining legislative framework stifles the ability to be strategic and can create a 'navel-gazing' culture rather than one of innovation and flexibility.

Finally, with the greater demands placed on modern fisheries management to ensure the sustainability of both fish stocks and the marine environment, SFCs need to be adequately and consistently funded, so they can attract and employ the appropriate mix of skills and utilise state-of-the-art technology and hardware. The Eastern SFC is the only SFC that has been able to keep pace with the resourcing needs of a modern inshore fisheries management body, with a staff complement and hardware that allows them to undertake the full range of fisheries management functions. The annual cost for this SFC is approximately £1.2 million. If the value of the commercial inshore fisheries of England and Wales is in excess of £130 million; the value of sea angling is £500 million (DEFRA 2004); and our coastal waters contain some of our greatest diversity of marine wildlife (English Nature 2004), it makes sense to have an adequately resourced organisation which is capable of maintaining and enhancing the economic and biological value of our inshore area.

9.7 Conclusion

In this analysis of Sea Fisheries Committees in England and Wales, I have explained their structure and functions, and their relationship with other fisheries management bodies, and I have evaluated their credentials of good governance and their effectiveness in managing inshore fisheries. My conclusion is that for over 100 years, SFCs have performed a valuable role in successfully managing inshore fisheries – the state of inshore stocks in comparison to those generally associated with the offshore provides testimony to this fact – but SFCs face many challenges which they are not fully equipped to deal with. First, a way has to be found of preventing some industry representatives from unduly influencing SFC decisions in their favour and for local authority representatives to be more aware and appreciative of the local fishing industry. Induction training for all new members so they fully understand and appreciate the role of the Committee would help (not least by stiffening local authority representatives' resistance to any inappropriate tendencies shown by their colleagues from the industry), while reimbursement of loss-of-earnings to working fishermen could encourage a wider representation from the industry. Second, increasing responsibility conferred on already over-worked SFCs for the protection of the marine environment must be matched by increased funding allowing them to invest in human and hardware resources that give them the capacity to broaden their capability and, as a result, their general outlook, as more than fish stock managers. Third, legislation is urgently required to give SFCs the flexibility they need to respond more speedily to local events to prevent them from turning into crises for the marine environment. Fourth, any proposal to merge SFCs into a nation-wide Marine Agency should be resisted, because it would undermine their essential characteristic – local autonomy.

References

DEFRA [Department for Environment, Food and Rural Affairs] (2004) *Review of Marine Fisheries and Environmental Enforcement,* London, DEFRA

English Nature (2004) *State of Nature Report – Maritime – Getting on to an Even Keel*, Peterborough, English Nature

MAFF [Ministry of Agriculture, Fisheries and Food] (2001) *Guidance Note on Nature Conservation and the Role of Sea Fisheries Committees*, London, DEFRA

Phillipson, J, and Symes, D (2001) 'Great Britain' in D Symes and J Phillipson (eds) *Inshore Fisheries Management*, Dordrecht, Kluwer

Symes, D (2002) *Inshore Fisheries Management in England and Wales: Facing up to the Challenges of the 21st Century*, Peterborough, English Nature

CHAPTER 10
A COMPARATIVE ANALYSIS OF TWO FORMS OF STAKEHOLDER PARTICIPATION IN EUROPEAN AQUACULTURE GOVERNANCE: SELF-REGULATION AND INTEGRATED COASTAL ZONE MANAGEMENT

SELINA M STEAD
School of Marine Science and Technology, Newcastle University, Ridley Building, Newcastle upon Tyne, NE1 7RU

Abstract

European aquaculture governance contains elements of the three main modes of governance: 1) hierarchical; 2) market; and 3) participative. This chapter focuses on the participative mode, both because it is the dominant mode, and because it offers a better prospect for the future of the aquaculture industry than either of the other two modes. There are two distinct forms of stakeholder participation: a) self-regulation, where participation is largely confined to the industry; and b) Integrated Coastal Zone Management (ICZM), where participation is (ideally) shared by all stakeholders. In this chapter, both forms of stakeholder participation are explained and evaluated, and the conclusion is drawn that the future of European aquaculture governance lies in strengthening the element of ICZM relative to the element of self-regulation.

10.1 Introduction

Aquaculture in Europe has been subjected to much less control by the European Commission than has the capture fisheries sector. This is partly because marine aquaculture generally takes place within Member States' territorial waters, and is therefore regulated mainly by the states themselves, and partly because aquaculture is a new industry, on a relatively small scale, and so has not raised many serious issues of competition between Member States in comparison with fisheries. As a result, the predominant mode of aquaculture governance in Europe is not the hierarchical mode (the top-down, centralised and coercive mode which predominates in the European catch fisheries in the form of the Common Fisheries Policy (CFP)), but what appears at first sight to be a modified version of the market mode, in which market forces of supply and demand are permitted to hold sway, subject to domestic legislation on planning, environmental protection, health and safety. However, scrutinising this modified market mode more closely, we can see that, notwithstanding its free market features, it contains a considerable amount of voluntary control by the aquaculture industry, laying down detailed guidelines and codes of conduct that all producers are virtually obliged to adopt. This form of governance, which has been termed 'self-regulation', thus embodies a 'thin' or partial form of stakeholder participation, in that the industry participates in decision-making, though other stakeholders are generally excluded. Accordingly, I have categorised it in the participative mode, rather than in the market mode, of fisheries governance.

However, there are increasing signs of a challenge to this self-regulating form of

T. S. Gray (ed.), Participation in Fisheries Governance, 179–192.

aquaculture governance in Europe, coming from two quarters. First, there are demands from other coastal resource users to participate in decision-making. Second, there is pressure from the European Union (EU) to shift from a single industry perspective to an eco-system approach, whereby aquaculture is governed in the context of the wider ecological environment in which it is located. The concept of Integrated Coastal Zone Management (ICZM) has arisen to satisfy these two aspirations, incorporating both a 'thick' or comprehensive form of stakeholder participation, and an ecosystem approach.

In this chapter, I examine each of these two forms of European aquaculture governance – self-regulation and ICZM – and show how the tide is gradually turning in favour of the latter. I conclude by arguing, however, that the best arrangement is where the two forms are combined, so that the industry retains its self-regulating capacity in spheres such as quality assurance, but that the whole coastal community is empowered to make decisions on such issues as the size and location of fish farms.

10.2 Development of aquaculture

Aquaculture is considered to be the fastest growing animal food production sector in the world, having increased at an average compounded rate of 9.2 per cent per year since 1970, compared with 1.4 per cent for capture fisheries and 2.8 per cent for terrestrial farmed meat production systems (FAO 2002). The marine aquaculture sector is dominated by high-value finfish, crustaceans, molluscs and aquatic plants. Finfish farming is the most important form of aquaculture in developed countries, having started commercially in the late 1970s/early 1980s, and having established itself as a successful alternative to fishing by the early 1990s. In western parts of the world, like Europe, the main opportunities for growth in the marine aquaculture sector lie in developing value-added products based on traditional farmed species, such as salmon (*Salmo salar* L.) and mussels (*Mytilus edulis*), and diversification into production of newer species such as cod (*Gadus* morhua) and haddock (*Melanogrammus aeglefinus*). Within the next 10 years, this sub-sector of aquaculture has been predicted to provide significant new employment. Organic fish farming is undertaken on a smaller scale, and the potential markets for organically farmed finfish, where a premium on price is paid, are more unpredictable, given that many consumers are more interested in competitively-priced products than in how fish are farmed.

Accordingly, despite some uncertainty, marine aquaculture is considered by many as a promising opportunity for diversification in coastal areas, especially in those areas that contain fisheries-dependent communities. Many human settlements are socially and economically dependent on unstable catch fishery resources, and, in some cases, aquaculture can offer an alternative sustainable livelihood, especially in rural areas where activities for income generation are limited. There are many examples where fishers have diversified into aquaculture successfully, meeting the need for employees skilled in working in and from a boat. Similarly, mollusc and cage culture provide additional revenue for fishers, who often perform them on a part-time basis. Indeed, in many parts of the world, fishing and aquaculture activities share similar coastal areas and services, and interaction between the two sectors is increasing. Moreover, offshore cage technology continues to advance and is becoming cheaper and increasingly viable, and again this could open up opportunities for diversification, relying on skilled boat users to operate the sites. Competition for space in coastal waters has meant that aquaculturists are also looking towards onshore marine aquaculture production units, such as those that rely on recirculation, which are becoming cheaper to operate and could offer an alternative year-round supply of fish to local processors.

However, aquaculture is not without its problems. For example, in addition to increasing competition for space, some farmed species, such as salmon, have faced a number of setbacks, including falling prices and negative media coverage. Furthermore, in developed nations, the aquaculture industry has undergone restructuring from many middle-sized production units into fewer and more efficient bigger units. This has occurred largely as a result of outside investment, and is a trend expected to increase, particularly in developing countries. As many aquaculture farms have become more technology-based, they have become less reliant on a large workforce, and some communities that are now dependent on aquaculture as a source of income, particularly in rural areas, have been forced to look for alternative employment opportunities. This has caused disappointment in some quarters, because the initial establishment of aquaculture production units was embraced largely on the understanding that this was an activity that would create jobs. Also, despite the commonly held view that aquaculture can easily become a suitable occupation for fishers seeking, or having, to leave the fishing industry, in practice, the two working environments - the culture of organisms ('gathering') and fishing ('hunting') - have important cultural differences, which may make it difficult to match individual skills in fishing with the labour requirements of aquaculture.

10.3 Participatory aquaculture governance

Although there are examples of both the hierarchical mode of aquaculture governance (for example, Thailand), and the market mode of aquaculture governance (for example, in Nigeria and Ukraine), the most common mode of aquaculture governance, at least in developed countries, is the participatory mode. There are, however, two forms of the participatory mode in aquaculture governance: a) self-regulation; and b) ICZM. In the remainder of this chapter, most of my analysis will focus on these two forms of participatory aquaculture governance.

Despite its rapid expansion and success, relatively little academic attention has been devoted to the participatory mode of aquaculture governance, by contrast to the considerable amount of interest in the participatory mode of governance of capture fisheries. The reasons why less attention has been paid to addressing stakeholder participation in aquaculture compared to other sectors, such as capture fisheries, include the following. First, aquaculture is a relatively young industry, whereas capture fisheries have a long cultural heritage associated with traditional coastal communities. Second, fish farms are often located in isolated and/or peripheral areas, where transport to and from can be time-consuming and costly. Third, people have less awareness of aquaculture, because they are less likely to come into contact with aquatic production units than with fishing vessels. Fourth, until recently, aquatic products available for sale to consumers, like finfish and shellfish in retail outlets, have not differentiated between farmed and wild origin. Fifth, negative media coverage of the aquaculture industry labels this sector more an 'abuser of the environment' than a 'victim of environmental change', which is sometimes how the fishing industry is perceived. These factors explain why so few social scientists have worked on aquaculture-related case studies of stakeholder participation, compared to those who focus on stakeholder participation in catch fisheries.

However, there are signs that this picture is changing, because more attention is slowly being directed towards promoting participation in decision-making processes that deal with aquaculture-related topics. This is partly because of the activities of organisations such as the Federation of European Aquaculture Producers, which are improving stakeholder involvement through initiatives that focus on self-regulatory measures. Similarly, the European Aquaculture Society organises workshops that bring together stakeholders, such as producers, scientists and policy makers, to debate current and emerging issues. Likewise, the European Strategy for the Sustainable Development of European Aquaculture (EC 2002b) highlights the potential role of ICZM (EC 2002a) as a type of participatory forum that could help to advance sustainable development of the aquaculture sector.

Let us now turn our attention to the two forms of the participatory mode in aquaculture governance, beginning with self-regulation.

10.3.1 SELF-REGULATION AS A PARTICIPATORY MODE OF AQUACULTURE GOVERNANCE

Although self-regulation is a thin form of participation, in that the main stakeholder to participate in governance decisions is the aquaculture industry, nevertheless, it is still a participative mode of governance. Self-regulation is distinguished from the hierarchical mode in that the government allows an extensive amount of self-government to the aquaculture industry; and it is distinguished from the free market mode in that the industry imposes upon its members a strict code of conduct, to prevent a free-for-all. The European Commission's Strategy for the Sustainable Development of European Aquaculture clearly endorsed the concept of self-regulation, when it proposed that "The industry should make more use of self-regulation and voluntary agreements" (EC 2002b).

Self-regulation is being encouraged by the Commission in order to address some of the problems experienced by fish farmers and legislative institutions, particularly in resolving distortions of competition between Member States. Self-regulation and associated codes are a less mandatory form of control than is control through licensing, and in Europe, Codes of Conduct, including Codes of Practice and Voluntary Codes, have been used extensively to regulate the aquaculture sector. Codes of Conduct, such as the voluntary code developed for Europe's aquaculture sector by the Federation of European Aquaculture Producers (FEAP) (Hough 2000), offer considerable benefits to farmers and to the industry as a whole. These benefits include increased consumer confidence through imparting knowledge that products adhere to high levels of product safety and are produced by environmentally-friendly farming methods. Linked to this is an improved image of aquacultural products, which can enlarge market demand and generate higher returns from sales. Incentives such as these help to promote wider compliance among producers and adherence to good farming practices, which are linked to eligibility for membership of quality assurance schemes such as the 'Label Rouge', used by French farmers, and, in Scotland, the 'Tartan Quality Salmon' label. The success of self-regulation, of course, relies on the implementation of good practices by producers, and then dissemination of the advantages to consumers.

It is probably true to say that the reason why self-regulation is the most widespread mode of governance in European aquaculture, is because, theoretically, at least, it takes into account not only the interests of the producers and consumers, but also the fish that are farmed, and the environment in which it operates (Hough 2000). The FEAP code is a voluntary and non-binding document, sometimes called soft law, drawn up in response to self-regulated sector development (Hough, 2000) and includes all species, types and scale of aquaculture. The code addresses the following topics among others:

- Guiding principles of the Code (expected conduct and attitudes by those involved);
- Husbandry;
- Environmental issues;
- Consumer issues;
- Social and economic considerations.

FEAP also promotes pro-active initiatives within the sector. For example, it encourages the development of Codes of Practice by Associations; Best Management Practices by Producer Groups and Co-operatives; and approved labelling schemes, such as "organic". As Hough (2000) notes, FEAP is particularly concerned to ensure the transparent management of aquaculture to the benefit of the consumer, the environment, and society. Self-publicity is also actively promoted. With this in mind, FEAP launched an initiative in 2001 – called 'Aqua-media' – designed to counter negative reports of the aquaculture industry in the media. This campaign was funded by the private sector, and its target audience included the general public, people in education (schools, colleges and universities), government and related institutions, the press, and consumer and special interest groups. FEAP released relevant information, using the Internet, multimedia Compact Discs, newsletters and brochures (Hough 2001). Similarly, some fish farmers actively try to engage with the local community by, for example, holding open days, participating in local fishery trusts and community council meetings, and sponsoring sporting events. In those areas where local coastal forums exist, representation of aquaculture interests is often found.

Outside Europe, we can find further examples of self-regulation in aquaculture governance. For instance, in Thailand, the hierarchical approach to aquaculture governance, which made use of command-and-control measures through legislation, proved inadequate in achieving sustainable development of the industry. Self-regulation as a form of stakeholder participation is now being pursued in preference to the earlier use of state regulation (Vandergeest *et al* 1999). State regulation failed to support the growth of aquaculture, because the laws and regulations quickly became outdated in relation to the rapidly developing industry (Lebel *et al* 2002) – dominated by shrimp. This legal inadequacy was compounded by a lack of support and involvement of relevant administration bodies; insufficient resources to enforce the regulations; and few incentives to promote compliance. This led to stakeholders working independently of one another and in an uncoordinated manner (Huitric *et al* 2002).

Codes of Conduct and Codes of Practice attracted considerable attention in Thailand in the 1990s in response to this shift of policy towards self-regulation. These voluntary and non-binding documents were viewed as tools that could help the Thai aquaculture industry to overcome difficulties experienced in marketing and production – especially in the marine shrimp sector, which is mainly export-driven. The Thai Code of Conduct developed for marine shrimp has been given a five-year implementation phase (2001-2006). By 2006, it is hoped that a majority of shrimp farms will be certified to the required standard outlined in the Code of Conduct. However, critics have complained that, in Thailand, such a voluntary code will not work because compliance with best management practices can only be attained through environmental laws and regulations. The Thai government is well aware of this criticism, and has attempted to reinforce the voluntary codes by legal tools. Whether this hybrid system will work, is difficult to tell: it is too early to evaluate the effectiveness of this form of self-regulation, and careful monitoring and evaluation are necessary in order to provide evidence for its retention beyond 2006.

However, even where self-regulation appears to work well within a loose governmental regulatory framework, it has been criticised for two deficiencies: first, for excluding all resource users other than the aquaculture industry (the democratic deficit); and second, for excluding consideration of wider ecological issues than the aquacultural (the ecosystem deficit). To address both these issues, the concept of ICZM has been devised.

10.3.2 ICZM AS A FORM OF PARTICIPATIVE AQUACULTURE GOVERNANCE

The Commission has enthusiastically embraced the concept of ICZM as a means of overcoming both the democratic deficit, and the ecosystem deficit, in self-regulatory aquaculture. With regard to the *democratic deficit*, the European Commission's Strategy for the Sustainable Development of European Aquaculture stated that stakeholder participation must be increased. Stakeholder participation was identified by the Commission as a way to include broader consultations in the process of policy formulation. This approach is viewed as necessary by the Commission in order to include more extensive information on economic, social and physical considerations, as opposed to concentrating on policies that are only production-orientated. Participation by stakeholders such as consumers, farmers, producer associations, researchers and special interest groups in the decision-making process, according to the Commission, will also help tackle the aquaculture sector's over-dependence on governments and the private sector.

Until recently, initiatives aimed at improving stakeholder participation in the context of aquaculture governance were largely focused on stakeholders who are directly or indirectly involved in the industry. But, as aquaculture has expanded at such a rapid rate, particularly in coastal areas where many complaints against aquaculture development reflect competition for space (EC 2002b), it has been deemed important to explore how aquaculture can also meet the needs of coastal governance. This is an opportune time for aquaculture – given that the industry is still in its infancy, and coastal policy is receiving a high level of attention by governments and the Commission to develop ways that will allow it to integrate with existing and emerging users of the same coastal resource. This is especially challenging, given the multifaceted and dynamic nature of coastal areas and their associated communities and industries. The

Commission's Demonstration Programme on ICZM (EC 1999) has shown that the best response to such complex situations is to adopt an integrated approach that both involves all the stakeholders, and addresses concurrently the many different problems an area faces.

With regard to the *ecosystem deficit* of aquaculture, participation is highlighted not only as a democratic objective, but also as a means of improving the ecological performance of the sector. The Commission recognised that insufficient involvement and consultation with relevant stakeholders, including members of the public, could lead to degradation and mismanagement of resources (EC 2000). Meaningful participation is, therefore, seen as a means of achieving sustainable management strategies (Kaiser and Stead 2002). The concept of ICZM, which received a lot of attention in Europe throughout the 1990s, was seen as a way of dealing with the severe environmental problems that exist in coastal areas, many of which impact directly on the aquaculture sector. These include, for example, ecological issues, such as the benthic impact of cage farming, sensitive ecology, and limited resources of fish stocks; social issues, such as a population shift from rural to urban areas and from the hinterland to the coast; and economic issues, such as declining traditional activities and limited employment opportunities. Concern over these issues led the Commission to launch a Demonstration Programme on ICZM in 1996, to examine the fate of coastal zones (EC 1999). The Commission was mindful of the facts that coastal problems often have a cross-national dimension, and may not be solvable by the Member States separately; that many EU policies (including fisheries and regional policies) impacted on the development of the coastal zones; and that there was a need for an exchange of experience and know-how in a field where successes are still rare, and where there is substantial public and political demand for the conservation of the coastal zones and their sustainable development.

The two main objectives of the ICZM Demonstration Programme were, therefore: 1) to address the democratic deficit, by stimulating a broad debate among stakeholders about issues of ICZM and the respective responsibilities of various actors; and 2) to address the ecosystem deficit, by providing technical information about sustainable management of coastal zones. This programme – 35 projects across Europe undertaken between 1996 and 1999 – sought to provide examples of good practice in ICZM in a range of socio-economic, cultural, administrative and physical conditions. The main conclusion from the findings of this initiative was that the sectoral approach to management (that is, self-regulation) does not meet the needs of managing complex issues in coastal areas. The ICZM initiative recommends that integrated planning management is the only way to solve problems in areas of intensive use and multiple pressures.

We can see these two dimensions of ICZM (addressing democratic and ecosystem deficits) in the very definition of ICZM. ICZM is defined as a dynamic, multi-disciplinary and iterative process that promotes sustainable management of coastal zones. ICZM brings together all those involved in the development, management and use of the coast within a framework that facilitates integration of their interests and responsibilities. Over the long-term, ICZM seeks to balance cultural, economic, environmental, recreational and social objectives in order to achieve common goals (adopted from definitions by Coastal Zone Canada Association (cited in Cordah 2001) and from EC 2000). The key notion is 'integration', which embraces four elements: democratic integration (that is, including all stakeholders); administrative integration (that is, involving all relevant authorities, to ensure joined-up policy-making); physical integration (that is, treating the coastal area, terrestrial and marine, as one interconnected whole); and disciplinary integration (that is, using a multi- and interdisciplinary methodology).

In my view, the first element – democratic integration – is the key factor. To explain: poor and fragmentary information can lead to uncertainty which is often at the root of the problem by preventing the main issues that need to be addressed in coastal management from being clearly identified. Given the combined complexities of coastal ecosystems and aquaculture, inter- and multi-disciplinary-sourced knowledge needs to be used in a co-ordinated manner, if relevant information is to be adequately integrated into the formulation of policy. One way to achieve this is by bringing together stakeholders from all relevant backgrounds and encouraging their involvement in the decision-making process.

According to chapter 2 of the Commission's ICZM Recommendation (EC 2002a), there are eight fundamental principles of ICZM: 1) broad perspective; 2) long-term perspective; 3) adaptive management; 4) local specificity; 5) ecosystem approach; 6) stakeholder inclusivity; 7) administrative co-ordination; and 8) extensive policy-making. The last three principles address the democratic deficit; the first five principles address the ecosystem deficit.

Let us look at these two groups of principles in turn, beginning with the last three principles. Clearly, the sixth principle – stakeholder inclusivity – directly addresses the democratic deficit. It demands the involvement of all parties: including aquaculturists, local authorities, organisations representing coastal interests, non-governmental organisations (NGOs), scientists, and the business sector. Such involvement is deemed essential to ensure that the perspectives held by all relevant stakeholders are factored into decision-making; that the whole community shares responsibility for setting priorities and committing itself to implementing policies of sustainable development; and that scientists conduct their work in the midst of a social debate about the use of coastal resources. As we shall see, much effort has been devoted to improving stakeholder participation in coastal management through mechanisms such as local coastal forums and workshops.

The seventh principle – administrative co-ordination – requires that all levels of administration (local, regional, national and inter-governmental) co-operate in finding common ground in approaching coastal issues. One way of ensuring this cooperation is to establish pro-active partnerships, linking the various administrative units. Such partnerships promote stakeholder engagement and thereby reduce the democratic deficit. The eighth principle – extensive policy-making – refers to the need to reach beyond the traditional approach to aquaculture governance (sectorally-oriented, dealing with single issues) and embrace the interests of all coastal resource users. Stakeholder input is also essential to the implementation of this principle.

Turning now to the five principles that address the ecosystem deficit, the fifth principle – the ecosystem approach – is obviously directly linked. The International Council for the Exploration of the Sea (ICES) has defined the ecosystem approach as the integrated management of human activities based in knowledge of ecosystem dynamics to achieve sustainable use of ecosystem goods and services and the maintenance of ecosystem integrity (RCEP 2004:170). DEFRA (2002) has described the ecosystem approach as a new and more strategic way of thinking than in past and current practice, emphasising the importance of maintaining a healthy ecosystem alongside appropriate use of the marine environment, for the benefit of current and future generations. In an ecosystem approach to aquaculture, full account is taken of the effects of fish farming on the marine ecosystem, and environmentally friendly farming methods are employed in order to protect, not only the ecological, but also the economic, social and cultural heritage of coasts.

The first four principles follow on from the fifth principle, in that they are each elements in the ecosystem approach. For example, the first principle – broad perspective – exemplifies the ecosystem approach, in that it requires that aquaculture policy makers consider the bigger picture of the whole coastal environment, taking account of ecological, technological, economic, social and cultural factors. It alerts us to the need to consider indirect and cumulative causes and effects of aquaculture, and the complex consequences of its onshore, inshore and offshore activities. The second principle – long term perspective – focuses on the needs of present and future generations of people in coastal areas, and entails adopting the precautionary principle to deal with uncertainties for the marine ecosystem, such as the impact of climate change, which could lead to a rise in water temperature, with significant consequences for species, including farmed species, in coastal waters.

The third principle – adaptive management – is also linked to the ecosystem approach, in that it urges policy makers to be flexible in the way in which they respond to coastal problems. Given the dynamic nature of coastal environments and systems, regular monitoring along with feedback of relevant data needs to be incorporated into management plans so that actions and/or supporting policies can be up-to-date and reflective of the current state of knowledge. Finally, the fourth principle – local specificity – reflects the fact that in diverse and sensitive coastal systems, one size does not fit all, and aquaculture activities must be adjusted to suit particular locations. It is important that a good understanding of the local characteristics of an area, including its strengths, weaknesses, opportunities and threats (SWOT analysis), are integrated into coastal management planning and strategic thinking. Local information can sometimes be difficult to obtain, particularly when language and cultural barriers need to be overcome, and time and resources should be made available to collect and apply informal-based information, such as indigenous knowledge, to decision-making processes.

Member States of the EU are being encouraged to implement the Commission's Recommendations for ICZM by 2006. It is important, therefore, that the aquaculture industry is closely involved in the implementation process. At present, Member States are drawing up national coastal strategies, and the UK recently completed a national stock-taking exercise where an analysis of major actors, laws and institutions that influence the management of the coastal zone is summarised (Atkins 2004). Participation in ICZM has been encouraged by the UK government, largely through the setting up of voluntary partnerships in the form of local coastal (and estuarine) forums. These have been established to persuade local stakeholders and the local community to participate in sustainable management (or 'wise use') of particular coastal areas. Scotland has been the most active in the UK in creating these local coastal forums, compared to its counterparts in England and Wales. Indeed, even in the 1970s, there were developments in Scotland that can be classed as examples of ICZM in its infancy. These include the Zetland County Council Act 1974; the Highland Region's Framework Plans for Fish Farming; and the Scottish Natural Heritage's Firths work on the Moray Firth, the Firth of Forth and the voluntary Marine Reserve at St. Abbs Head (Burbridge 2001). In the 1990s, many more voluntary coastal management initiatives were developed throughout Scotland, including the Forth Estuary Forum, the Tay Estuary Forum, the Moray Firth Partnership, the Cromarty Firth Liaison Group, the Fair Isle Marine Environment Tourism Initiative, the Minch Project, the Firth of Clyde Forum, the Loch Ryan Advisory Management Forum, and the Solway Firth Partnership (Burbridge, 2001). Further illustrations of partnerships and coastal management initiatives are described in Atkins (2004) – some including stakeholder representation from the aquaculture sector.

An example of a partnership that specifically addresses the needs of aquaculture by building consensus between different resource users of the same coastal areas, is the Tripartite Working Group (TWG). Members of the TWG represent fish farming and local wild fisheries interests, and it was set up in 1999 by the Scottish Executive to help establish Area Management Agreements (AMAs) in Scotland. The creation of an AMA is a voluntary process, the main aim of which is to develop practical measures to promote the maintenance and management of wild and farmed fish stocks, largely salmon (*Salmo salar* L.) at the local level. There are currently nine AMAs in Scotland. The TWG, and the AMAs that have resulted from this initiative, supports many of the principles advocated by ICZM, and they have aided co-operative working to achieve mutual objectives, through facilitating dialogue between the participants involved. This is a good example of positive action taken to address the need for improved communication and understanding between two groups of coastal stakeholders (wild salmon fisheries and the farmed salmon industry), which have traditionally worked against each other in an environment of mistrust. The advantage of AMAs is that they can be flexible and tailored to specific local needs. It is important to retain this local specificity (the fourth ICZM principle), which is considered an important contribution to the success of AMAs, and is also credited as the ICZM principle which has been implemented most effectively in comparison to the other ICZM principles in the UK (Atkins 2004). There are further opportunities to extend the benefits offered by the existing AMAs to other resource users, thereby implementing the sixth ICZM principle.

The TWG process and associated AMAs are considered to be a more effective tool than the regulatory system alone, which is typical of the hierarchical mode of governance. However, success from these initiatives, in terms of developing effective ICZM models, does not occur overnight, and it is now well known that in order for such developments to attain substantive integrated management and planning, between 5 and 10 years is required (Humphrey and Burbridge 1999). This is because participation must be gradual, whatever approach is used, to build trust, understanding and confidence, (Kaiser and Stead 2002). Although there are many examples of positive experiences, such as some of the outcomes from the Local Coastal Forums in Scotland (McGlashan 2002), in practice, initiatives based on stakeholder participation usually underestimate the time and resources required for maintaining success in the longer term. Moreover, success is more likely to come from pro-active, rather than reactive, responses. For example, the TWG initiative was a reactive response to the increasing level of conflict between the various interested parties from the two stakeholder groups, but it is better if partnership-based projects arise through anticipatory work rather than waiting until a problem has arisen or deteriorated.

To date, the success of ICZM initiatives in the UK has been variable. Although there are many examples of good practice in ICZM that the aquaculture industry can draw upon, it should be noted that some ICZM initiatives have failed to achieve success. Four key reasons for failure have been revealed by the extensive study conducted in the UK on ICZM. First, there is a lack of long-term funding. Participative initiatives cost money, and little progress can be expected if there is a dearth of funding provided. Second, and partly because of limited funding, there has been a lack of involvement with stakeholders. Despite the rapid increase in initiatives that have centred on promoting involvement of stakeholders, many have had limited success in making a long term difference in practice, especially in factoring information derived from stakeholders into the formulation of policy. In particular, there has been a failure to fully engage local communities (Fletcher 2003; McGlashan and Barker 2004).

From my own experience, there needs to be more effort focused on developing methodologies that not only promote the initial involvement of participants, but also sustain stakeholder participation. The participatory mechanisms used so far, such as local coastal forums, may not be sufficient to achieve this sustained commitment. The aquaculture sector needs to recognise this shortfall, and to try and improve the continuation of stakeholder participation. Clear and achievable goals are one way to help retain interest, along with a demonstration of transparent and democratic methods on issues such as membership composition and rationale for set-up. Another suggestion is the use of pilot case studies in which stakeholder participation plays a central part in determining which form of governance best supports coastal aquaculture activity along with the needs of other resource use and users. Also, consensus conferences (Kaiser and Stead 2002) and stakeholder-led dialogue initiatives, focusing on the characteristics of an area, its people, associated aquaculture activity and other resource users, could be more widely implemented. Past experiences of consensus conferences have attracted good media coverage, which can offer the opportunity to portray the industry in a positive light. However, this can only be achieved with support from resources at the local, national and international levels, along with the involvement of key actors and players.

Third, there lacks a suitable legal framework for Member States to establish participatory ICZM. For one thing, although the European Commission has issued ICZM recommendations (which they hope to be implemented by Member States by 2006), thereisnolegalrequirementforMemberStates to set them up. As a result of this 'soft-law'approach,thereis little incentive for governments to prioritise ICZM over otherpolicyinitiatives If ICZMwereput on a statutory basis, we would see much more urgentgovernmental action to establish and support them. Moreover, without statutory status,participatoryformsof ICZM such as local coastal forums in the UK are weak in the face of developers. Let me explain this point. Each of the local coastal forums must have produced a management plan outlining the approach they wish to adopt to implement, and the majority of the members of these forums have usually agreed (at least in principle) to the content of the plan. However, as highlighted by McGlashan and Barker (2004), because these management plans carry no statutory recognition, if a proposed development (for example, a business venture that could constrain aquaculture production) conflicts with the management plan, the local authority would find it very difficult to refuse permission for the development, and may even be open to a legal challenge if it rejected the plan. It is this lack of statutory powers, either forcing (or encouraging) all of the relevant stakeholders to participate, combined with the frustration of trying to implement a voluntary management plan with no specific funds, and a lack of sustainable funding, that is leading to low morale and high staff turnover in coastal forums in the UK. This is contributing to a negative perception about ICZM.

In the view of some critics, the problems with European marine aquaculture governance are largely attributable to a lack of coherent and specific EU legislation for aquaculture, combined with no overarching legislation for ICZM. These deficiencies arise because many aquaculture issues are regulated by national legislation, which is influenced by a number of horizontal Community Directives that can lead to competition distortions among producers from different Member States (EC 2002b). The 'Strategy for the Sustainable Development of European Aquaculture' (EC 2002b) compounds the problem, in that Member States are invited to promote stakeholders' participation in the process of policy planning for aquaculture, yet it leaves incoherence in aquaculture policies at the inter-governmental level. On the one hand, the Commission is pushing the problem back to the industry and its Member States to be solved; yet, on the other hand, the Commission is failing to clarify the legalities at the European level. It is clear there is a need to simplify and better integrate the complex body of legislation that influences aquaculture activities at the inter-governmental level, especially in relation to the formulation, implementation and monitoring of different policies, and to ensure that these policies complement those that exist, or are evolving, at the regional, national and local levels (Stead 2003).

Fourth, there is an assumption, always implicit, and sometimes explicit, in ICZM theory that stakeholder participation and the ecosystem approach are consistent with each other, and indeed, that the one is a pre-condition of the other. This assumption is based on the argument that the more people who are involved in decision-making about the use of a resource, the more likely it is that that resource will be well-protected. However, is this argument convincing? What is the empirical evidence that increasing the extent of stakeholder participation will ensure that the ecosystem approach is adopted?

10.4 Conclusion

In European aquaculture governance, there are elements of all three main modes of governance: 1) the hierarchical mode, evident in the limited and somewhat incoherent framework laid down at EU and Member State levels; 2) the free market mode, evident in the forces of supply and demand which drive aquacultural producers; and 3) the participative mode, evident in two forms: a) self-regulation; and b) ICZM. Because of the deficiencies of each of the first two modes – the hierarchical mode suffers from over-centralisation and inflexibility, while the free market mode suffers from a lack of accountability – the participative mode is the best option to ensure sustainable development of aquaculture. However, neither of the two forms of the participative mode, self-regulation and ICZM, is adequate on its own. Self-regulation, while excellent at the role of quality control over the industry, fails to take into account either the claims of other marine resource users, or the ecological impact of aquaculture on the ecosystem. For its part, ICZM, while potentially designed to remedy both these failures (the democratic deficit and the ecosystem deficit), has little commercial discipline to offer the industry. My conclusion is that both self-regulation and ICZM are needed for the effective governance of the aquaculture industry. Since, at present, self-regulation is far more prevalent than is ICZM, the proper balance between them requires a great deal more investment in ICZM during the next few years. Unless this occurs, there is a danger that the EU will step in and introduce a much more severe strain of the hierarchical mode of governance.

References

Atkins, WS (2004) *ICZM in the UK: a Stocktake*, London, HMSO

Burbridge, PR (2001) *Lessons Learned from Local Coastal Management Partnerships*, The Scottish Coastal Forum, the Scottish Executive, Countryside and Natural Heritage Unit, Edinburgh, Scottish Natural Heritage

Cordah (2001) *Indicators to Monitor Progress of Integrated Coastal Zone Management: a Review of Worldwide Practice,* Countryside and Natural Heritage Research Programme Research Findings No. 8, Scottish Executive Central Research Unit, Edinburgh, Stationery Office

DEFRA [Department for Environment, Food and Rural Affairs] (2002) *Safeguarding our Seas*, London, DEFRA

EC [European Community] (1999) Towards *a European Integrated Coastal Zone (ICZM) Strategy: General Principles and Policy Options – a Reflection Paper*, Luxemburg, Office for Official Publications of the European Communities, European Commission

EC [European Community] (2000) *Communication from the Commission to the Council and the European Parliament on Integrated Coastal Zone Management: a Strategy for Europe*, COM (2000) 547 final, Brussels, European Commission

EC [European Community] (2002a) *Recommendation of the European Parliament and of the Council (2002/413/EC of 30 May 2002) concerning the implementation of Integrated Coastal Zone Management in Europe*, Official Journal (L148/24 of 6 June 2002), European Commission

EC [European Community] (2002b) *Communication from the Commission to the Council and European Parliament – a Strategy for the Sustainable Development of European Aquaculture*, COM (2002) 511 final, Brussels, European Commission, 19 September

FAO [Food and Agriculture Organisation of the United Nations] (2002) *The State of World Fisheries and Aquaculture*, Rome, FAO

Fletcher, S (2003) 'Stakeholder representation and the democratic nature of coastal partnerships in the UK' *Marine Policy* 27:229-240

Hough, C (2000) 'Codes of conduct and aquaculture' in *Responsible Aquaculture for the New Millennium,* European Aquaculture Society Special Publication 28, AQUA 2000, Nice, France, 2-6 May, ISBN 90-71625-16-8

Hough, C (2001) 'Aqua-media: a new initiative for the promotion and support of European fish farming' in E Kjørsvik and S Stead (eds) *New Species: New Technologies,* European Aquaculture Society, Trondheim, Norway, 4-7 August

Humphrey, SL and Burbridge, PR (1999) *Thematic Analysis of Sectoral and Territorial Co-operation in the EC ICZM Demonstration Programme*, Brussels, European Commission

Huitric, M, Folke, C, and Kautsky, N (2002) 'Development and government policies of the shrimp farming industry in Thailand in relation to mangrove ecosystems' *Ecological Economics* 40:441-455

Kaiser, M, and Stead, SM (2002) 'Uncertainties and values in European aquaculture: communication, management and policy issues in times of "changing public perceptions"' *Aquaculture International* 10:469-490

Lebel, L, Tri, NH, Saengnoree, A, Pasong S, Buatama, U and Thoa, LK (2002) 'Industrial transformation and shrimp aquaculture in Thailand and Vietnam: pathways to ecological, social and economic sustainability' *Ambio* 31(4):311-323

McGlashan, DJ (2002) 'Coastal management and economic development in developed nations: the Forth Estuary Forum' *Coastal Management* 30:221-236

McGlashan, DJ and Barker, N (2004) 'The partnership approach to integrated coastal management' in JS Potts and H Smith (eds) *The British Sea: Towards a Sustainable Future*, London, Routledge

RCEP [Royal Commission on Environmental Pollution] (2004) *Turning the Tide - Addressing the Impact of Fisheries on the Marine Environment*, London, RCEP

Stead, SM (2003) 'The legislative puzzle – aid or barrier to integration?' in T Chopin and H Reinertsen (eds) *Beyond Monoculture,* European Aquaculture Society Special Publication, August 2003

Vandergeest, P, Flaherty, M, and Miller, P (1999) 'A political ecology of shrimp aquaculture in Thailand' *Rural Sociology* 64(4):73-596

CHAPTER 11
THE ROLE OF UK STATUTORY NATURE CONSERVATION AGENCIES IN THE ENVIRONMENTAL GOVERNANCE OF FISHERIES

N CLARE ENO AND MARK GRAY

Senior Fisheries Policy Officer and Sea Fisheries Liaison Officer, Countryside Council for Wales, Bangor, Gwynedd, LL57 2LQ

Abstract

The subject of this chapter is the increasing role of statutory nature conservation agencies (NCAs) in fisheries governance in the UK. There are three main sections: in the first section, we set out the powers and responsibilities of UK NCAs in relation to the designation of marine sites and the potential for them to be protected from fishing activity. In the second section, we explain the wider strategic role of NCAs in helping to shape future fisheries policy at European, UK, national and local levels. In the third section, we discuss how effective NCAs are in fulfilling each of these roles, and what the main obstacles are to improving their effectiveness.

11.1 Introduction

There is no doubt that the statutory nature conservation agencies (NCAs) are playing an increasingly important role in UK fisheries governance, and, in this chapter, we explain and evaluate this role. The Country Agencies – Countryside Council for Wales (CCW), Scottish Natural Heritage (SNH) and English Nature (EN) – are the statutory wildlife advisers to national governments: they deliver their statutory responsibilities for Great Britain as a whole, and internationally, through the Joint Nature Conservation Committee (JNCC) and are collectively known as the nature conservation agencies (NCAs). CCW and EN are empowered by three main pieces of legislation: the National Parks and Access to the Countryside Act 1949; the Wildlife and Countryside Act 1981 (amended by the Countryside and Rights of Way (CroW) Act 2000); and The Conservation (Natural Habitats, &c.) Regulations 1994. SNH is empowered by the Natural Heritage (Scotland) Act 1991 and The Nature Conservation (Scotland) Act 2004.

We divide the influence on fisheries governance of the work of NCAs into two categories: **statutory** and **strategic**. The statutory work consists mainly in advice on selecting, designating and managing marine sites, and on the effect which activities could have upon the environment. This work can impact directly on fishing activities. NCAs do not have powers to manage fisheries, but their designating authority does give them substantial leverage over the way that other bodies manage fisheries. In section two, we explain this statutory work, giving illustrations, and pointing out certain difficulties faced by NCAs in carrying out these duties.

NCAs also give advice to governments (at international, national, and sub-national levels); to the fisheries regulators; and to others on wildlife conservation including on the potential environmental impacts of fishing activities. This is discussed in section

T. S. Gray (ed.), Participation in Fisheries Governance, 193–207.

three, where we explain the strategic work of NCAs, which is essentially their advocacy role by which they seek to influence fisheries and environmental policy makers, including politicians.

In section four, we evaluate how effective NCAs are in carrying out these two roles, and we examine four ways in which their performance could be improved by removing legal and political obstacles, and by plugging information deficits. We conclude by summarising the findings of the chapter, and discussing five further issues that NCAs might address in their role as marine environmental stewards.

11.2 Statutory role: Influence over fisheries governance in marine sites

The statutory work of NCAs that affects fisheries lies essentially in their powers in relation to the designation of national – SSSIs (Sites of Special Scientific Interest) and MNRs (Marine Nature Reserves) – and European marine sites – SACs (Special Areas of Conservation) and SPAs (Special Protected Areas). Let us consider these sites in turn.

11.2.1 SITES OF SPECIAL SCIENTIFIC INTEREST (SSSIs)

The NCAs can designate SSSIs on land and along the intertidal zone for "special interest by reason of its flora, fauna or geological or physiographical features" (JNCC 1996). Each individual site notification contains a list of activities "likely to damage the special interest" (potentially damaging operations), and these include fishing activity, fisheries management, marine life collection or alterations to fishery management practice. SSSIs may encompass inter-tidal fisheries such as shellfish cultivation, mechanical and hand gathering of shellfish (for example, cockles and mussels) and netting for finfish. The amendment of the Wildlife and Countryside Act 1981 by the Countryside Rights of Way Act 2000 (CRoW Act) (or The Nature Conservation (Scotland) Act 2004) provides an assessment process that fisheries management authorities have to undertake before permitting an operation that is likely to damage a feature of the SSSI.

However, whilst the CroW Act requires a fisheries authority to seek assent from a NCA before permitting a potentially damaging operation in a SSSI, a NCA cannot stop a fishing activity within an SSSI unless it can serve notice on an owner/occupier. The protection of SSSIs is limited, therefore, because management agreements/notices do not apply to third parties and there is no provision to hold a fisheries authority accountable for allowing third parties to act. An offence could only be committed by a third party (such as a fisherman) if damage to a SSSI feature was intentional and the feature was known to be within the SSSI. The difficulty of prosecuting third party activity is further compounded by the public right to fish, because a fisherman could argue he was exercising his right to fish.[1]

11.2.2 MARINE NATURE RESERVES (MNRs)

Seven MNRs were originally proposed by the NCAs in Great Britain (in the 1980s

[1] In common law, the public has a right to fish in the sea within the territorial waters of the UK unless an individual has acquired exclusive rights or Parliament has restricted the common law rights of the public. The public right extends to taking fish from the foreshore – the land between high and low water (CCW 2002).

following enactment of the Wildlife and Countryside Act 1981) of which only 2 were designated. The first to be designated was around Lundy Island in the Bristol Channel in 1986, and another around Skomer Island off the coast of South Wales in 1990. A further proposal for Strangford Lough in Northern Ireland was added and designated in 1995. However, the designation of these three MNRs took over 15 years because there was an undertaking in Parliament that everyone would have to agree regardless of the nature of the objection (the proposed Menai Strait MNR stalled for many years due to objections from recreational anglers), and there was neither the political will nor the means to overcome the objections received. A further hindrance to progressing the establishment and management of the sites was the fact that no other authorities beyond the NCAs held any responsibilities for promoting them. There was some control of fisheries in MNRs, but only through fisheries legislation/byelaws instigated through the Sea Fisheries Committees (SFCs).

11.2.3 EUROPEAN MARINE SITES (EMSs) – SPECIAL AREAS OF CONSERVATION (SACs) AND SPECIAL PROTECTED AREAS (SPAs)

In 1992, the European Community adopted the Habitats Directive (EC 1992) to ensure the conservation of habitats and wildlife by European Member States. One of the key objectives of the Directive is the creation of a network of SACs across all the land and sea areas of the European Union (EU). This 'Natura 2000' network of sites would enable the conservation of a diverse range of habitats and species and would incorporate the SPAs for bird species, designated under the 1979 Birds Directive (EC 1979). The Conservation (Natural Habitats, &c.) Regulations 1994 (HMSO 1994) is a Statutory Instrument that transposed the Habitats and Birds Directive into UK law. The NCAs have been given authority under these regulations to designate SACs and SPAs in UK waters.

While there have been many reviews of the impact of fishing on the ecosystem, the NCAs commissioned the first work assessing the potential implications of fishing on features of the 1992 EU Habitats Directive (Gubbay and Knapman 1999). By doing so, they laid the way for considering how fishing might be governed in those sites in future.

Moves towards designating 45 EMSs under the Natura 2000 series have greatly expanded the area of protection for marine conservation purposes in the UK as well as in the rest of the EU (EC 2000; Owen 2004; Hernandez-Aguilar 2004). The recent ruling following a case in the Waddenzee (ECJ 2002) has given legal backing to the interpretation of fishing as a 'plan or project', which means that any fishery likely to have a significant effect on the integrity of a site must be subject to appropriate assessment. The implications for fisheries governance could be quite considerable, as a precedent has been set for a precautionary stance.

While it is not the role of the NCAs to manage the fisheries taking place in the EMSs, as the statutory advisors to Government and others on the implications for wildlife of activities in the marine environment, their opinions must be acted upon by the competent authorities. In the UK, a variety of fisheries regulators act as competent authorities according to their jurisdiction.[2] It is the responsibility of a competent authority to undertake an assessment appropriate to a "plan or project" being proposed

[2] 'Competent authority' means any Minister, government department, public or statutory undertaker, public body or person holding a public office that exercises statutory powers (EN 1998).

within a Natura 2000 site. This entails assessment of all new fisheries or changes to current practices, including issuing permits or licenses when a fishery is reopened. The competent authority has to take account of the advice of the NCAs and to ensure that no significant deterioration of the favourable conservation status of the site will result. The competent authority can impose restrictions on the practice to ensure that but the integrity of the site is maintained.

11.2.4 CASE STUDIES

Since 2002, several Ministerial Orders have been requested by NCAs to prohibit particular inshore fisheries in EMSs in England.[3] For example, EN made a formal request to the UK Department for Environment, Food and Rural Affairs (DEFRA) that an Order be put in place to protect Eelgrass beds in the Solent EMS from cockle dredging. EN presented evidence that the recent low level use of cockle pump scoop dredges was causing damage to the beds (which are, in themselves, a feature of the SAC, as well as providing an important food source for Brent Geese – a feature of the SPA (DEFRA 2004). The prohibition (the Solent European Marine Site (Prohibition of Method of Dredging) Order 2004) was thought to affect about five vessels using pump scoop dredges. Inadequate fisheries management led EN to recommend this Ministerial Order in the Solent. A SFC byelaw could be used to prohibit shellfish dredging in the Solent, but because the byelaw making process can take a year to complete, a Ministerial Order was required to provide necessary protection (DEFRA 2004). A prohibition Order could be temporary and allow time for a SFC or the Environment Agency (EA) to develop their own byelaw or allow an appropriate assessment to take place.

In Wales, the Shellfish (Specified Sea Area) (Prohibition of Fishing Methods) (Wales) Order 2003, under the Sea Fish (Conservation) Act 1967, was introduced to prohibit hydraulic dredging for bivalve molluscs in Carmarthen Bay, off the South Wales coast. The Order was introduced following a reasoned opinion by the European Commission against the UK in 2002 in relation to the non-designation of Carmarthen Bay as an SPA for the Common Scoter duck, and against the South Wales SFC (SWSFC) for not carrying out an appropriate assessment before authorising hydraulic dredging for razor clams (a third party was believed to have complained to the Commission). The Order followed a long-running dispute between CCW and the SWSFC over the assessment of the impact of hydraulic dredging for razor clams (a new fishery) in Carmarthen Bay. Uncertainties and difficulties of implementing the requirements of the Habitats Directive to ensure that hydraulic dredging did not have a significant impact on the features of the Carmarthen Bay SPA, may have led to the reluctance of the SWSFC to introduce further restrictions or undertake a costly impact assessment. The lack of suitable local control led the Welsh Assembly Government to introduce an Order to prohibit hydraulic dredging.

CCW could agree to this Order being rescinded if an adequate environmental assessment took place and measures to control fishing were introduced to ensure that

---

[3] Ministerial Orders can be made under section 5 of the Sea Fish (Conservation) Act 1967 in England and Wales (section 5A of the Act permits Orders to be made for marine environmental purposes), and sections 3 and 15(3).

the integrity of the site is maintained. (Note the Bay was designated an SPA for non-breeding Common Scoter in 2003). However, in most fisheries in the UK there is an inability to control effort or intensity of fishing effort, other than vessel size restrictions and through opening or closing a fishery. This is an example where a feature of a site may comfortably withstand light fishing intensity, the integrity of the site might be compromised if fishing effort increases. Another example concerned scallop dredging in an SAC in North Wales (Pen Llyn a'r Sarnau) where CCW advised the North Western and North Wales SFC (NWNWSFC) to prohibit scallop dredging from a bay where, although a low level of fishing would have been acceptable, the SFC were unable to guarantee that only a low level of fishing would ensue.

SWSFC have sought advice on the implications of the Habitats Directive from both DEFRA and the previous Government department, the Ministry of Agriculture, Fisheries and Food (MAFF) since 1998, including answers to the following questions:

- Who pays for an assessment? (Developers (fishermen), or regulators (government department /SFC)?)
- How far should an assessment go in the collection of primary ecological information to support site designation? (This question reflects the dearth of ecological information collected by statutory conservation bodies, despite the sites having been submitted to the Commission for designation.)

The SWSFC were informed that the Environment Minister, Elliot Morley, was aware that the rules were not as clear as they might be, but he asked for some consideration, on grounds that this was a developing policy area. He hoped that more definitive guidance would be available in the near future (SWSFC 2002).

The cost of, and responsibility for, marine impact assessments is not unique to the fishing industry (PMSU 2004). Other marine industries, such as aggregate dredging, oil and gas and wind farms, abide by the polluter-pays principle and the developer provides the necessary information for the competent authority to decide whether the development can proceed. Yet, in the case of the fishing industry, the lack of ownership or exclusive rights to fish stocks or an area of seabed deters fishermen from paying for an assessment, if others who have not paid are likely to benefit.

Offshore (beyond 12 miles), the NCAs have been instrumental in identifying conservation sites. The first such site is the Darwin Mounds. The UK has been able to secure protection of these sites, which require special provision in relation to the European Common Fisheries Policy (CFP) (Clorley 2004). Offshore habitats regulations are currently being drafted.

It can be seen that, while the NCAs do not govern any fisheries, through the introduction of the Habitats Directive, and its subsequent EU interpretation and the Habitats regulations, they have an increasing influence on how fisheries are managed within sites of European marine nature conservation importance. Indeed, there is a growing obligation to manage fisheries to accord with nature conservation interests (Eno 2004). The fishing industry and regulators need to be aware that third party complaints to the Commission can lead to quick and decisive action (such as the hydraulic dredging prohibition in Carmarthen Bay). The implications of not preventing damage to site integrity might also be considerable (for example, the UK government is currently subject to pre-infraction proceedings by the European Commission as a result of the loss of the majority of the biogenic horse mussel reef communities within Strangford Lough SAC, proceedings that may result in a very substantial fine. In this case, the fisheries department is the competent authority.)

11.3 Strategic role: Influence over fisheries policies and initiatives of government, the fishing industry and other agencies

Turning now to the strategic work of the NCAs achieved through advocacy, we examine the influence that NCAs exert on fisheries policy at three levels of decision-making: European and UK; devolved administrations; and local.

11.3.1 EUROPEAN AND UK LEVEL

At the European level, the NCAs have, increasingly over the last ten years, taken on a greater advocacy role regarding fisheries and their governance, reflecting the growing prominence in the EU of marine conservation and protection of marine wildlife and habitats. For example, a considerable amount of time was spent influencing the 2002 reform of the CFP, by lobbying, replying to consultations, and engaging in key meetings. A number of reports were commissioned to help stimulate different approaches to fisheries governance (Symes 1998; Pope and Symes 2000; Symes and Pope 2000; Symes *et al* 2002). The objective of the new CFP Regulation (EC 2002) has moved towards managing the whole of the marine ecosystem, rather than just one component (fish), acknowledging the impact of fisheries on the marine environment; and resolving to control those damaging activities (Clorley 2004). This is known as the ecosystem-based approach (EBA), the elements of which are set out authoritatively in Table 11.1.

Many of the greening influences over EU fisheries policy that have been claimed by environmental non-governmental organisations (ENGOs) resulted from a combination of pressures exerted by both statutory and voluntary environmental bodies, involving changes in public opinion and political will. The NCAs have certainly invested heavily in influencing the development of more integrated fisheries governance (integrating both stakeholder participation and environmental policies). For instance, EN commissioned work on the EBA to fisheries management (Pope and Symes 2000; Symes and Pope 2000), to determine how the EBA might be implemented. For its part, CCW has worked with stakeholders from around the Irish Sea to develop the concept of integrated fisheries management, by organising workshops and meetings. This work has contributed to the development of Regional Advisory Councils (RACs) and particularly to progressing the development of sub-areas within the North Western Waters RAC. While the NCAs will not be members of RACs as such, they will act as observers and monitor the RACs, and advise governments on whether they are applying an ecosystem approach. NCAs may also play a more active role in the working groups and sub-area discussions.

Table 11.1. Twelve principles of the EBA (UN 2003)

	UN Convention on Biological Diversity's twelve principles of the EBA
1.	The objectives of management of land, water and living resources are a matter of societal choice.
2.	Management should be decentralised to the lowest appropriate level.
3.	The Ecosystem Approach should be undertaken at the appropriate spatial and temporal scales.
4.	Recognising the varying temporal scales and lag-effects that characterise ecosystem process, objectives for ecosystem management should be set for the long-term.
5.	Ecosystem managers should consider the effects (actual or potential) of their activities on adjacent and other ecosystems.
6.	Recognising potential gains from management, there is usually a need to understand and manage the ecosystem in an economic context. Any such ecosystem-management programme should: • reduce those market distortions that adversely affect biological diversity; • align incentives to promote biodiversity conservation and sustainable use; and • internalise costs and benefits in the given ecosystem to the extent feasible.
7.	Conservation of ecosystem structure and functioning, in order to maintain ecosystem services, should be a priority target of the Ecosystem Approach.
8.	Ecosystems must be managed within the limits of their functioning.
9.	Management must recognise that change is inevitable.
10.	The Ecosystem Approach should seek the appropriate balance between, and integration of, conservation and use of biological diversity.
11.	The Ecosystem Approach should consider all forms of relevant information including scientific and indigenous and local knowledge, innovations and practices.
12.	The Ecosystem Approach should involve all relevant sectors of society and scientific disciplines.

In 2003, following representation from the fishing industry, the British Prime Minister instructed the Cabinet Office's Strategy Unit to prepare a medium to long term fisheries strategy for the UK. In 2004, the Strategy Unit published a report entitled *Net Benefits: A Sustainable and Profitable Future for UK Fishing* (PMSU 2004). In addition to a long-term strategy, the report emphasised the importance of bringing the industry and other stakeholders into a partnership with government over management decisions. It is noteworthy that officers from the NCAs were seconded into the PMSU's core group to channel and provide nature conservation input. A series of working groups were organised following the publication of the report to agree on how the recommendations could be implemented. The NCAs are represented on, and make a significant contribution to, all of these groups, as well as to a full stakeholder group, the workings of which will influence how the DEFRA Sustainable Fisheries Programme draws up the government's response to *Net Benefits*, which is likely to determine how fisheries governance is to change in the UK.

Also, together with representatives from Government, the Association of SFCs, and a scientific expert on bycatch, the NCAs sat on the UK Small Cetacean Response Strategy Group set up in 2001. The aim of the strategy was to identify measures that could be introduced to reduce small cetacean bycatch to below the target (1.7 per cent of the population) set by the ASCOBANS Meeting of the Parties in 2000.[4] A consultation paper was produced in 2003 but its recommendations were superseded by an EC (2004) regulation which came into force in July 2004, and which included the mandatory use of acoustic devices ('pingers') in specified bottom net fisheries.

[4] ASCOBANS – the international Agreement on the Conservation of Small Cetaceans of the Baltic and North Seas – is an annex of the Bonn Convention.

11.3.2 DEVOLVED ADMINISTRATIONS: SCOTTISH EXECUTIVE ENVIRONMENT AND RURAL AFFAIRS DEPARTMENT (SEERAD); WELSH ASSEMBLY GOVERNMENT (WAG)

The second level of fishery policy decision-making at which NCAs exercise influence is that of the so-called 'devolved administrations' in Scotland and Wales. In the case of Scotland, improved markets for shellfish, the decline in offshore fisheries, and greater consideration of the marine environment, have led to a review of the inshore sector. SNH are members of the Scottish Inshore Fisheries Advisory Group (SIFAG), which has been asked by SEERAD to develop a strategy for inshore fisheries, to include an element of stakeholder management.[5]

In the case of Wales, in order to ensure a sustainable future for all Welsh fisheries, the WAG intends to create an over-arching Welsh Fisheries Strategy that will cover commercial sea and inland fisheries, aquaculture, and recreational fisheries. This strategy will involve a statutory steering group of which CCW is a member, and a stakeholder Advisory Group (comprising representatives of the sea fisheries sector). CCW sit on the Welsh Fisheries and Aquaculture Strategy Groups (FASG)[6], which has already produced a component strategic action plan for the development of the Welsh Fisheries and Aquaculture sector, the focus of which was "to develop a profitable and sustainable Welsh fisheries industry" (WDA 2003). CCW also had input to the group developing of *A Strategy for the Recreational Fisheries of Wales*' (WAG 2003). CCW thus ensures that wildlife conservation elements are considered in Wales, and it particularly promotes those policies contained in its Sea Fisheries Policy (CCW 2003).

11.3.3 LOCAL LEVEL

The third level of fishery policy decision-making influenced by NCAs is the sub-national or local level. Within Wales and England, this means influencing the SFCs. For example, in 2002, the Skomer Advisory Committee[7] instructed CCW and SWSFC to investigate the desirability of establishing 'no-take-zones' (NTZs) within the Reserve, where no removal of living material would be permitted. In addition to nature conservation, CCW also recognised the potential benefits to fisheries of such no-take-zones, and, since 2002, CCW and SWSFC have consulted widely with the local community, fishermen and anglers. A draft byelaw was submitted to the SWSFC in January 2004, which recommended a limited commercial pot fishery, and restriction of boat and shore angling to specific areas.

This initiative followed two recent no-take-zones in England: St Agnes in North Cornwall led by local shell-fishermen to test whether the viability of their fishery could be improved; and Lundy MNR that resulted from a joint development between the Devon SFC and EN for both fisheries and nature conservation purposes (Phil Coates,

---

[5] SIFAG provides advice on inshore fisheries matters to the Scottish Executive, and consists of 16 organisations, including 11 fishermen's organizations, local authority representation, Scottish Natural Heritage, Scottish Environmental Link, Seafish and Highlands and Islands Enterprise, as well as two Executive agencies (Fisheries Research Services and the Scottish Fisheries Protection Agency).

[6] FASG comprised industry representatives from the catch, aquaculture, processing and retail sectors, SFCs, CCW and NGOs. It has just split into two component groups with overlapping membership.

[7] Skomer Advisory Committee represents commercial, recreational, fisheries and educational interests, and comprises about 60 members.

SWSFC pers comm 2002). However, there has been a mixed reaction to a project to develop a network of no-take zones around the Cornish coast, which is being driven by Cornwall County Council and is backed by EN, the EA, sea fisheries bodies and other sectors. The zones, inside the six-mile limit, are intended to maintain marine biodiversity and improve fish stocks, and they could be either legally enforceable or voluntary. Moreover, there was almost unanimous opposition to the setting up of a NTZ at Whitsand Bay at a public meeting in November 2004 (*Fishing News* 2004:19).

11.4 Discussion: Evaluation of the effectiveness of NCAs in their environmental influence on fisheries governance

We divide our discussion of NCAs' effectiveness into their two main roles: statutory and strategic.

11.4.1 STATUTORY

How effective are the NCAs in their statutory interventions with fisheries activities through designation of protected marine sites? In our view, they are having an increasingly marked impact on fisheries governance in the UK. An important element in ensuring the effectiveness of the NCAs in relation to fisheries governance is in their ability to work collectively. This was facilitated through the JNCC, whose chair at the time was Lord Selborne (who had chaired the House of Lords enquiry into the mid-term review of the CFP), aided by the formation of an inter-agency Marine Fisheries Working Group. This working group developed a work programme that was accompanied by the appointment of dedicated fishery officers. It was the first time that the NCAs had made a concerted effort to influence sea fisheries policies, and it was undertaken as fishing was the most widespread activity in the marine environment, with many fishing practices seen as unsustainable. The group's work was a combination of identifying the needs of marine wildlife, undertaking research on the effects of fishing (for instance through EU funded studies on ghost netting and potting work – Eno *et al* 2001; Bullimore *et al* 2000; Kaiser *et al* 1996) and developing their reputation with the fisheries sector from a scientific and policy perspective. The work of the NCAs' group gradually became more policy-focussed pending the reform of the CFP, as their written and oral evidence was sought in response to fisheries consultations. At the same time, the work of the individual agencies is being increasingly felt in their designation of marine sites.

11.4.2 STRATEGIC

How effective are NCAs in influencing fisheries policy by their advocacy activity? This is a more difficult question to answer, because it is notoriously hard to demonstrate cause and effect relations in policy arenas, and in this particular policy arena, it is especially hard to separate the effectiveness of NCAs from that of ENGOs. However, it seems clear that NCAs have contributed significantly to the contemporary shift in direction in European fisheries governance towards an EBA, and the integration of environmental objectives into fisheries policy. Evidence to support this claim comes from the CFP 2002 reform process which was heavily influenced by NCAs, along with ENGOs, and which marked a turning point in the way in which European fisheries are governed, from a single stock management strategy, to an EBA strategy. NCAs have also influenced thinking on the new RACs. Some of the points from the CCW-commissioned study on integrated regional management of the Irish Sea (Symes *et al* 2002) have featured in Commission guidance on setting up RACs. Domestically, UK decision makers are keen to involve the NCAs – indeed, they are sometimes the only 'environmental' organisation involved in fisheries strategy development (for instance in DEFRA's small cetacean bycatch strategy group). NCAs have also been closely engaged in developing and implementing the national fisheries strategies of SEERAD and WAG.

The UK political climate towards marine conservation has dramatically changed in recent years as the UK has signed up to international agreements (such as the FAO Code of Responsible Fishing) and conformed to new European fisheries (as a result of the reformed CFP) and environmental framework legislation. The NCAs are now formally part of UK and National Administrations' fisheries governance strategies as members of government decision-making groups, and they are formally consulted on changes to technical regulations (including Statutory Instruments). There is a general acceptance that fisheries management should take an EBA (fundamental to the reformed CFP: EC 2002), and that the precautionary principle should be used where necessary. The NCAs advise government on how the EBA can be applied to fisheries and the implications of it, and when and where the precautionary principle should be applied. Fisheries regulators appear keen to engage the NCAs to ensure that environmental issues are considered adequately, not only as a matter of formality, but as a matter of political necessity.

11.4.3 HOW COULD THE NCAs BE MORE EFFECTIVE?

There are four ways in which NCAs could be made more effective: by the removal of legal constraints; the resolution of political obstacles; the availability of better marine environmental information; and by a strengthening of the European network of fisheries and nature conservation advisors.

11.4.3.1 *The Removal of Legal Constraints*

One of the most frustrating features of NCAs' work in relation to fisheries is that they often have to recommend drastic action – such as closing fisheries – to protect habitats or species, because the competent authorities are legally unable to cut fishing effort. There is an urgent need for a change in the law to enable SFCs to reduce fishing effort, which would facilitate a win-win situation. The legal problem of how to deal with third party violations of SSSI agreements must also be addressed. At present, third parties cannot be prosecuted for damaging an SSSI, unless that damage was shown to be deliberate, an anomaly that is compounded by confusion over the extent of the public's right to fish. Notifying the fishing industry of SSSI features susceptible to damage from fishing operations could help to address such enforcement difficulties. For example, following advice from EN, information on SSSI features susceptible to damage from cockle hand gathering in Morecambe Bay was included in a NWNWSFC's cockle and mussel hand gathering permit scheme in 2003.

There is an urgent need for legal clarification on Natura 2000 site issues. For instance, can Member States implement measures to protect Natura 2000 sites from fisheries without the need for Commission involvement? It has been argued that Member States' obligations to manage fisheries to meet obligations under the Habitats Directive do not fall within the remit of fisheries management, but under the environmental part of the treaty (Owen 2004). If this is the case, then Member States could restrict fisheries in Natura 2000 sites without the agreement of the Commission (Clorley 2004; Owen 2004). Another Natura 2000 issue arises out of the application of Article 6(2) of the Habitats Directive (EC 1992), which is a general duty of care imposed upon competent authorities to ensure that the activities they authorise do not threaten the integrity of a Natura 2000 site. How will competent authorities monitor potential impacts of fisheries in order to determine whether they threaten the integrity of the sites? In other words, who is responsible for funding the required impact assessments?

The work of NCAs is also seriously hampered by the lack of legal ownership over, or exclusive rights to use, marine resources, and the resulting difficulty of finding relevant responsible parties against whom to take action for harm to the marine environment. This is a major issue, currently being grappled by the PMSU and resultant Sustainable Fisheries Programme, which is in favour of more property rights being established over marine resources, with concomitant responsibilities, but the issue is complicated by the public right to fish.

11.4.3.2 *The Resolution of Political Obstacles*

Until 2002, when the PMSU began to take an interest in fisheries management, NCAs had long felt weakened because environmental issues in relation to fisheries management had not been at the top of the UK political agenda. The feeling was that, especially in Scotland, the government was more concerned about the fate of the fishing industry than it was about the fate of the marine environment. However, whether the PMSU initiative will shift the balance significantly in the environmental direction remains to be seen, because the primary emphasis in the PMSU report (2004) seems to be on ensuring a profitable industry (unlike the report from the Royal Commission on Environmental Pollution, which categorically prioritises the heath of the marine ecosystem over that of the fishing industry (RCEP 2004: para 10.78)). Another political obstacle lies in the attitudes of other EU countries towards environmental issues of fishing. For example, because of the political strength of fishing interests in Spain and France, UK attempts to regulate offshore fisheries for marine conservation purposes, such as reducing dolphin fatalities, face tough opposition. However, in the case of Darwin Mounds, pressure came via the Habitats Directive, which trumps objections from individual Member States.

11.4.3.3 *The Availability of Better Marine Environmental Information*

It is no secret that there is an information deficit with regard to the marine environment. The knowledge that we possess about the way the marine ecosystem functions is very patchy. Although we know quite a lot about some individual commercially valuable species, we are especially ignorant about the population size and dynamics of many protected species, such as cetaceans and basking sharks. We also lack information on the distribution of habitats and species of key conservation importance (such as elasmobranchs), and on the marine ecosystem and trophic webs: knowledge of which would help us to determine how important commercially exploited fish (target and non-target species) are to protected predators. The critical question is who will pay for all the research necessary to plug these information gaps? Researchers at ICES (the International Council for the Exploration of the Sea) are already fully stretched, and without a massive increase in their resources, they would be incapable of undertaking the vast amount of increased work that would be required to fully understand the marine ecosystem. But without this understanding, the attempts by NCAs to advance the EBA in fisheries governance will be difficult.

11.4.3.4 *The Strengthening of the European Network of Fisheries And Nature Conservation Advisors*

While there is a European network of nature conservation agencies – the European Environment Advisory Council (EEAC) – there are very few examples of equivalent agencies to the NCAs throughout Europe who advise on the effects of fisheries on marine nature conservation interests. The UK NCAs established a European Nature Conservation and Fisheries Advisory Network (ENCFAN), but many of the members are not from equivalent bodies and include academics, independent consultants and representatives from fisheries departments. ENCFAN has been used to share information and ideas on European policy issues, and the UK NCAs have often facilitated this exchange including the commissioning of a series of workshops organised by the Institute for European Environmental Policy (IEEP) in the run-up to the reform of the CFP (see Coffey, this volume). A strengthening of this network would help facilitate exchange, and possibly coordination, of advice and policy messages across Europe. Interfacing with RACs is an obvious area where this could bring benefits.

11.5 Conclusion

Our conclusion has two parts. First, in summarising the findings of the chapter, we find that the work of NCAs in influencing fisheries governance in an environmental direction, takes two forms: statutory, which consists in designating protected sites, thereby, where appropriate, restricting fishing activity immediately; and strategic, which consists in influencing fisheries policy, thereby, where appropriate, restricting fishing activity in the future. Both roles are important, but the latter is more difficult to evaluate. We also identified several impediments, which prevent NCAs from playing an even more effective role in fisheries governance, some of which can be addressed speedily by governments.

Second, we suggest five issues that NCAs themselves might address. One issue is the implementation of the EBA and the role of NCAs. The NCAs need to convince the fishing industry and regulators that the EBA provides a fundamental delivery mechanism for progress towards sustainable development rather than preservationism – the latter being a commonly held view within industry. This has led CCW, NWNWSFC and representatives of the fishing industry to develop a sustainable fisheries project in North Wales to trial the application of the EBA to fisheries management. In addition to an analysis of the economic, environmental and social elements of three fisheries (mussel cultivation, potting and scallop dredging) there will be a comparative investigation of the fisheries against the internationally agreed (United Nations Convention on Biological Diversity (CBD)) twelve principles of the EBA (Table 11.1). There are also moves to develop more integrated, multi-species analysis (under a new project 'Science for sustainable marine bioresources' being sponsored by the Natural Environment Research Council (NERC), DEFRA and SEERAD), and efforts to identify meaningful indicators will help with predictive models and monitoring.

Another issue is the potential tension (or even conflict of interest) that NCAs face in simultaneously serving as judge of, and collaborator with, the fishing industry. Their statutory role of designating protected sites pulls them in the direction of judge, identifying the environmental harm done by fishing, but their strategic role of advocacy pulls them in the direction of collaborator, working alongside fishers to arrive at a better environmental outcome. Conflict could arise between these two roles, if, for example, in its collaborative role, the NCA encouraged the industry to think that they were meeting environmental requirements, but subsequently, in its statutory role, it informed them that they were not. Such tension is faced equally by other bodies, such as the Environment Agency, and it is likely to be resolved only by the relevant senior management deciding on the organisation's priority between coercion and cooperation.

The third issue is the involvement of NCAs in the development of marine spatial planning. In a speech at the Coastal Futures Conference in London in January 2005, The Minister for the Environment, Elliot Morley indicated that the proposed Marine Bill would cover marine spatial planning to aid integrated management of all activities in our coastal waters. The Government's first Marine Stewardship Report, *Safeguarding our Seas*, published in May 2002, contained a commitment to explore the role of marine spatial planning. In March 2004, the Government's response to its Marine Stewardship follow-up consultation paper, *Seas of Change*, proposed an investigation into how a marine spatial plan could be prepared and used. The NCAs are now contributing to the development of marine spatial planning through, for example:

- A DEFRA led project research into marine spatial planning using the Irish Sea as an example: due to report in June 2005;
- A JNCC project mapping marine regulations and policy implications on marine activities in the UK sectors of the Irish Sea;
- The formation of an internal NCA Marine Spatial Planning group to contribute to Government-led initiatives, to ensure consistency between NCAs, to develop strategic thinking and to share the work-load.

The fourth issue is the relationship between NCAs and ENGOs. On the one hand, they are quite different types of organisations: NCAs are public servants; ENGOs are independent of government. On the other hand, NCAs and ENGOs share many common objectives in relation to fisheries governance: in particular, they both want to integrate environmental policies (particularly the EBA) into fisheries management, and there is sometimes a fine line between the advocacy role of NCAs and the campaigning role of ENGOs. A dilemma NCAs face is maintaining a proper distance from ENGOs, yet embracing them as fellow workers in the field. The latter response could have significant cost saving implications for both organisations, by sharing the work load.

The fifth issue is how to influence the protection of offshore nature conservation interests from fishing. The rapid extension of statutory marine protected sites in recent years, from intertidal SSSIs to offshore Natura 2000 sites (eg Darwin Mounds), has raised a number of issues that the NCAs are having to consider. These include the legal basis for measures to restrict fishing, for example through the CFP or through individual Member States, monitoring the condition of the sites and enforcement. The internal NCA Habitats Group is currently considering these issues and offshore habitats regulations are currently being drafted.

Regardless of how the advice is delivered, a vital component part of fisheries governance is that statutory nature conservation responsibilities are acted upon. The NCA will therefore continue to play a role in this important objective, thereby contributing significantly to the achievement of fisheries which are sustainable from all perspectives.

References

Bullimore, BA, Newman, PB, Kaiser, MJ, Gilbert, SE and Lock, KM (2001) 'A study of catches in a fleet of "ghost-fishing" pots' *Fisheries Bulletin* 99:247-253

CCW [Countryside Council of Wales] (2002) *Legal advice to CCW on the Right to Fish versus Wildlife and Countryside Act 1981, as Amended*, Bangor, CCW, June

CCW [Countryside Council of Wales] (2003) *Sea Fisheries Policy,* Bangor, CCW

Clorley, J (2004) 'Relationship of the Common Fisheries Policy and the Habitats and Birds Directives' in *Marine Protected Areas and Fisheries*, Proceedings of the International Expert Workshop, Isle of Vilm, Germany 28 June-2 July

Coates, Phil (2002) personal communication

DEFRA [Department for Environment, Food and Rural Affairs] (2004) Regulatory Impact Assessment, Explanatory Memorandum to the Solent European Marine Site (Prohibition of Method of Dredging) Order, London, DEFRA

EC [European Community] (1979) *Council Directive of 2 April 1979 on the conservation of wild birds (79/409/EEC)*, OJ L 103, Office for Official Publications of the European Communities 25 April

EC [European Community] (1992) *Council Directive 92/43/EEC of 21 May 1992 on the conservation of natural habitats and of wild fauna and flora* (The Habitats Directive), Office for Official Publications of the European Communities, 22nd July

EC [European Community] (2000) *Managing Natura 2000 Sites: the provisions of Article 6 of the 'Habitats' Directive 92/43/EEC,* European Commission, Luxembourg, Office for Official Publications of the European Communities

EC [European Community] (2002) *Council Regulation (EC) No 2371/2002 of 20 December 2002 on the conservation and sustainable exploitation of fisheries resources under the Common Fisheries Policy*, Office for Official Publications of the European Communities

EC [European Community] (2004) *Council Regulation (EC) No 812/2004 of 26.4.2004 laying down measures concerning incidental catches of cetaceans in fisheries and amending Regulation (EC) No 88/98*, Office for Official Publications of the European Communities

ECJ [European Court of Justice] (2002) *European Court of Justice (ECJ) ruling on the Waddenzee Cockle fishery Case*, C-127/02

EN [English Nature] (1998) *Natura 2000: European Marine Sites: Guidelines Relating to Statutory Conservation Objectives and Operations Which May Cause Deterioration or Disturbance*, Peterborough, English Nature

Eno, NC, MacDonald, DS, Kinnear, JAM, Amos, SC, Chapman, CJ, Clark, RA, Bunker, F St PD and Munro, C (2001) 'Effects of crustacean traps on benthic fauna' *ICES Journal of Marine Science* 58: 1-20

Eno, NC (2004) 'Fishing in European Marine Sites in Welsh waters – science underpinning regulation' in *Marine Protected Areas and Fisheries*, Proceedings of the International Expert Workshop, Isle of Vilm, Germany 28 June-2 July

Fishing News (2004) Article, London, Informa

Gubbay, S and Knapman, PA (1999) *A Review of the Effects of Fishing within UK European Marine Sites,* UK Marine SACs Project, Peterborough, English Nature

Hernandez-Aguilar, P (2004) 'Nature protection in the marine environment: views on the relation between fisheries and the Birds and Habitats Directives' *Marine Protected Areas and Fisheries*, Proceedings of the International Expert Workshop, Isle of Vilm, Germany 28 June-2 July

HMSO [Her Majesty's Stationery Office] (1994) *The Conservation (Natural Habitats, &c.) Regulations 1994*, Statutory Instrument No. 2716, HMSO

JNCC [Joint Nature Conservation Committee] (1996) *Guidelines for Selection of Biological SSSIs: Intertidal Marine Habitats and Saline Lagoons*, Peterborough, JNCC

Kaiser, MJ, Bullimore, BA, Newman, PB, Lock, KM and Gilbert, SE (1996) 'Catches in 'ghost-fishing' set nets' *Marine Ecology Progress Series* 145:11-16

Owen, D (2004) *Interaction between the EU Common Fisheries Policy and the Habitats and Birds*

Directives, IEEP Policy Briefing, London, Institute for European Environmental Policy

PMSU [Prime Minister's Strategy Unit] (2004) *Net Benefits: a Sustainable and Profitable Future for UK Fishing*, London, Cabinet Office

Pope, JG and Symes D (2000) *An Ecosystem Based Approach to the Common Fisheries Policy: Defining the Goals*, a report to English Nature, Peterborough, English Nature

RCEP [Royal Commission on Environmental Pollution] (2004) Turning the Tide: Addressing the Impact of Fisheries on the Marine Environment, London, RCEP

SWSFC [South West Sea Fisheries Committee] (2002) SFC committee meeting minutes, October

Symes, D and Pope, JG (2000) *An Ecosystem Based Approach to the Common Fisheries Policy: Achieving the Objectives*, a report to English Nature, Peterborough, English Nature

Symes, D (1998) *The Integration of Fisheries Management and Marine Wildlife Conservation*, JNCC report no. 287, Peterborough, JNCC

Symes, D, Ridgway, S and Crean K (2002) *Integration of Regional Fisheries Management in the Irish Sea: a Feasibility Study*, a report to the Countryside Council for Wales, Bangor, CCW

UN [United Nations] (2003) *Convention on Biological Diversity*, Rio de Janeiro, June

WAG [Welsh Assembly Governmemt] (2003) *A Strategy for the Recreational Fisheries of Wales*, Recreational Fisheries Strategy Group

WDA [Welsh Development Agency] (2003) *Development of the Welsh Fisheries and Aquaculture Sector: a Strategic Action Plan*, a report for the Welsh Assembly Government, Cardiff, WDA

CHAPTER 12
THE ROLE OF ENVIRONMENTAL NGOs IN FISHERIES GOVERNANCE

EUAN DUNN
Head of Marine Policy, Royal Society for the Protection of Birds (RSPB), The Lodge, Sandy, Bedfordshire SG19 2DL

Abstract

This chapter explores the evolving role played by environmental NGOs (ENGOs) in UK fisheries governance in recent years. This role has grown exponentially, as the environmental dimension of fisheries activity has become increasingly understood and accepted by regulators and even by fishermen. There are three sections in the chapter: the first section explains how ENGOs have moved on from 'problem identification' to embrace 'problem solving'; the second section provides five illustrations of ENGO engagement in problem solving; and the third section addresses the challenges of the future, focusing on ENGO participation in the new European Regional Advisory Councils (RACs), and discussing ways in which ENGOs can fulfil their vastly increased workload in relation to fisheries governance.

12.1 Introduction

During the last twenty years, the context of fisheries governance has changed dramatically. From a time when the focus was almost exclusively upon managing the exploitation of fish stocks in a way that would maximise long term commercial catches, the focus today is increasingly upon protecting marine ecosystems, so that fishing activity can only take place when and where it is consistent with ecosystem health. Environmental NGOs (ENGOs) have played a central part in bringing about this change, and they continue their efforts to consolidate and implement it. However, they are finding that it is committing them to a major allocation of resources, and they have to find ways of coping with this inflated workload. In a sense, they are the victims of their own success.

In the first section of this chapter, I explain the changing role of ENGOs in relation to fisheries governance – from campaigning for 'problem identification' to collaborating in 'problem solving'. This shift is a familiar pattern for ENGOs who are working in other areas of environmental governance, but in the case of fisheries, it has been a particularly rapid process. This is not to say that ENGOs have abandoned their campaigning for problem identification – far from it; there are many environmental issues about fisheries that still have not been properly recognised, and ENGOs continue to mount campaigns to highlight them (for example, the Royal Society for the Protection of Birds (RSPB) long-standing campaign against industrial fishing, and the Deep Sea Conservation Coalition (DSCC)'s current high profile campaign for a moratorium on deep water trawling on the high seas). Nor is it to suggest that collaboration in problem solving has sorted out all the problems that have been identified. There is still a lot of resistance among resource users to environmental constraints on their activities, and ENGOs have to continue to push hard if they are to make headway. However, at least some progress

T. S. Gray (ed.), Participation in Fisheries Governance, 209–218.

is being made in moving beyond issue recognition to doing something about the issues, and ENGOs are contributing significantly to that progress.

In the second section, I examine five instances where ENGOs have been engaged in problem solving, and evaluate how effective they have been. The first case is the Shetland sandeel fisheries management partnership, which was a pathfinding example of cooperation between fishers, regulators and ENGOs (the RSPB) to find a way of protecting seabirds' supply of sandeel prey, while balancing the needs of the commercial exploitation of the sandeel stock. The second case is ENGO participation in the meetings of the influential North Sea Conferences, which materially pushed the agenda forward on the issues of the ecosystem-based approach (EBA) and the precautionary principle (PP), now widely accepted rules of fisheries management. The third case is the 2002 reform process of the European Union (EU) Common Fisheries Policy (CFP), to which the ENGOs contributed, helping to push the CFP in the direction of EBA and PP. The fourth case is more diffuse – the forging of various bilateral links with the fishing industry: some ENGOs have succeeded in working with fishers to bring in environmental measures that they can accept. The fifth case is the ENGOs' recent admission to the European Commission's Advisory Committee on Fisheries and Aquaculture (ACFA), which may stiffen the committee's resolve on environmental impacts of fishing.

In the third section, I discuss the future prospects for ENGOs in European fisheries governance, concentrating on the opportunities provided by their participation in the newly established seven Regional Advisory Councils (RACs). A significant concern here is with the added burden that this and other opportunities entail for ENGOs. One strategy that they can employ to deal with this workload is to work more closely with each other to share the burden. The fact that the ENGOs have such a problem is, however, a welcome sign of their success in stamping an environmental signature on fisheries governance.

12.2 The nature of the change

The role of ENGOs in fisheries governance has developed significantly in recent years, reflecting a transition from the phase of 'problem identification' to include also 'problem solving'. The 'problem identification' phase is the ENGO's classic 'whistleblower' role where it capitalises on its communication resources to raise awareness about an identified conflict in attempts to galvanise the political and practical action which characterises the problem solving phase. Problem identification was the hallmark of the relatively limited ENGO activity in fisheries in the 1980s – 'limited' because fisheries was scarcely perceived as an environmental issue for ENGOs. In addition, there was little progress towards getting involved in problem solving because there was no access route for ENGOs to the levers of fisheries governance in those days. All things considered, we could characterise the ENGOs' role in the early days as 'occasional whistleblower'.

As the ENGOs strove to enter the closed shop of fisheries, raise the environmental issues they had identified, and be acknowledged as a stakeholder, they were generally viewed with suspicion, especially by the fishing industry. Twenty years ago, the fishing industry itself was less well organised as a lobbying force than it is now (one civil servant described it to me then as “a broad church but with very narrow aisles”). As a result, the industry perceived the ENGOs, with their well-oiled publicity machines, as capable of drowning out their own voice. With overfishing the widespread accusation, the industry saw in the ENGOs only a ‘green’ threat exerting yet further pressure on their beleaguered industry. As a result there was lack of mutual trust and dialogue.

12.3 Evidence of effectiveness

We can chart the factors that began to involve the ENGOs with fisheries governance as a series of opportunities for deeper engagement which helped develop their relationship with other players. Five illustrations can be given of this effective engagement: the Shetland sandeel fisheries partnership; the North Sea Conferences; the CFP 2002 Reform; the bilateral links with the fishing industry; and access to the European Commission’s ACFA.

12.3.1 THE SHETLAND SANDEEL FISHERY PARTNERSHIP

As far as the relationship between the ENGOs and the industry was concerned, trust-building had to precede a more constructive dialogue. In the case of the RSPB, a turning point was its engagement with the Shetland sandeel fishery in the 1980s. The collapse of the local sandeel stock at that time and the resulting widespread breeding failure of sandeel-dependent seabirds brought fishermen, fishery managers (initially DAFS: Department of Agriculture and Fisheries Scotland), the country agency (Scottish Natural Heritage) and the RSPB into sharp conflict. It was a steep learning curve for everyone. The RSPB invoked the precautionary principle (probably for the first time in a UK fisheries issue). Engaging seriously with the fisheries science, such as the niceties of Virtual Population Analysis (VPA), we challenged the inadequacy of the scientific basis for managing the fishery. Recognition by the other parties of this comprehensive approach helped to legitimise our place at the negotiating table although it was the eventual collapse of the sandeel stock which actually catalysed most consensus.

In Shetland, there followed an unprecedented level of cooperative research, dialogue and transparency about the best way to move forward, a process greatly assisted in this case by the small size of the Shetland community in which individuals could easily meet on a regular basis and get to know one another. From this debate evolved the current management regime, which balances fishing and environmental interests, with the RSPB having an active voice in governance.

While the Shetland case study was a pioneering application of an ‘ecosystem approach’ (Dunn 1998), the interpretation and operational integration of the approach into the bigger picture of UK and European fisheries management remains a key challenge. Nevertheless, the experience gained from the small, inshore Shetland sandeel fishery did help to facilitate the RSPB’s advocacy for the creation in 2000 of the closed area for the much larger offshore sandeel fishery (prosecuted mainly by Denmark) off the east coast of Scotland and northeast England. The closure, which remains in place to this day, was taken in response to the declining productivity and population of the kittiwake *Rissatridactyla* on the adjoining coast and represents the first major management measure for a Community fishery taken in the interests of wider biodiversity. In the preparation of the relevant advice to the European Commission in 1999, the RSPB participated on an *ad hoc* basis in the ICES Study Group on Effects of Sandeel Fishing, another precedent in the ENGOs' developing stakeholder role in fisheries governance.

12.3.2 THE NORTH SEA CONFERENCES

In retrospect, another watershed for ENGO engagement was the Fourth (1995) International Conference on the Protection of the North Sea in Esbjerg. Welcoming and recognising the growing stakeholder involvement of environmental NGOs at Esbjerg, the high-profile Danish Environment Minister, Svend Auken, said in the Plenary meeting that there was a time when people wondered what such organisations were doing in a serious debate. "On the contrary", he continued, "it is these organisations that make the debate serious".

This was a generous declaration of support but it had a lot to do with the febrile atmosphere at that conference generated by the controversy over the dumping of Shell's 'Brent Spar' oil installation in the North-East Atlantic. Arguably, it marked the start of serious engagement by the ENGOs with European fishing policy issues. The then UK Environment Minister, John Gummer, who had much to teach the ENGOs about grabbing the limelight, was furious with the Danes' threat to censure the UK over dumping oil installations at sea. Gummer accused the Danes of using 'Brent Spar' as a device to divert attention from "raping the North Sea" with their industrial fishing. Interestingly he also accused Greenpeace (the 'Brent Spar' protagonists) of ducking the fishing issue because, he said, "you can't raise money from slithery creatures like fish."

Esbjerg spawned the Bergen Intermediate Ministerial Meeting (IMM) on the Integration of Fisheries and Environmental Issues (IMM 1997): a key conference which enabled the ENGOs to lobby on a wide variety of fisheries-related issues, and not just the industrial-fishing gauntlet thrown down by John Gummer. The ENGOs were particularly instrumental in lobbying for the basic principles that had to underlie fisheries governance in the new millennium, namely the precautionary approach, the ecosystem-based approach, and the overarching goal of sustainable development. Later, they were also influential in helping to shape the European Commission's Fisheries Biodiversity Action Plan.

12.3.3 REFORM OF THE EU COMMON FISHERIES POLICY (CFP)

The North Sea Conference developments, in turn, gave the ENGOs a platform for the looming reform of the CFP, which came to a head in 2002. Key issues for most of the ENGOs were the reduction of fleet capacity, the removal of public subsidies and the implementation of an ecosystem-based approach. At least in the early stages of the reform consultation, the ENGOs put most of their energy into influencing the European Commission, less so the European Parliament whose governance status in EU fisheries policy is currently 'consultation' rather than 'co-decision'.[1]

---

[1]This distribution of ENGO effort may conceivably change in future: the Convention on the Future of Europe, a high-level forum of politicians from across Europe, produced a draft Constitution for Europe in July 2003.

The ENGOs appeared to be moderately successful in influencing the European Commission, so much so that the Commission's 2001 'Green Paper' was seen as literally too 'green' by much of the industry (EC 2001). However, at the same time, the Commission arguably had their own agenda, which coincided with parts of the ENGOs' manifesto, notably on subsidies and the ecosystem-based approach, so this created synergy and a win-win situation. Put another way, the Commission might be said to have used the ENGOs as a bridgehead for putting controversial and ambitious proposals to industry.

While campaigning on the big CFP reform issues such as the ecosystem-based approach and subsidy reduction, the RSPB and BirdLife International also decided to focus on a very specific issue, which reflected their traditional concern with inshore fisheries and the importance of inshore waters for biodiversity. Accordingly, we commissioned a report (Coffey and Dwyer 2000) from the Institute for European Environmental Policy (IEEP), which presented a compelling case for a new European inshore fisheries management regime. Among the report's recommendations, the case was made for giving Member States the right to manage all fishing activities within their 12nm territorial waters, including also 'foreign' vessels fishing under historic rights agreements. The CFP reform (EC 2002) did secure this very right for Member States, though the extent to which the RSPB's advocacy actually assisted this change is, as always, hard to judge.

Overall, the ENGOs appeared to have played a greater part in shaping the CFP during this reform period than ever before, but the Green Paper and the Commission's proposals may be described as the high water mark of the ENGOs' aspirations, because opposition hardened thereafter. Compared with their earlier inroads, the ENGOs failed to engage effectively enough with the critical final hurdle, the Fisheries Council. While all stakeholders find the Council particularly intractable, there were other reasons for this shortfall. In particular, the ENGOs are well resourced and organised for fisheries advocacy in northern Europe but, with some notable exceptions, less so in the southern European Member States where – with a sharp divide in the goals of CFP reform – little progress was made in lobbying these so-called 'Friends of Fishing' nations. The environmental agenda became increasingly marginalised as the sense of crisis management deepened. As December 2002 approached, ICES were calling for a cod ban in the North Sea, and the southern Member States became increasingly opposed to reduction of their fishing fleets and subsidies.

12.3.4 BILATERAL LINKS WITH THE FISHING INDUSTRY

At country and regional level, however, the CFP reform presented opportunities for a deeper engagement between ENGOs and the fishing industry. To an increasing extent, both could rally under the win-win banner of seeking 'sustainable fisheries'. There was recognition from the industry that the ENGOs – geared up as they are for communication – can function as grit in the oyster to help raise the profile of issues and

This Constitution would give the European Parliament co-decision powers in the area of fisheries policy. From October 2003, the text was discussed by government representatives at the Inter-Governmental Conference (IGC). The IGC came forward with a final Constitution for Europe in Spring 2004, and it was signed in October 2004. It is currently uncertain whether government representatives will agree to an extension of the Parliament's powers on fisheries.

create pressure for mutually advantageous change. Such alliances were also facilitated by the perception that, in the catastrophic failure of the pre-2002 CFP, the industry was the victim of external (especially Commission and Ministerial) mismanagement.

There are numerous recent examples of close and constructive involvement between ENGOs and industry. Apart from the pioneering engagement on the Shetland sandeel fishery (above), the RSPB has a long tradition of working with the Sea Fisheries Committees in England and Wales. In 1999, we collaborated with the Cornwall Sea Fisheries Committee to draft a byelaw (the first ever environmental byelaw for a UK inshore fishery and the first exercising of the 1995 Environment Act by a Sea Fisheries Committee) to mitigate seabird by-catch in the St Ives Bay gillnet fishery. We have also cooperated closely with the Scottish Fishermen's Federation on inshore fisheries, and hopefully this helped to secure the new access restrictions adopted in the CFP reform. Again, the World Wide Fund for Nature (WWF) has worked closely with the fishermen's organisations north and south of the England/Scotland border to lobby for regional management and for short-term investment in the industry to secure long-term sustainable fisheries.

12.3.5 ACCESS TO ACFA

To the extent that the ENGOs have been successful in shaping the growing mountain of fisheries rhetoric with which we all struggle now, the ENGOs have influenced the governance of fisheries. But what of their stakeholder role in the decision-making process? The European Commission's consultation process on the reform of the CFP gave the ENGOs unprecedented access. As part of this, in 1999, towards the end of her term as Fisheries Commissioner, Emma Bonino was instrumental in opening up the Commission's influential ACFA to Environmental (and Development) NGOs. However, while the ENGOs continue to participate in this, their perception is that proceedings and outcomes are dominated by the traditional strength of the offshore producer sector, and this creates the impression of ENGO involvement being 'greenwashing'. Although this deters some ENGOs from engagement with ACFA, others find particular value in the dissemination of information from the Commission that such engagement brings: for example, early indication of intentions, strategies and proposals, enabling the participant ENGOs to be as up to speed as the fishing sector with the Commission's thinking on the evolving CFP.

12.4 Future challenges

I now turn to the future, where I see two challenges for ENGO engagement in fisheries governance. The first challenge is specific, relating to ENGO involvement in the RACs newly established by the European Commission. The second challenge is more general (exemplified by ENGO's membership of the RACs): that of the greatly increasing workload falling on the shoulders of ENGOs who have taken up the opportunities now available to participate in fisheries governance.

12.4.1 ENGAGING WITH REGIONAL ADVISORY COUNCILS (RACs)

Despite its shortcomings, ENGO access to ACFA has helped to consolidate the stakeholder role of ENGOs in fisheries governance, and this, in turn, creates a precedent for the ENGOs' stakeholder role in decentralised management. This has gained considerable momentum with the growing emergence of RACs, seven of which are in various stages of creation following the revision of the CFP, which gave them a mandate. As the IEEP Workshop on Regional Advisory Councils (Anon 2003) demonstrated, there are understandably different views on, and expectations of, RACs. For example, fishermen see RACs primarily as a means for achieving more active participation in the management process as well as building bridges with scientists, and they envision the management guidance role of the RACs as a transitional step to a stronger hand on the tiller in the future. This aspiration was spelled out as early as 2000 in a joint proposal on 'Zonal Management' (that is, management of regional seas) by the National Federation of Fishermen's Organisations (NFFO) and the Scottish Fishermen's Federation (SFF): "Initially the committees would advise the European Fisheries Council but it is expected they would be empowered to make management decisions for their respective zones after a suitable induction period" (NFFO/SFF 2000). For their part, the recreational anglers see an opportunity in the RACs to be recognised as important resource users and contributors to national economies, which qualifies them to have a significant say in management. Environmental interests hope RACs will facilitate ecosystem-based management. The extent to which these respective stakeholder ambitions will be realised will depend on a number of factors, not least mutual trust, respect and openness.

Resolution of the different aspirations for RACs will also be influenced, however, by their **structure** and **composition**. The framework for this is set out in the 'Council Decision of 19 July 2004 establishing Regional Advisory Councils under the Common Fisheries Policy (EC 2004). Each RAC will have a General Assembly of all interested parties which, in turn, appoints an Executive Committee of up to 24 members, comprising two-thirds from the fishing sector and a third from the other interest groups affected by the CFP. In effect, this means that in an Executive Committee of 24 members, the environmental NGOs need to try and secure representation among the eight seats available.

The first RAC to get up and running was the North Sea RAC, which had its inaugural meeting, effectively the first meeting of the General Assembly, in Edinburgh in November 2004. This had been preceded by several meetings of the 'shadow' North Sea RAC, which served to shape the final proposal for the RAC to Member States the European Commission. The ENGOs helped to ensure that the principle of working towards sustainable management of North Sea fisheries was set in a wider context of environmental integration. In this regard, the RAC's objective (NSRAC 2004) to provide stakeholder-led advice includes the following qualifying statement: "This will be done within the general aim of attaining the sustainable management of fisheries, incorporating an ecosystem based approach and based on the precautionary principle." In an initiative led by the fishing sector, the holistic nature of this goal would have been unimaginable ten years ago. During these meetings of the 'shadow' RAC, the broad composition of the RAC's Executive Committee was also formulated. This resulted in those ENGOs which had a Europe-wide network being offered three seats on the Executive Committee elected at the Edinburgh meeting. As UK Partner of BirdLife International, the RSPB took up one of these seats with the others being occupied by WWF and Seas At Risk.

Each RAC can set up as many subsidiary working groups as it needs to formulate the advice it needs and, in the case of the North Sea RAC, there are three at the time of writing: one for Demersal fisheries, another for Flatfish and most recently one on Spatial Planning, incorporating closed areas and other Marine Protected Areas (MPAs). Each working group should, as far as possible, mirror the balance of representation in the Executive Committee, so the environmental NGOs are represented on all three working groups. The RSPB/BirdLife International representative currently has the chair of the Spatial Planning group, which in itself is indicative of how the RACs can pay more than lip service to the breadth of stakeholder representation. The work of the group is likely to include the impending network of marine Natura 2000 sites and other MPAs (as laid down by OSPAR – Commission for the Protection of the Marine Environment of the North-East Atlantic) and their relatedness to any closed areas set up as a fisheries technical measure.

For the ENGOs, spatial planning is important for effective delivery of an ecosystem-based approach to fisheries management. For the fishermen's part, their willingness to approve such a working group should not be taken to signal a sudden conversion to the ecosystem-based approach, but rather the growing concern they feel for being buffeted and squeezed out of the North Sea and other Community Waters by the increasing plethora of other human activities, notably oil and gas licensing, and the marine renewables industry (especially wind farms). To this end, fishermen are increasingly alive to the opportunity such developments offer them in adopting a more hands-on stewardship role by using their own vessels and experience to contribute to spatial planning by providing data on the distribution of stocks, spawning and nursery areas. For the fisheries sector, marine spatial planning is thus regarded as a potential guarantor of fishing space and fishing rights. They see the growing demand for closing fishing grounds and are aware of the need to enter the spatial planning debate armed with strategic thinking if they are to carry influence in Brussels.

The RACS are only in their infancy, but from their brief engagement with the North Sea RAC, the experience of the ENGOs is of a more interactive and rewarding relationship with the fishing sector and other stakeholders than takes place in the ACFA. There is more dialogue with the sector in the RAC, whereas debate in the ACFA tends to be characterised by set-piece exchanges. In the end, however, the process matters less than the result, and it remains to be seen to what extent the Commission will take on board the advice of the RACs. Certainly, the greater the consensus among the RAC's stakeholders on any particular issue, the more united a front the advice to the Commission will represent, and the more persuasive it is likely to be. This puts genuine pressure on the stakeholders to compromise in order to offer tractable, usable advice to the Commission and thus prevent the RAC degenerating into a talking shop. The North Sea RAC is still finding its way in relatively uncharted waters, but it already has a strong sense of purpose and is keen to assert itself, set its own agenda, and not let the Commission lead it by the nose.

12.4.2 STAKEHOLDER FATIGUE

Several years ago, environmental NGOs would neither have expected to be stakeholders in governance bodies nor would they have had the opportunity. So, significant progress is being made. However, the price of involvement in governance is also a huge amount of time and resources in preparation for, and engagement with, these institutions, which presents a daunting challenge for all stakeholders, not just the ENGOs. With seven RACs likely to be up and running by 2006, and each having a number of working groups to service, the prospect of stakeholder fatigue is real. No single development in fisheries has so challenged the ENGOs to respond; having sought for years to influence the management of fisheries, the door has swung open and failure to step in now will send out a damaging signal about commitment and resolve to see the ecosystem-based approach made operational.

This pressure to engage is not coming, however, just from the emergence of RACs. UK devolution alone (which created the Scottish Parliament and the Welsh Assembly) has generated an almost exponential demand on ENGOs for expertise, consultation and direct participation on Government and other institutional advisory committees and steering groups. In recent months, this has been highlighted by the consultation process following the publication in March 2004 of the Prime Minister's Strategy Unit report '*Net benefits: A Sustainable and Profitable Future for UK Fishing*' (PMSU 2004). The Strategy Unit's Stakeholder Advisory Group set up a series of sub-groups dealing with issues such as environmental incentives and MPAs, which gave the ENGOs centre stage for inputting into a high level decision-making process, but at the same time, stretched their capacity to do so.

One outcome of this pressure on individual ENGOs to make an effective stakeholder presence at meetings on fisheries governance, has been growing co-ordination between ENGOs to arrive at common positions and negotiating strategies, so that they can effectively time-share their involvement at the exponentially expanding fora which invite them. While individual ENGOs have retained their specialist areas of expertise (for example the RSPB on industrial fisheries), this cross-ENGO fertilisation has produced hybrid vigour, and helped to sharpen and streamline advocacy messages. Thus, concerted action by ENGOs has significantly increased both in the UK and in Brussels.

It is clear that the ENGOs have an increasingly central role to play in fisheries governance. Nevertheless, they sometimes feel they have an identity crisis, caught as they are between the conflicting pressures to play the inclusive governance game on the one hand, and, on the other, to maintain their traditional external lobbying role. In essence, they have to try and do both, thereby aspiring to maintain the untenable condition of being partly pregnant.

12.5 Conclusion

In this chapter, I have, first, charted the evolution of the ENGOs' role in fisheries governance, showing how it has changed dramatically in the last twenty years from a relatively isolated position to an active stakeholder. Twenty years ago, fisheries were not widely perceived as an environmental issue and were not, therefore, a high priority focus of ENGO activity. With little access to the fishing industry and its decision-making machinery, the ENGOs effectively had only an occasional 'whistle-blowing' role. In this stand-off, the ENGOs were generally regarded with suspicion by the industry.

Second, I have examined the effectiveness of ENGOs in playing their new role of 'problem solving', by looking at five cases of fisheries governance in which they have been closely involved. In the first case, I have shown how it took the RSPB an intense encounter over the Shetland sandeel fishery for barriers to break down and for a more inclusive approach to develop towards conflict resolution; mutual trust also developed as a result. In the second case, I explained how the recognition of fisheries as a major environmental issue by the North Sea Conferences also assisted the ENGOs towards greater participation in governance, not least by helping them to contribute to the agenda for CFP reform. In the third case, I discussed how, in the CFP reform process, the ENGOs generally appeared to be successful in influencing European Commission proposals, but less successful at influencing the Fisheries Council, partly because the ENGOs' advocacy is poorer in the southern European Member States. In the fourth case, the CFP reform process fostered new lobbying alliances between ENGOs and the UK fishing industry. In the fifth case, I noted that, in the run-up to CFP reform, the ENGOs were granted access to the ACFA, although the ENGOs appreciate that their actual influence is generally limited, encouraging the perception that their involvement is merely 'tokenism'.

Third, I have looked at the future challenges for the ENGOs in the sphere of fisheries governance, focusing on their participation in the RACs, and on their capacity to cope with increasingly heavy demands on their resources. I drew attention to the fact that the pressure to meet growing stakeholder demand has resulted in more concerted action between ENGOs, enabling them to pool their resources and lobby better as a collective set of organisations. But this changing advocacy arena highlights an ongoing debate within ENGOs about how to be most effective, weighing up the relative benefits of external lobbying and a more hands-on stakeholder role in governance. Perhaps ENGOs are becoming the victims of their own success?

References

Anon (2003) *Report of the IEEP Workshop on CFP Regional Advisory Councils* (The workshop was held at Aberdeen, 2-3 June 2003), London, Institute for European Environmental Policy

Coffey, C & Dwyer, J (2000) *Managing EC Inshore Fisheries: Time for Change*, Royal Society for the Protection of Birds, Sandy, UK

Dunn, E (1998) 'The Shetland sandeel fishery: the ecosystem approach in action' *El Anzuelo* (IEEP newsletter on fisheries and the environment) 1:4-5

EC [European Community] (2001) *Green Paper on the Future of the Common Fisheries Policy*, COM(2001) 135 final, Brussels, European Commission

EC [European Community] (2002) 'Council Regulation (EC) No 2371/2002 of 20 December 2002 on the conservation and sustainable exploitation of fisheries resources under the Common Fisheries Policy' *Official Journal of the European Communities* L358:0059-0080, European Council of Ministers, 31 December

EC [European Community] (2004) '2004/585/EC: Council Decision of 19 July 2004 establishing Regional Advisory Councils under the Common Fisheries Policy' *Official Journal of the European Communities* L256:0017-0022, European Council of Ministers, 3 August

IMM [Intermediate Ministerial Meetings] (1997) *Statement of Conclusions from the Intermediate Ministerial Meeting on the Integration of Fisheries and Environmental Issues*, Bergen, Norway , 13/14 March

NFFO/SFF [National Federation of Fishermen's Organisations/Scottish Fishermen's Federation] (2000) *Zonal Management*, Grimsby, NFFO

NSRAC [North Sea Regional Advisory Council] (2004) 'Submission to Member States & the European Commission for the Establishment of a Regional Advisory Council for the North Sea', submitted September/October, www.nsrac.org

PMSU [Prime Minister's Strategy Unit] (2004) *Net Benefits: a Sustainable and Profitable Future for UK Fishing*, London, Cabinet Office

CHAPTER 13
THE PARTICIPATORY ROLE OF THE MEDIA IN FISHERIES GOVERNANCE

TIM OLIVER
Editor, Fishing News, T&F Informa, Telephone House, 3rd Floor, 69-77 Paul Street, London ECA 4LQ

Abstract

This chapter considers the role of the British media in European and UK national fisheries governance politics. It is divided into three parts. First, I shall argue that the national print and television media has influenced, and continues to influence, fisheries policy, but that this is largely in the direction of environmentalism and marine conservation, at the expense of fishing. Second, I shall argue that, by contrast, regional newspapers sometimes influence fisheries policy in favour of fishermen. Third, I shall argue that the direct influence exerted by the specialist fishing trade press over fisheries policy is negligible, but that it plays an important role in speaking up for the fishing industry and fisheries-dependent communities.

13.1 Introduction

One of the most notable features of European Union (EU) fisheries policy in recent years has been a marked shift away from concern primarily with the interests and well-being of fishermen, and towards an overwhelming concern with conservation and marine eco-system preservation. This shift in priorities has been driven by the environmental movement, helped by a friendly media. It reflects the growth in political power of the environmental movement and a corresponding decline in the political influence of the fishing industry.

One of the most significant results of this shift has been a gradual change in the public perception of fishermen. Until recent times, the public, in general, regarded fishermen with a mixture of respect and admiration. They were almost heroic figures, who braved the elements and did a physically arduous and dangerous job under difficult conditions to put high protein food on people's tables. A visit to the harbour to see fishing vessels landing their catches, and to take in the general atmosphere of fishing, was a central feature of seaside holidays. But in a relatively short time, this perception has changed dramatically. Many people, and especially young people, now see fishermen as greedy, self interested pirates who plunder the oceans with powerful and technically sophisticated vessels and equipment, without a thought for the marine environment or for future generations of fishermen. This perception is particularly prevalent in urban regions where people have little knowledge of the sea and do not identify with rural or coastal communities and ways of life.

This change has occurred largely because of the environmental movement's relentless portrayal of the world's oceans as barren deserts due to over-fishing. This message has

T. S. Gray (ed.), Participation in Fisheries Governance, 219–229.

been relayed via a print and TV media that is hugely receptive to such a portrayal. This has, in turn, led to public support for the environmental cause, and persuaded politicians to take its side in pushing for curbs on fishing to protect fish stocks and the marine environment. There are many votes in being seen to be 'green', but few votes for backing the fishing industry. Even in Scotland, fishing represents only about 2.5 per cent of the total economy, and in the UK as a whole it is only 0.03 per cent of GDP. The number of fishermen in the UK is only 12,000, with another 22,000 employed in the fish processing industry (PMSU 2004:27). This compares with a figure of well over 2 million members of environmental NGOs (Rawcliffe 1998:74).

In the next section (section 2) of this chapter, I show how the national media has presented this environmentalist agenda, at the expense of fishing interests. In section 3, I explain how the picture is rather different at the regional level, where news media are more sympathetic to the fishermen's cause, and sometimes succeed in their campaigns for policy shifts in their direction. In section 4, I focus on the specialist fishing trade press, which, while having little effect on fisheries policies, nevertheless serves a valuable function as the voice of the beleaguered industry facing bankruptcy, and the champion of fisheries-dependent communities facing decline.

13.2 National media

The national print and television media, in relaying the concerns of the public and the environmental movement over the marine environment, has significantly influenced fisheries governance. Journalism is nowadays very much the province of young people, who tend, rightly, to be idealistic and to want to change the world. Many young journalists are environmentally inclined, and thus are susceptible to accept sympathetically, although often uncritically, the stance of the environmental movement in general, not just the marine environmentalists. They tend to be 'eco-left' politically, and their instincts lean towards the environmental cause and against the commercial fishing industry.

An example from my own experience graphically illustrates this tendency. Some time ago, when the current 'cod crisis' was beginning to build up, I received a call from a journalist putting together an item on fishing and fish stocks for a TV news slot. She asked me if I would comment on the 'fish stocks crisis'. I pointed out to her that not all stocks were in crisis and that some were very healthy such as pelagic stocks (herring and mackerel), prawns and haddock, among others. I also pointed out that not all cod stocks were in trouble and that cod was in good shape in waters around Iceland, Norway and the Faeroes. She then became very hesitant and said that that was 'not really the line I was looking for', and that she would call me back. Needless to say, she did not.

It has become noticeable in recent years that articles in the national broadsheets on marine environmental issues have failed to include even a perspective from the fishing industry. Articles often appear following press releases from environmental organisations such as the World Wide Fund for Nature (WWF), English Nature, the Institute for European Environmental Policy (IEEP), Greenpeace, the Marine Conservation Society (MCS) and the Marine Stewardship Council (MSC) and several others. The topics of these articles are issues such as seals, cetacean by-catches, damage to the marine environment, sustainable fishing and eco-system fisheries management. While the spokespersons for these organisations are quoted and their point of view recorded, there is often no view from the fishing industry. Let us look at the unbalanced coverage of five of these issues in the national media.

13.2.1 SEALS

First, the case of seals is a very good example of the influence of the environmental movement and the prominence accorded to its views in the media, illustrating well the government's sensitivity to environmental issues and the extent to which it responds to public concern, but ignores fishing industry concerns.

Fishermen have for some years been expressing great alarm at the explosion of seal stocks (seals consume a large amount of fish), and have called for seal populations to be managed. Such concern has been expressed by fishermen in Norway and Canada as well as in the UK and EU. The response from the UK government has been nil. But in 2001, a virus began to kill common seals, beginning in the Baltic and spreading to Scotland and the North Sea. It was widely reported in the media, and the Scottish Executive Environment and Rural Affairs Department (SEERAD) – which is also responsible for fisheries, although this is not evident from its title – perceived that there was public concern over the seal deaths. It called a press conference and announced a £30,000 fund to finance a vaccination programme for seals. This showed starkly where the government's priorities lay – protecting the welfare of seals, but ignoring the fishermen's concerns about the expanding seal population's impact on fish stocks.

13.2.2 DOLPHINS

The second example of the environmental bias of the national media is the case of dolphins. The issue here, which is given great prominence by the Department for Environment, Food and Rural Affairs (DEFRA – again no mention of fishing in the department responsible for fisheries in England and Wales), is that of by-catches of dolphins by big trawlers that tow one net between two boats, primarily fishing for bass in the Channel and Western Approaches. Environmentalists have argued for some time, rightly, for action to be taken to stop this cetacean by-catch from occurring.

DEFRA has been carrying out trials with certain types of grids that theoretically allow the dolphins to escape before the net is hauled and the dolphins drowned. It recently made an announcement on the progress of these trials, and the story was widely reported in the national print media. As editor of the leading weekly trade newspaper, *Fishing News,* I received a call from a senior press officer in DEFRA to check that I had received the release – something that I do not recall ever happening in relation to fisheries press releases. (Significantly, even more recently, I had a similar call to check that I had received a notice about a press call involving new fisheries minister Ben Bradshaw and moves to ban 'shark finning' – a practice that does not even occur in UK fisheries, as the Shark Trust that is campaigning against the practice has acknowledged).

Yet at the end of 2002, when the Council of Ministers met in December to finalise no less an issue than the reform of the Common Fisheries Policy (CFP) that would set the scene for fisheries in EU waters for the foreseeable future, there were no phone calls and no press conferences from either DEFRA or SEERAD.

13.2.3 DARWIN MOUNDS

The third illustration of the environmental movement's media influence is the Darwin Mounds. DEFRA triumphantly announced in 2003 that it had persuaded the European Commission to impose emergency measures to close to commercial fishing the area of the so-called Darwin Mounds in deep water off the west coast of Scotland. This is an area of rare, slow-growing cold-water reefs that were discovered only comparatively recently. The government and the Commission were responding to an intensive campaign by the environmental movement to protect these reefs. It is, of course, right that the Darwin Mounds should be preserved, but the comparative speed with which the ban was put in place – with no consultation with the fishing industry and using emergency powers – is another telling example of the high priority afforded to marine environmental concerns and the low priority to the opinions and interests of the fishing industry.

13.2.4 COD CRISIS

The fourth example of environmental dominance in the national media is the so-called 'cod crisis'. The media reporting of the cod crisis has reflected an unquestioning acceptance of the position of the scientific and fisheries management establishment that North Sea and other cod stocks are on the verge of collapse through over-fishing. Although concerns are expressed in the media over the effects of pollution and, occasionally, of industrial fishing (the capture with very fine mesh nets of species such as sandeels and pout specifically for fish meal and oil), the media reports overfishing as the primary cause of the cod crisis.

However, many fishermen believe that the decline in cod stocks is due to environmental changes, and, in particular, to higher sea temperatures that are strongly correlated with poor spawning and recruitment of cod. This, they believe, has pushed the cod further north, so that there are now few cod in the southern and central North Sea but plenty in the Faroe Islands waters. The fishermen, therefore, argue that EU fisheries policy is fundamentally wrong, because it is based on an obsession with trying to achieve an unachievable goal. Cod in EU waters cannot be 'saved' by any amount of restriction on cod fishing, because its decline is due to environmental changes that are outside man's control. Yet the entire whitefish sector is being sacrificed in pursuit of this one unattainable objective.

It is noticeable, however, that in reports on this issue in the broadsheet papers, there is an increasing tendency for the views of environmental groups to be reported at length, with reporters not even bothering to ask for a statement from the fishermen or their organisations. There seems now to be an almost unconscious assumption of 'conservation good, fishermen bad' among journalists. For instance, tabloid headlines such as 'Has cod had its chips?' suggest to an uninformed public that cod everywhere is in crisis, whereas the reality is that cod stocks in the Faeoese and Icelandic waters are particularly healthy, and in Norwegian waters they are well within safe limits.

The picture painted by the media of cod stocks being at risk throughout the entire North Atlantic, only helps to reinforce the Commission's determination to stick to its policies in pursuit of the Holy Grail of 'saving the cod'. The ruinous effects of this policy on the fishing industry can safely be ignored because the industry is economically and politically weak and can do little to retaliate, and because the public – heavily influenced by the media – is behind the Commission and against the fishermen.

13.2.5 DEEP-WATER SPECIES

This knee jerk reaction by the Commission to a perceived environmental imperative is also evident in the fifth example – that of deep-water species. In this case, the charge is that, in its haste to respond to the environmental lobby's doom-laden predictions of the over-fishing of slow growing species such as orange roughy and grenadier, the Commission overnight imposed total allowable catches (TACs) and quotas on the deepwater fisheries west of Scotland and Ireland that had previously been unregulated. This action resulted in the virtual loss of this entire fishery to UK fishermen, and in its handing over to the French and, to a lesser extent, Spain.

This decision was taken on the very day after Commissioner Franz Fischler had assured UK fishermen's leaders in London that the Commission would not impose a TAC regime on the fishery. The decision was also in direct contradiction to the Commission's statements, repeated with ever increasing frequency over the last two or three years, that TACs and quotas do not work as conservation tools. This acknowledgement is, in fact, the central reason why the Commission is now fighting with all the means at its command to move towards effort control as its main fisheries management tool.

These examples illustrate starkly that the interests of the commercial fishing industry are very far down the list of priorities of the government departments responsible for fishing, but that politicians are extremely sensitive to marine environmental concerns, and they court the media assiduously on such issues.

13.3 Regional media

Let us now turn to the regional media, where a rather different picture emerges. As the fishing industry has become smaller, there is less interest shown in its affairs by the national media. There may be a flurry of reporting at the end of each year when the European Commission and Council of Ministers set the TACs for the following year, but then the industry is forgotten again. This relative indifference to the industry's well-being is not, however, manifested in regional newspapers that have fishing ports in their catchment areas. Such papers, which have a fishing industry readership, tend to be more supportive of the fishing industry, and to blame foreign fishermen, particularly the Spanish, for the fishing industry's problems. They take a 'jingoistic' approach, and portray foreign fishermen as rapacious invaders of what should be UK fishing grounds. 'They have killed their own stocks and now they want ours' is a general theme that runs through much of the coverage by regional papers.

Such newspapers are also very critical of the CFP, as the instrument by which foreign fishermen have gained access to UK grounds. The CFP is portrayed as an instrument of oppression against UK fishermen that simultaneously supports foreign fishermen with generous grants to build new vessels. This is a criticism that is not wholly without foundation. However, it is not the CFP alone that has caused problems for British fishermen; it is also the lack of support by successive British governments for the UK fishing industry, and the extent to which they have taken up, or, more accurately, failed to take up, the support that is available via the CFP.

13.3.1 SCOTTISH TRAWLERMEN'S COMPENSATION

There are two outstanding examples in the UK of regional newspapers exerting influence on behalf of fishermen – in Scotland and the southwest of England. In Scotland, fishing has a much higher political importance than elsewhere in the UK. Scottish MPs, Members of the Scottish Parliament and Scottish Euro MPs have much more to say on fishing issues than have those south of the border, even in fishing port constituencies, with the possible exception of those in the south west of England. Naturally, therefore, press releases and statements from Scottish politicians in response to industry issues, and as part of the general debate about the future of the industry, flow much more prolifically from them, than from their counterparts south of the border. In Scotland, the national papers such as the *Scotsman* and the *Herald*, and the northeast of Scotland-based *Press and Journal,* give significant coverage to fishing industry issues, reflecting the much higher profile the industry has there. This reflects the fact that there are many more votes in fishing in Scotland than elsewhere, although even in Scotland, fishing is still very much more of an issue in the ports and coastal regions than in the Glasgow/Edinburgh 'middle belt' where the mass of the Scottish population lives.

The clearest evidence of the influence of the Scottish press in championing the cause of fishermen relates not to current regulation of fishing opportunities, but rather to the issue of compensation for lost fishing opportunities in the past. This is the campaign mounted by the *Press and Journal* in Scotland, based in Aberdeen and with a wide readership in northeast Scotland, for compensation for redundant trawlermen who lost their jobs as a result of the extension of fishing limits to 200 miles during the 1970s, and specifically the 200-mile limit imposed by Iceland in 1976.

After some 20 years of campaigning by an organisation called the British Fishermen's Association (BFA), the Labour government, after it came to power in 1997, eventually set up the 'Icelandic Trawlermen's Compensation Scheme'. The BFA had branches in Hull, Grimsby and Fleetwood, where almost all of the now vanished British distant water trawler fleet had been based. It had no branch in Aberdeen, although there were a few trawlers there that occasionally made trips to Iceland, and many Aberdeen fishermen sailed on distant water trawlers from Hull, Grimsby and Fleetwood. But former fishermen in Aberdeen felt that they were not getting their share of the compensation, and the *Press and Journal* launched a major campaign on their behalf. This newspaper campaign, which won an award from the *Media Week* magazine, resulted in the Department of Trade and Industry (DTI) - the government body that was administering the compensation scheme - deciding to look again at its eligibility requirements, and DTI officials travelled to Aberdeen to talk to the ex-fishermen and their supporters. A lot of hitherto unknown evidence and information about, for instance, which Aberdeen vessels had made trips to Iceland, and when they had made them, became available as a result of the *Press and Journal*'s campaign. The outcome was that

Aberdeen-based trawlermen obtained compensation on the same terms as their counterparts elsewhere in the UK.

13.3.2 THE AREA VII MONKFISH TAC

It is in the southwest of England, however, that we find the clearest example of a regional newspaper contributing to a change in government policy in favour of fishermen rather than environmentalists. This occurred in 2003 over the EU's regulations governing the catching of monkfish (often known as anglerfish). Fishermen in the southwest found a huge imbalance between the amounts of cod and monkfish they were catching, and the quotas they had for these species. Because EU regulations forbid the **landing** of over-quota fish – fishermen cannot avoid **catching** them if they are on the grounds – the fishermen were forced to discard large amounts of valuable over-quota fish to stay within their quota limits. This fish is, however, effectively dead once it has been caught. Therefore, while dumping this over-quota monkfish did nothing for conservation, it robbed fishermen of many thousands of pounds of income and consumers of a first class food, simply to 'balance the quota books'. Fishermen have repeatedly pointed out the waste of a quota system for individual species in mixed fisheries, where they cannot avoid catching fish for which they have little or no quota when they fish for other species for which they do have quota.

The southwestern fishermen set about publicising this waste and the imbalance between the realities of the stocks they were catching and the quotas they had, to embarrass the Commission and the politicians, with the aid of the *Western Morning News*. They made a video of perfectly good fish being discarded, which was sent to EU Fisheries Commissioner Franz Fischler, Prime Minister Tony Blair and UK Fisheries minister Ben Bradshaw. The *Western Morning News* gave this video maximum coverage, and it was picked up to some extent by the national TV media, in particular, the Sunday morning *Country File* programme.

The outcome of this coverage was that the UK government sent fisheries scientists down to the southwest to make trips to sea in order to verify the fishermen's catches. A formal approach was then made to the Commission, backed up with catch data, for an increase in the TAC of monkfish. Some time later, the Commission did, in fact, increase the TAC of monkfish in ICES Area VII by 30 per cent.

Although the fishermen were lobbying hard for an increase in their own right, it is doubtful if they would have swayed the Commission without the sustained press campaign in their support, which embarrassed the fishery managers. It was, of course, covered in the trade press, but again this alone would not have influenced policy. Significantly, there was no increase in the monkfish quota in ICES Area VI, the west coast of Scotland, where a similar problem exists with monkfish, despite intensive lobbying by the Scottish fishing industry, but for which there was no similar media campaign. It may also be a significant factor in this differential treatment that the UK fisheries minister, Ben Bradshaw, is MP for a southwest constituency, Exeter. The minister tends to concern himself primarily with the industry in England and Wales, leaving Scottish issues to his Scottish counterpart, Ross Finnie. Fishing is a devolved issue under Scottish devolution, but the UK minister remains the lead minister in negotiations with Brussels.

However, the above two examples are noteworthy not only for their relative success, but also because of their rarity. Campaigns on behalf of any group of fishermen – as opposed to simply reporting events and complaints – are very rare even in regional newspapers that have a fishing 'constituency'. It is significant that even this level of publicity and embarrassment over the Area VII monkfish TAC did not cause the UK government to put the case for an increase in the cod TAC, as it is now received wisdom that all cod stocks are in trouble, despite compelling evidence to the contrary, in the southwest at any rate.

13.4 Specialist fishing trade press

Turning to the specialist press such as my own paper, *Fishing News*, I would say that in so far as having any direct effect on policy is concerned, its impact is zero. It is the voice of a weak and politically powerless industry and, like the industry itself, can safely be ignored by policy makers. Through its pages, for year after year after year, fishermen and their representatives have railed against the iniquities of the CFP, against the waste and ineffectiveness of the quota system, against flag ships, against industrial fishing, against the politicisation of fisheries management, against bureaucracy, and against flawed science. Their voice has been variously angry, proud, defiant, despairing, fearful, and contemptuous – all to no avail. The system under which they have to work, grinds on relentlessly, the bureaucrats loftily indifferent to their pleas.

In the columns of the paper and in other fora, fishermen have asked repeatedly to be given a genuine role in managing the fisheries upon which they depend for their livelihoods. Yet, despite their intimate knowledge of the condition of the stocks and what management techniques will and will not work, until recently, they have been totally excluded from any meaningful role in fisheries management. It is true, however, that this situation is now beginning to change. The EU has recently begun to conduct formal annual surveys of fishermen in each Member State, asking for their opinions on the state of certain key stocks such as cod, haddock, plaice and hake. This information is passed on to the scientists who advise the Commission on TACs (total allowable catches) each year, for incorporation into the scientific advice on which the fisheries management regime is based. But to what extent the scientists actually use this information in making their stock assessments and formulating their advice, is open to question.

Another development that fishermen hope will give them a genuine voice and role in fisheries management is the establishment in 2004 of Regional Advisory Councils (RACs). These are being set up as part of the reform of the CFP, agreed at the end of 2002. Five RACS will advise the Commission on management in distinct regions such as the North Sea, Irish Sea, West of Scotland, plus a sixth separate RAC for pelagic (for example, mackerel, herring and sprats) fisheries, pelagic fish being migratory and pursued by specialised vessels. The RACs will comprise representatives of the 'stakeholders' in the fisheries of that region, including fishermen, fish processors, sea anglers, and NGOs, plus scientists and officials.

Fishermen hope that the RACs will give them a genuine management role, but there are also fears that they will become only 'talking shops', with the clash of conflicts of interest preventing a unified view from being formulated. The hope is that they will lead to effective 'decentralisation' of the CFP, but the fear is that they may just create five 'mini CFPs'. It remains to be seen which view will be proved correct.

In the meantime, fishermen and their leaders can only lobby the scientists, officials and politicians as best they can, to get across the industry's views, and can only sit around helplessly in hotel lobbies and Brussels corridors as the bureaucrats and politicians take the decisions that can make or break fishermen and their communities. A spectacular example of this helplessness occurred at the 2003 December Fisheries Council in Brussels. The Commission and EU ministers agreed on a set of complex and highly restrictive measures in the North Sea involving restrictions on fishing time, a 'cod recovery area' and a special 'permit area' where haddock could be fished in reasonable quantities. These measures applied only to the UK, and they so outraged the UK industry, that the Scottish Fishermen's Federation sought advice on mounting a legal challenge to the regulations.

Following an intensive lobbying campaign by the Scottish whitefish sector, involving expensive weekly visits to Brussels, some of these restrictions were, unusually, modified during 2004. The iniquity and folly of these regulations was extensively reported in *Fishing News*. However, the modest changes that were eventually made to the regulations, were due entirely to the huge pressure exerted by the industry leadership during 2004, belatedly supported by the Scottish Executive as it gradually realised the extent of the blunder to which the UK fisheries ministers had agreed.

This event illustrates the fact that the specialist press can only be influential in affecting policy-making in direct proportion to the weight and influence of the industry it represents. As has already been noted, the fishing industry has little weight or leverage in the UK, so therefore neither does its press. The fishing press may be more influential in countries such as Spain or Norway, where there are powerful fishing lobbies. Even in those countries, however, it is probably the fishing lobbies themselves that bring influence to bear, rather than the press itself.

In so far as the fishing press does have a role at all in fisheries governance, it is rather one of expressing industry opinion, or, more accurately, opinions, than of exerting direct influence on the policy makers. It is a forum for dialogue and debate, not just between fishermen themselves, but also between their leaders, scientists, environmental groups and even, on occasions, ministers. This role should not be underestimated. While it is common to refer to 'the industry', as if it is a homogenous whole, the reality is that the fishing industry is composed of scores of individual 'industries' all around the coast, in hundreds of separate ports, harbours, coves and beaches. Each location, indeed, each boat almost, is an industry in itself. There is also a strong element of competition and rivalry between many boats, ports, regions and methods of fishing. The fishing industry is, by its very nature, diverse and scattered and hugely varied, ranging from the £15m pelagic super trawlers in Shetland to 16ft cove boats in Cornwall; from distant water freezers - two – on the Humber to cockle dredgers in the Thames and the Wash. Fishing methods include trawling, beam trawling, pelagic trawling, pair trawling, seine netting, gill netting, tangle netting, potting, long lining, suction dredging and one or two more minor methods such as fyke netting for eels and drift netting for salmon and herring.

It is perhaps no exaggeration to say that *Fishing News* is the only single unifying thread that draws together this enormous range of diversity and interests, and creates the illusion of 'an industry'. It is also the case that the trade press is a 'stakeholder' in the industry, just as much as any other business that exists to provide a service to the industry, and relies on the industry for its profitability. The fortunes of the trade press rise and fall with those of the industry, just as much as do the fortunes of all its other associated industries. Advertising revenue is the lifeblood of most trade newspapers, both display and classified. When times are good, fishermen invest in new boats and equipment; companies undertake research and develop new products and services; the second-hand boat market is buoyant and, as a result, advertising revenues are healthy. But when the industry is depressed and confidence ebbs, investment in new boats and equipment dries up, supply companies divert their resources to developing new markets and outlets, and advertising revenues are squeezed.

The trade press, therefore, cannot be insulated from the economic fortunes of the fishing industry. Accordingly, as well as being 'the voice of the fishermen', and promoting the industry's cause at a purely editorial level, the fishing press has a vested interest in the well-being of the industry and doing everything it can to promote that cause. This is perhaps another reason why the trade press has so little influence on fisheries governance; it is perceived as simply just another part of the industry, echoing the views the fishery managers hear repeatedly from the industry itself.

It could even be argued that the fishing press contributes to the problem for fisheries managers of ever increasing technical efficiency. So-called 'technical creep' has been calculated to account for an average increase in efficiency of about 3 per cent a year, meaning that fishing effort should be cut by this amount each year just to stand still, as it were. As a primary medium for the dissemination of new technical information that leads to ever greater efficiency and thus pressure on stocks, the fishing press perhaps helps to make the difficult task of fisheries governance even more challenging. This process is inevitable, because the fishing press survives on the advertising of companies that are marketing these new products and services, and which rely on the trade press as the primary medium by which they can reach their target market.

13.5 Conclusion

To sum up, I believe that in evaluating the influence of the media in fisheries governance, we can distinguish between three kinds of media. First, the national media, which tends to be sympathetic to the politically powerful and internationally well-organised environmental movement and exerts influence on fisheries governance, by highlighting the adverse environmental effects of commercial fishing, to the detriment of the fishing industry. Second, the regional media, which tends to favour the interests of its regional fishing industries and has enjoyed some success in obtaining relief for local fishermen in a few policy areas. Third, the specialist fishing trade press, which strongly supports the fishing industry as a whole and has little or no influence on the formulation of fisheries policy. Fisheries policy is made and implemented largely at EU level, while the specialist fishing press is national and therefore fragmented on a European scale. Nevertheless, the trade press serves an important social function as the mouthpiece of small and politically weak industries.

References

PMSU [Prime Minister's Strategy Unit] (2004) *Net Benefits – a Sustainable and Profitable Future for UK Fishing*, London, Cabinet Office

Rawcliffe, Peter (1998) *Environmental Pressure Groups in Transition*, Manchester, Manchester University Press

CHAPTER 14
THE ROLE OF MARINE SCIENCE IN PARTICIPATORY FISHERIES GOVERNANCE

CHRIS FRID
School of Biological Sciences, University of Liverpool, Crown Street, Liverpool, L69 7ZB, UK

Abstract

In the North Atlantic, formal international agreements to provide a co-ordinated response to the requirement for marine science to underpin fisheries governance have been in place for over 100 years. In this chapter, I consider how marine science has been used during that period, and the extent to which failures in fisheries management result from deficiencies or misuse of the science. My analysis of our failures in the past leads me to a consideration of ways to avoid such failures in the future, including an account of the possible role for marine science in an objective-based management regime, such as the 'ecosystem-based approach to fisheries management'. This role will include a significant element of prediction of the ecosystem effects of management scenarios, and also much greater dialogue with industry stakeholders and society to allow the informed selection of management objectives. The traditional fisheries science sector is inadequately prepared for this task, and much greater use of the wider marine science community will be required. In addition to the scientific challenges, the development of effective communication mechanisms between marine scientists and fisheries scientists, and between the science sector and society, must be acknowledged as necessary conditions for the success of these initiatives.

14.1 Introduction

Marine science as a discipline is often traced back to the oceanic voyages of exploration in the late nineteenth century. However, 'science' has been involved in fisheries governance for probably as long as there has been fisheries governance. In early times, this input came from 'advisors' who included stakeholders, such as fishers, resource 'owners' such as the Crown, and learned men. With the development of what we would now describe as the 'scientific approach', there was scope for its application to fisheries problems. This was seen most in freshwaters where scientific investigation led to great advances in our understanding of, for example, salmon lifecycles. However, the continued increase year on year in marine catches, and the extent and richness of life in the oceans as revealed by the early research surveys, meant that most people, including many early scientists, believed that the oceans' bounty was so vast that it could not be impacted by anything man could do. As late as 1884, Thomas Henry Huxley, President of the Royal Society, stated "The cod fishery, the herring fishery, the pilchard fishery, the mackerel fishery, and probably all the great sea fisheries, are inexhaustible". However, by the turn of the last century this view was being challenged and it was recognised that a coordinated international programme of research was required. It was this that led to the establishment of the International Council for the Exploration of the

T. S. Gray (ed.), Participation in Fisheries Governance, 231–247.

Sea (ICES) in 1901, which brought together marine science in the North Atlantic. Similar agreements were established subsequently, covering the Baltic, Mediterranean, North Pacific and other regions.

The aim of this chapter is to provide an analysis of the role of science in current governance, using the NE Atlantic as a case study, and to consider the science needs of an alternative governance framework – that of ecosystem management.

14.2 The role of science in fisheries governance

Science is defined by the Oxford English Dictionary as the "systematic study of the structure and behaviour of the physical and natural world through observation and experiment". It therefore provides a body of knowledge and a mechanism for answering questions posed by society. In the context of fisheries, these questions are likely to concern the possible configurations of the exploited system and how to manage the system in order to arrive at the condition desired by society and expressed by their democratic choices. This leads us to answer questions such as:

- How many fish are in the sea?
- How many fish can be removed without compromising the stock (i.e. what is the Total Allowable Catch (TAC))?
- What impact will this removal have on the habitat and other parts of the ecosystem?
- What is the level of annual recruitment, and how does it relate to size/age and number of adults?
- What should be the minimum size/age of the fish we catch?
- What are the options for technical measures to limit exploitation (such as restrictions on fishing methods or the imposition of marine protected areas (MPAs))?
- What impact will these options have on fishers?
- What impact will the measures have on the exploited stocks?

It is immediately apparent, then, that the central question for fisheries science is 'how many fish can we harvest without impacting on the ability of those that remain to maintain the population'? Population models would seem to offer a solution, as they can allow us to predict future population size, based on a limited number of measures of the current situation and some knowledge of the biology of the organisms. Fish, like all living organisms, have the capacity to produce an excess of offspring. That is to say, each pair of parents can produce more than one pair of offspring. In a stable population, disease, predation and other natural processes, mean that within a generation each pair of parents replaces itself. However, if we remove a proportion of the offspring produced, we reduce the amount of competition for food, and the size of the population may show no effect of the harvesting. In other words, the number of individuals we took as a harvest would have died later of natural causes anyway. Given that an adult fish like a cod can produce in excess of two million eggs each year, and may have a reproductive life of ten or more years, there would seem to be a massive scope for the harvesting of the excess production.

As it has long been known that mortality varies between individuals in a population, in large part due to their age/size, fisheries models try to model the population using just three basic input parameters; recruitment, growth and mortality. Recruitment provides an estimate of the number of young fish entering the adult stock that year; this subsumes all the mortality and losses at the egg and larval stages. This obviates the need for much data gathering and the need to model the full life cycle. Growth functions are then required to allow the movement of individuals from one size group to another to be modelled, while mortality terms describe the natural and fishing mortality on each size class.

For many years, the optimum harvest was seen as the maximum yield that could be removed without reducing the population in subsequent generations. This is known as the Maximum Sustainable Yield (MSY). The approach used in many contemporary fisheries management schemes to assess the state of stocks and provide advice on the amount of harvest, is Virtual Population Analysis (VPA). This seeks to estimate fishing mortality (F) and the numbers at age in a stock from catch at age data and estimates of natural mortality only.

14.2.1 VIRTUAL POPULATION ANALYSIS (VPA OR COHORT ANALYSIS)

In order to estimate fishing mortality (F) and the numbers at age in a stock from only catch data, VPA must assume that natural mortality (M) is constant at any given age. For example, consider the total number of five-year old fish in the stock, which is the number of fish reaching the age of four the previous year minus the number of four year olds dying (total mortality Z_4). The total mortality of four-year old fish is composed of fishing mortality (F_4) and natural mortality (M_4). We assume natural mortality is constant over years but can vary for different age groups. The number of fish of a particular age class caught is a proportion of the total number dying. These two relationships can be combined into a single equation containing natural mortality (M), numbers at age plus 1, and catch and fishing mortality at age. If M, numbers at age plus 1, and catch at age, are known, then F can be calculated. The actual equation is complex, so solving for F is done by computer iteration. The iterations begin with the oldest cohort (as the total mortality is then 1, none survives to the next year) and assume a value for F for the oldest cohort. This iteration can then be used to calculate the numbers in the cohort one year younger. The procedure can then be repeated, using the appropriate catch data, to get a value of F for the next cohort and so on until all the cohorts have been modelled.

A big assumption of VPA is that M is constant across years and is known. M can be calculated if total mortality and fishing mortality are known, but if F is known then VPA is unnecessary! It is relatively easy to include age specific rather than constant natural mortalities into the model but it is not easy to incorporate inter-annual variability in M or density dependence. Density dependence is the term used to describe when the values of a parameter vary depending on the population size; more fish are likely to starve when the population is large than when it is small, for example.

Unfortunately, in using VPA as a basis for management measures, the least reliable estimates of F are those for the older age classes, that is, those parts of the stock most heavily exploited. Considerable effort has gone into developing more accurate VPA, including multi-species VPA (MSVPA) and more recently the 4M package (Multispecies, Multi-fleet, Multi-area Modelling-package). These multi-species models contain inter-specific interactions such as large cod predation on small whiting.

The MSVPA has its origins in a multi-compartment production model of the North Sea (Anderson and Ursin 1977). However, this model was too complex, containing too many inestimable parameters to be useful in fisheries management. ICES, therefore, developed a simpler model that focused on the predatory interactions between the commercially exploited fish stocks. For those stocks for which catch-at-age data were available, and assuming constant, instead of food dependent, individual food intake and growth, it was possible to construct a multispecies model, MSVPA, with only three equations. These were the catch and stock number equations of the single species VPA plus an equation describing how predation mortality, M2, depends on the biomass of the prey and the total food intake of the predator.

In order to gather information on predation and diet for the model, a major international programme of fish stomach sampling was carried out. The so-called 'Year of the Stomach', 1981, saw co-ordinated sampling by research cruises across the North Sea during which approximately 60,000 stomachs from five commercially exploited fish species (cod, haddock, whiting, saithe and mackerel) were collected. These five species were assumed to be the major fish predators in the North Sea. The stomach contents were analysed to provide estimates of the average food composition and total weight of stomach content by predator age, prey age and quarter of the year. In 1984, the first quarterly North Sea MSVPA was produced. The model was further refined and additional fish stomachs were collected in 1985, 1986 and 1987 for some of the predators. In 1991, the second 'Year of the Stomach' saw additional food composition data collected for all of the MSVPA predators as well as for a suite of other predators expected to prey on commercially important fish species. The total food composition database for the North Sea now contains the results from analysing approximately 200,000 fish stomachs (Greenstreet *et al* 1997).

Over the period from 1984 to 1997, ICES has performed sensitivity analyses of the MSVPA, examined the assumptions, the difference between single and multi-species, long- and short-term predictions of effort and mesh changes, added additional 'other predators', developed alternative, simpler models and tried to reduce the parameters of the model describing the food selection of the major fish predators (Pope 1991; ICES 2002a:82). The major conclusion of the work is that natural mortality is much larger for the younger ages of the species exploited for human consumption than previously assumed. The MSVPA was found to be quite robust to changes in input parameters. Most importantly, it was found that the long-term predictions arising from a multi-species approach differed significantly from single species predictions.

Outside the North Sea, the MSVPA has been applied in the Baltic, in the Barents Sea, Georges Bank and in the Bering Sea. The development of MSVPA centred on biological interactions between fish, their prey and their predators. However, from a management point of view, technical interactions between fleets and species are also important. This has prompted the development of the, so-called, 4M model (Multi-species, Multi-fleet, Multi-area Modelling-package). Within the 4M model the impact of technical interactions can be evaluated. However a lack of data disaggregated by fleet, has so far prevented it being used operationally.

Table 14.1. List of stocks used in the analysis of the effectiveness of ICES fishery advice

Fishery	*Species*	*Area*
Flatfish	Plaice	Division VIId (Eastern Channel)
		Division VIIe (Western Channel)
		Sub-area IV (North Sea)
	Sole	VIId (Eastern Channel)
		Division VIIe (Western Channel)
		Sub-area IV (North Sea)
Industrial	Norway pout	Sub-area IV (North Sea) and Division IIIa (Skagerrak – Kattegat)
	Sandeel	Sub-area IV
Pelagic	Herring	Divisions VIa (South) and VIIb,c
		Sub-area IV Division VIId and Division IIIa (autumn spawners)
	Mackerel	combined Southern, Western & N.Sea spawn.comp.
Roundfish	Cod	Division VIa (West of Scotland)
		Sub-area IV (North Sea), Division VIId (Eastern English Channel) and Division IIIa (Skagerrak)
	Haddock	Division VIa (West of Scotland)
		Division VIb (Rockall)
		Sub-area IV (North Sea) and Division IIIa (Skagerrak – Kattegat)
	Saithe	Sub-area IV (North Sea), Division IIIa (Skagerrak) and Sub-area VI (West of Scotland and Rockall)
	Whiting	Sub-area IV (North Sea) and Division VIIId (Eastern Channel)

14.3 Performance of science advice in North Sea fisheries governance

The current management objective for fish stocks (Table 14.1) is to keep the spawning stock within 'safe biological limits' – that is, where there is a high likelihood that the stock will not suffer a catastrophic decline and that sufficient fish are available to replace losses. However, ICES produce advice on the status and levels of exploitation for only a limited number of species and stocks. There were many other species exist in the North Sea. A number of these are fished commercially, but they are not assessed, and estimates of spawning stock biomass (SSB) are not available for these. It is therefore impossible to ascertain if these stocks are within safe biological limits.

Management advice is given in a precautionary framework and with respect to the desired biomass of fish in the sea (B_{pa}) and the level of fishing mortality that matches this biomass (F_{pa}). Three criteria can be used to determine whether a stock is within these limits and hence whether the objective was met:

- SSB was above the desired level ($SSB>B_{pa}$);
- F was below the desired level ($F<F_{pa}$);
- Both of the above ($SSB>B_{pa}$ and $F<F_{pa}$).

In order to evaluate the performance of science advice to fisheries managers, ICES carried out an evaluation of its past advice (ICES 2003). For each stock for which advice was produced, both the actual annual management advice given and the action taken was assessed, using the observations tabulated in the "Catch Data" and assessment output tables from the 2002 round of fisheries advice (ICES 2002b). The evaluation identified four possible scenarios:

1. **Stock does not meet the objective, correct advice**: the estimate of SSB and/or F *in the assessment year* led to advice to reduce catch, when the estimate of SSB and/or F *in the 2002 assessment* now indicates that the stock did not meet the objective (i.e. respectively SSB < B_{pa} , F > F_{pa} or SSB < B_{pa} and F > F_{pa})
2. **Stock does not meet the objective, *incorrect* advice**: the estimate of SSB and/or F *in the assessment year* led to advice for status quo or increased total allowable catch (TAC), when the estimate of SSB and/or F *in the 2002 assessment* now indicates that the stock did not meet the objective.
3. **Stock meets the objective, *incorrect* advice**: the estimate of SSB and/or F *in the assessment year* led to advice to reduce catch, when the estimate of SSB and/or F *in the 2002 assessment* now indicates that the stock met its objective
4. **Stock meets the objective, correct advice**: advice for *status quo* or increased TAC, when the estimate of SSB and/or F *in the 2002 assessment* now indicates that the stock did meet its objective

Signal theory was applied to these scenarios to determine the proportion of Hits (1 and 4), Misses (2) and False Alarms (3) per year as the proportion of the stocks for which the respective scenarios applied. If the analysis shows a high Hit rate and low rates of Misses and False Alarms, it is support for the view that precautionary reference points are a robust basis for fisheries management advice, generally advising managers to take actions that would move the stock in the proper direction. High Miss rates would suggest that precautionary reference points, as currently used, do not lead to advice that is sufficiently restrictive to ensure stocks remain within safe biological limits. High False Alarm rates would indicate that precautionary reference points, as currently used, lead to overly intrusive management advice. The actual performance of B_{pa} and F_{pa} as objectives and as guides to fisheries management is presented in Table 2.

Overall, the main difference between the criteria used, is that using only F, will result in relatively low False Alarm rates, but high Miss rates. Using only SSB, results in a decrease in Misses but a higher proportion of False Alarms. The best results were achieved using both criteria with a 53 per cent Hit rate, 23 per cent Miss rate and 24 per cent False Alarms.

Tables 3 and 4 give a quantitative indication of the true impact of the advice depending on the scenario, not just what advice was provided, but how management actually responded to the advice and the indicator. This also shows that in general the advice was appropriate. If the objective was not met, a reduction in TAC of about 18 per cent was suggested in case of a correct advice (Hit), whereas there was an increase of TAC averaging between 10 per cent (SSB) and 15 per cent (F) in case of a Miss. If the objective was met, correct advice resulted in a suggested increase of the TAC between 26 per cent (SSB) and 16 per cent (F), whereas in the case of a False Alarm, the TAC was suggested to decrease between 9 per cent and 18 per cent. Overall, the advice using SSB appears more appropriate with relatively small changes in case of a Miss and False Alarm but relatively higher changes of TAC in case of Hits.

Table 14.2. Proportion (%) of Hit, Miss or False Alarm depending on the criteria used (i.e. respectively SSB > B_{pa}, F < F_{pa} or SSB > B_{pa} and F < F_{pa}) and the type of fishery (from ICES 2003)

Criteria	*Fishery*	*Hit*	*Miss*	*False Alarm*
F	All	49	44	7
F	Flatfish	52	43	5
F	Pelagic	27	63	10
F	Roundfish	52	40	8
SSB	All	51	25	24
SSB	Flatfish	52	18	30
SSB	Industrial	16	21	63
SSB	Pelagic	57	26	17
SSB	Roundfish	56	29	15
F & SSB	All	53	23	24
F & SSB	Flatfish	50	17	33
F & SSB	Pelagic	57	26	17
F & SSB	Roundfish	55	25	20

Table 14.3. The average change of the TAC (%) that was actually implemented for various scenarios: i.e. the objective (SSB> B_{pa}) is met (1) or not met (0) and advice is correct (1) or incorrect (0) (from ICES 2003)

Scenario	*The objective*	*Advice*	*Flatfish*	*Industrial*	*Pelagic*	*Roundfish*	*Total*
1	0	1	-11.8		-37.7	-21.5	-18.8
2	0	0	9.6	0.0	8.7	12.2	10.3
3	1	0	-6.0	-2.4	-13.1	-17.7	-9.4
4	1	1	23.9	11.4	10.8	38.9	26.1

Table 14.4. The average change of the TAC (%) that was implemented for various scenarios: i.e. the objective (F<F_{pa}) is met (1) or not met (0) and advice is correct (1) or incorrect (0) (from ICES 2003)

Scenario	*The objective*	*Advice*	*Flatfish*	*Industrial*	*Pelagic*	*Roundfish*	*Total*
1	0	1	-11.7		-34.5	-20.9	-18.1
2	0	0	11.5		10.2	20.8	15.5
3	1	0	-7.3		-4.3	-26.4	-17.5
4	1	1	17.0		5.6	17.4	16.3

14.4 Marine science and fisheries management tools

There are a number of approaches to fisheries management, involving control of fishing effort and/or catches, and a wide range of other technical measures. Traditionally, the approach has focused on conservation of the target stocks with measures such as catch quotas and controls on size/age at capture (mesh size controls). In recent years, the effectiveness of much of the management effort has been questioned – after almost 100 years of fishery management in the North Sea, more stocks were listed as endangered (outside 'safe biological limits') than ever before! This, along with growing recognition of the need to manage fisheries in the context of the wider ecosystem, has led to developments of other management approaches including effort controls, technological changes to provide better selection of the target species, and closed seasons/areas to minimise habitat damage.

14.4.1 CATCH QUOTAS

The most simple and intuitive means of ensuring sustainability of fisheries is to limit the catch to a level that removes only the surplus production. This is the basis of most fisheries regulations worldwide, and also the source of many of the problems of over exploitation. However, there are two major problems in applying this approach: first, the difficulties of calculating, in real time, the levels of this surplus production; and second, how to match the effort of the fleet to this level of production in space and time.

Science is the key tool in estimating the size of stock a year in advance through modelling the population (see above). The scientists are then able to advise on both the TAC and also the amount of fishing activity likely to result in this take. However, it is managers that use this information and convert it into management measures. TACs are usually set to reduce the total fishing effort on a stock in order to limit the rate of fishing mortality. TACs are then translated into quotas that restrict landings of individual fishers. Such restriction of landings is supposed to restrict fishing effort, but the link is not direct. For example, fishers can continue to fish, but discard excess catches, and this is recognised as one of the problems of managing using TACs and quotas. In theory, however, there is no fundamental difference between control of effort and control of catches.

The managers have to balance the biological and social aspects of the fishery in setting the quota. Fishers' livelihoods, their families' welfare, the investment in infrastructure, both in fish capture but also post-capture processing, have to be considered. This results in reductions in quotas often being less than those recommended by scientists. If the quota set is so small that the quota for each vessel fails to provide enough legitimate catch for the fisher, then either the fisher will behave illegally or the economics of the whole fleet will collapse.

Fishers in the European Union (EU) have learned to live with a quota-based system, but, generally, they regard it as 'unfair' (Hatchard, this volume). The quotas recommended are regularly challenged and are always set too low, according to the industry. Fishers complain that the TAC is not allocated into quota according to the 'best' algorithm, and they denounce the need in mixed fisheries to discard species for which the quota has been filled, while still pursuing other species.

14.4.2 FISHING EFFORT REDUCTION

Currently, fishing vessel and fleet capacity are widely considered to be excessive and not in balance with the resources (ICES 2002b; FAO 1998). The relationship between the fleet size and its ability to impose a particular level of mortality on a stock is a complex one. Efforts to develop models such as the 4M, which incorporate these relationships, are severely handicapped by a lack on data on these relationships.

The European Commission has in place a programme, whereby fishing vessel de-commissioning targets are set for each Member State, in an attempt to reduce the overall fishing effort in EU waters, thus reducing the pressure on stocks. However, vessel de-commissioning is likely to be more attractive to those skippers with the least economically successful vessels. For instance, in a UK shrimp fishery consisting of 98 vessels, seven of these vessels were responsible for 49 per cent of the total fleet effort, and the combined effort of a further fifty-six vessels was estimated to be only 6 per cent of the total effort (Revill 1996). This implies that a programme that cut the fishing effort, in terms of the number of vessels, by around 55 per cent, might only actually reduce effort on the stock by 6 per cent!

Such programmes also fail to deal with the issue of 'technological creep'. It is estimated that the ability of the fleet to catch fish (the efficiency of the fleet) increases by 7-8 per cent every year as a result of new vessels replacing old ones and changes in technology on the existing vessels. This implies that even if the fleet size were reduced by 40 per cent immediately, there would be a need to make a further cut, of the same magnitude, in the fleet every 12.5 years! Thus science can advise not only on the need and extent of effort reductions, but how these can be targeted effectively. However, lack of information about the scale of the vessel, usually withheld in order to provide anonymity, has prevented this in most cases.

In the EU, an additional restriction has been applied, which limits effort during periods of low stock size and hence low quotas. Known as the 'days-at-sea' regulations, they restrict the number of days per month an individual vessel can fish without removing the vessel permanently from the fleet. These measures have been a key element of the European Commission stock recovery plans, but they are unpopular with fishers, whose costs remain the same, but their opportunity to earn is restricted. Days-at-sea restrictions also force vessels to sea in poor weather if quota is not used up and there are days remaining in that month. Moreover, they encourage the race to fish, and lead to further high grading (and black fish landings) on the days when the vessel is at sea.

14.4.3 SIZE/AGE AT CAPTURE RESTRICTIONS

Technical conservation measures such as minimum mesh sizes affect the composition of the catch, but do not restrict the total quantity of fish caught. For this, direct conservation measures such as catch or effort controls are required. The simplest technical measure to apply is one based on the size (age) at which individuals are caught – a Minimum Landing Size (MLS). This can be set to ensure that all individuals spawn at least once before becoming available for capture. Selection may occur post-capture with small individuals being returned to the wild or, in net fisheries, through alterations in the gear, such as the mesh size of the net.

In fisheries targeting individuals, it is easy to enforce minimum landing sizes. In net capture fisheries, it is usual to apply a mesh size to the net; fishers still have to comply with a MLS but the number of small fish caught that have to be sorted and then discarded is reduced. The mesh size used should ensure that most of the time, individuals below the desired size escape. However, selection will never be 100 per cent: as a net fills with fish then the fish in the net 'blind' the mesh openings, preventing the escape of smaller individuals, while a heavy catch being towed through the water will stretch the net causing the mesh to deform.

When fishing for a variety of species that differ in morphology (round fish versus flatfish) and life history (early breeders versus late developers), the selection of an appropriate mesh size is a compromise. The small individuals in the catch, below the species' MLS are discarded, but this is of little use if by that time they are dead. There are now efforts directed at developing gears that sort the species in the water by criteria other than size.

14.4.4 SORTING THE CATCH BY SELECTION IN THE WATER

Fishing gears are size selective. However, fishing gears can be made more selective towards the target species by altering the geometry of the gear. This would reduce post-capture discarding. The mesh sizes of the netting in fishing gear, and in particular the cod-end mesh size, is influential in determining the selective properties of towed demersal fishing gear (Anon 1996). The minimum cod-end mesh size is widely legislated for in EU waters and is specific to each target fishery. In many fisheries, however, the demersal fishing grounds are multi-species in nature and by-catch and discarding are common resultant features, due to poor cod-end selection (Evans *et al* 1994). The twine diameter (thickness) of the meshes in the cod-end is also known to affect the selection process in the cod end, as are seasonal processes such as spawning status that may affect the overall shape of the fish.

Research into the incorporation of selectivity devices such as square mesh panels, funnels (sieve nets) and separator grids into towed fishing gears to enhance their overall selectivity is becoming more widespread, and, as a result, their use within fishing gears is gaining acceptance as a management tool. Legislation requiring the use of selectivity devices is being implemented in many instances; for example to allow turtles to escape. However, while such technical measures as mesh size and sorter nets go some way to addressing the problem of by-catch, they will never completely solve it.

14.4.5 CLOSED AREAS

Technical conservation measures such as closed areas, which prohibit or restrict fishing activity from an area, are also common, but need to be supported by direct management limiting catch or effort. In some cases, these spatial restrictions may be related to the need to protect military, oil and gas installations or sites of special scientific or historical interest (Rogers 1997). In other cases, fishing is restricted or prohibited in order to protect the fish stocks (usually juveniles) themselves from over-exploitation. For example, in the North Sea, much of the North Yorkshire (NE England) inshore coastal waters (inside 3 miles) are closed to all towed forms of fishing in order to protect the juvenile codling and other gadoid species that aggregate in these waters (Rogers 1997). In other areas, the marine protected areas (MPAs) may be established to segregate recreational activities, including tourism, from fisheries.

14.4.5.1 *The 'Plaice Box'*

One of the largest restricted fishing areas in the North Sea is the 'plaice box'. This is a nursery ground for large numbers of juvenile commercially important flatfish species, such as plaice and sole (Anon 1994). The plaice box is closed to fishing vessels with engine powers above 300 horsepower, and therefore excludes the large whitefish beam trawling fleets from accessing these grounds, and thereby inflicting mortality by discarding juveniles. The plaice box has been in existence for over ten years, but only recently has been closed for the entire year. However, few beneficial effects on the flatfish stocks have been identified, though it is postulated that environmental factors (such as climate change) may be affecting the structure of the fish stocks and overshadowing the beneficial effect resulting from the existence of the 'plaice box'. In addition, the efficacy of the closed area may be compromised by the continued use of the region by small (below 300 horsepower) vessels.

14.4.5.2 *The 'Cod Box'*

The 'cod box' was a temporary closure imposed because the North Sea cod stock was considered by ICES to be outside safe biological limits and at serious risk of collapse (ICES 2001). On 14 February 2001, an area of more than 40,000 square miles of the North Sea, almost a fifth of its entire area, was closed for 75 days to fisheries likely to catch cod (Figure 14.1). The areas closed included some of the main fishing grounds for the international North Sea beam and otter trawl fleets. The aim of the emergency closure was to reduce fishing mortality on spawning cod, but the wider consequences of this closure were not considered at the outset.

Since the beam trawl fleet was allowed to continue fishing during the period of the closure, but could not fish in the closed area (the activities of the beam trawl fleet could be effectively monitored and enforced through the satellite Vessel Monitoring System (VMS)), the fleet sought alternate fishing grounds. Many of the grounds to which the vessels were displaced were not the grounds that the fleet usually fished (Rijnsdorp *et al* 2001). Modelling suggests that the closure led to a different spatial distribution of trawl effort than in normal years, with slightly greater cumulative impacts on the production of sea floor living animals. This effect occurred because the effects of a given trawl impact are relatively greater when habitats are impacted the first time, than when they are fished frequently. Organisms that are less vulnerable to impacts will inhabit an area that is regularly impacted. Some of the trawling effort was displaced to areas that had never been trawled before, and recovery of the seafloor communities in these areas was expected to take decades. Thus protection of spawning cod for 75 days leads to impacts on other ecosystem components that may persist for several decades.

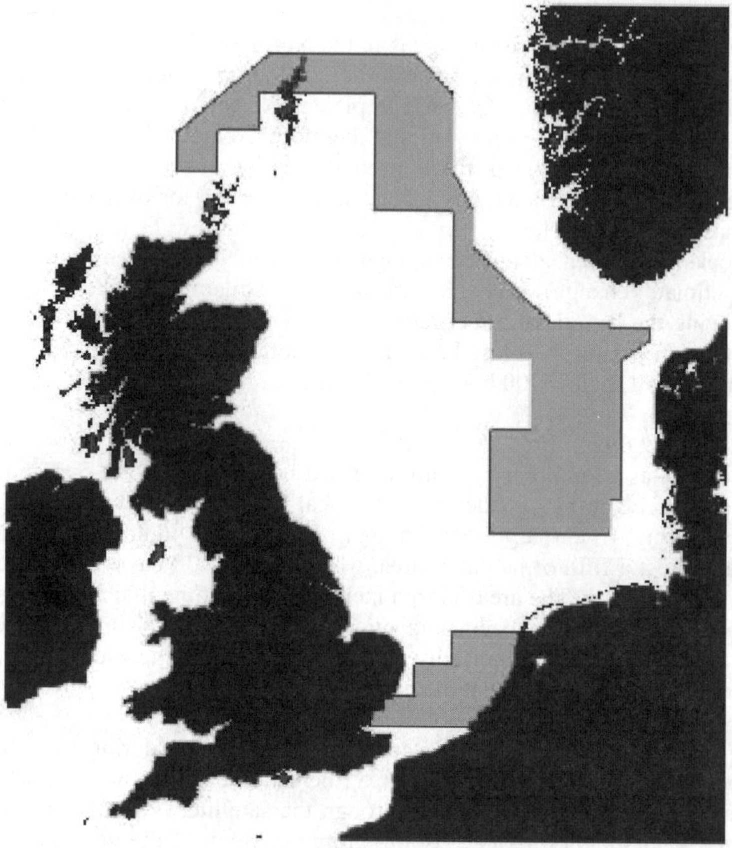

Fig. 14.1. The area of the cod box closure from 14 February - 31 April 2001: stage 1 of the North Sea cod recovery plan.

14.5 Ecosystem-based management

As noted above, to date, fisheries management has focused on providing a sustainable stock of fish. However, various international agreements, including the Convention on Biological Diversity (CBD), now require protection of the ecosystem. Ecosystem management schemes are in their infancy, and considerable effort is being directed at developing appropriate measures for ecosystem status (health) and function.

One example of an ecosystem level management scheme that has been implemented is the sandeel fishery off the east coast of Scotland and NE England. A number of internationally important seabird colonies occur in this area, including the Isle of May and the Farne Islands. The Isle of May alone hosts around 70,000 pairs of breeding seabirds per year. While these birds range far and wide and take a variety of prey outside the breeding season, sandeels are a very important component of the diet of adults and young during breeding. During the breeding season, the birds' foraging is also restricted to sites relatively close to the breeding grounds. In the 1980s, a number of inshore areas were exploited for the first time by industrial fisheries targeting the sandeels. At this time, many spectacular breeding failures by the seabirds. For example, 4,300 pairs of kittiwakes in the Isle of May in 1998 raised less than 200 young (a pair normally raises 1 or 2 chicks from a clutch of 3 eggs). While the evidence of a fishery-seabird interaction is only circumstantial, it was sufficient to prompt a precautionary response. Industrial fishing in the 'sandeel box' (which covers the inshore area from eastern Scotland down to NE England) is closed if the breeding success of kittiwakes in the nearby colonies falls below 0.5 chicks per pair for 3 successive years. The fishery does not reopen until breeding success has been above 0.7 for 3 consecutive years. Thus management of this fishery is based on an ecosystem objective (seabird population health), is precautionary (the link is not yet proven) and uses the kittiwake breeding success as a biological indicator of the ecosystem effects of the fishery.

14.5.1 THE ECOSYSTEM – THE EMERGING CHALLENGE

With the adoption of the Convention on Biological Diversity (1992), managing the environment in an ecologically sustainable manner has shifted from being an option to a legal necessity – sustainability is now the goal of management policy. Given that reproduction and adaptability are fundamental biological attributes, the real challenges for managing the system are two-fold: first, determining the key limits – that is, what are the ways and rates which can be sustained; and second, setting in place policies to obtain society's goals for the marine ecosystem. The latter is a socio-political issue, while the former is very much a scientific issue and may be the greatest challenge facing ecologists in the third millennium.

14.5.2 THE ECOSYSTEM APPROACH TO ENVIRONMENTAL MANAGEMENT

The ecosystem approach has been defined as:

> (T)he comprehensive integrated management of human activities based on best available scientific knowledge about the ecosystem and its dynamics, in order to identify and take action on influences which are critical to the health of the marine ecosystems, thereby achieving sustainable use of ecosystem goods and services and maintenance of ecosystem integrity. (Danish Presidency 2002)

The ecosystem approach is seen as requiring the setting of clear objectives covering ecological, social and economic goals (Koge Conference 2002). From these objectives, it is possible to develop appropriate metrics of each class of objective and to develop management measures that aim to ensure the objectives are met.

14.5.3 THE MARINE SCIENCE REQUIREMENT FOR AN ECOSYSTEM APPROACH TO FISHERIES MANAGEMENT

Science has to contribute to this process in two distinct ways: first, in clearly communicating with all stakeholders about possible configurations of the ecosystem – the **educator** role; and second, the provision of clear advice to managers – the **advisory** role.

The educator role will essentially focus on informing stakeholders of 'What it is feasible to wish for?' This will involve explicitly predicting possible ecosystem scenarios – for example, this many seals in the North Sea will mean a maximum catch of this many salmon and this much cod, and will also mean this many birds. Or, if we can catch this many cod and this many sandeels, we would then expect this many porpoise to be killed each year in our nets and only this many birds to breed. This task requires a major shift in the attitudes and behaviour of scientists. The fisheries science community is not used to communicating directly with society. This approach would also necessitate a massive advance in our predictive capabilities. There are presently no models that can do for the ecosystem what MSVP/4M models do for the ten or so species of fish and their predators modelled. Given the complexity of multispecies systems and the recognised importance of climatic variations in driving marine productivity, ecosystems modelling is a massive undertaking. Also, we need to identify aspects of the ecosystem which can be used as measures of the success of a management scheme in achieving a particular configuration.

After society has been informed by the science through the educator role, it will be expected to express a preference for the state of the ecosystem. This will lead to the setting of clear objectives. Science now has to fulfil its advisor role in advising managers on the steps to be taken to meet the objectives, and in monitoring the system, to continue to provide advice in response to the observed status. This role is similar to that currently fulfilled in fisheries management, although the broader, ecosystem, basis of the management objectives presents greater challenges.

14.6 Marine science and governance models

The most obvious way of governing or managing social activity is through government regulation and enforcement - what is known as 'hierarchical governance'. But there are at least two other ways of governing social activity: 'market governance' and 'participatory governance' (Gray 2003). In the case of fisheries, market governance could mean a system of Individual Transferable Quotas (ITQs), while participatory governance could mean a system of co-management. In practice, we are unlikely to find a pure form of any of these three alternative modes of governance – virtually every kind of fisheries management system is likely to have some elements of all three modes. What differentiates one system from another, therefore, is the proportion of the three elements that they respectively embody.

Most commentators would probably regard current EU fisheries governance as being dominated by the hierarchical model. Science input to the governance process tends to be at a high level in this hierarchy. I would suggest that this is historic. ICES was established as an intergovernmental organisation, and so ICES advice flows to governments (and supra-governments such as the EU). One must acknowledge that certain individual scientists (such as Shepherd 1993:19), and, increasingly, scientific institutions, have made great efforts to inform the industry of the science behind the advice. So would a different governance regime require different science?

14.6.1 IS MARINE SCIENCE USED DIFFERENTLY DEPENDING ON THE GOVERNANCE REGIME?

In many ways, the science required to answer the questions posed at the beginning of this chapter is probably required by whatever governance regime is being used, although the importance of the various questions may vary. However, it appears to me that the adoption of a true 'ecosystem approach', as opposed to merely carrying on as before but paying lip service to the ecosystem, is inherently linked to more participatory governance. This will involve marine science in new roles. For some scientists playing the role of informed advocates, and entering into a debate about possible objectives and management schemes, may be an uncomfortable experience.

What is clear is that participatory management involves much greater 'education/communication' and much greater openness in terms of data exchange, both from scientists to stakeholders and from stakeholders to scientists (for instance, on fishers' behaviour in response to management measures). An 'ecosystem based approach' also requires a much wider range of science than traditional fisheries management, hence the title of this chapter. It is no longer **fisheries** science but **marine** science that needs to inform fisheries management.

14.7 Conclusion

Technological advances are in part responsible for the perilous state of our fisheries. Improvements in vessel design, gear efficiency, gear handling, catch processing and navigation have all helped us to impose a greater mortality on fish stocks than ever before while using fewer vessels and fishers. Technology does not provide a solution to this problem, but the priority action must be to adjust national fleets so that with the available, and still developing, fish catching technology, the level of exploitation reflects the biological reality of fish stock production. This is the challenge for politicians and policy makers.

In realising the potential sustainable yield from fisheries, we must also have regard for the sustainability of the ecosystem, both because it ultimately supports the fish stocks, and because of society's desire to maintain healthy and natural ecosystems. Closed areas are a very efficient means of protecting key habitats or vulnerable species, but as the scales applied to date they are unable to provide an effective mitigation against the direct mortality of fishing. Given that closed areas often, in reality, merely redirect effort into open areas, which then suffer higher levels of fishing, their role in ecosystem management is really restricted to protection of key species or habitat features. Rather than closing areas to fishing, we should seek ways of catching fish that do not do collateral damage to non-target species or ecosystem/habitat features. This will involve a move to more selective and lighter gears and possibly a return to static traps in place of towed gears. This may lead to financial hardships for the fishers in the short term and society should be willing to pay compensation to fishers for playing a stewardship role. Development of such 'ecologically friendly' fishing gears is the challenge for technologists.

We know much about fish biology, but predicting the size of a stock, even a couple of years in advance, remains difficult. A reduction in effort will make year-to-year fluctuations in stock size (and catch and market price) less marked, but it is still important that we develop a better understanding of the relationship between the environment and fish stocks and between fish stocks and the rest of the ecosystem. This is needed to underpin any attempt to provide a holistic ecosystem approach to coastal management. This is the challenge for marine scientists.

For marine scientists, the tools that need to be developed include both better ecosystem models and ways of predicting and incorporating the role of extrinsic drivers, such as climate into our predictions. These may seem to reflect the views expressed by some of the stakeholders. Trade papers and meetings frequently feature the assertion that the perilous state of the fish stocks is the result of external factors – the climate, and seal predation, being two that are widely cited. These extrinsic and ecosystem effects can be important, particularly when stock sizes are low so that there is little buffering capacity. However, focusing on them as the cause of the state of fish stocks ignores two important facts: first, fishing effort/mortality on the stocks is at an all time high; and second, we cannot manage the climate or the ecosystem, but we can manage fisheries. If we wish to rebuild stocks, then it can only be achieved through management measures imposed on fishing. Our poor record over the last 100 years of fisheries science and management, and the current need to incorporate ecosystem issues, including predators and climatic drivers, into our management, argue very powerfully that we need a new approach.

Marine science should fulfil two important roles in this new, participatory approach – an educational and an advisory role. To do so, requires some radical thinking within fisheries governance institutions and a redirection of resources by government and other advisory customers.

Acknowledgements

I would like to thank Tim Gray for inviting me to be part of the workshop and in particular for explaining to me what 'governance' meant! The ideas presented here have benefited from discussions with many colleagues, in particular Tom Catchpole, Simon Greenstreet, Jenny Hatchard, Steve Hall, Odette Paramor, Gerjan Piet, Jake Rice, Leonie Robinson, Stuart Rogers, Catherine Scott, and Simon Thrush. I am particularly grateful to Stuart Rogers, Odette Paramor and Catherine Scott who read a draft of the manuscript. This work was in part supported by the European Commission through study contract Q5RS-2001—1685 *European Fisheries Ecosystem Plans: Northern Seas.*

References

Anderson, KP and Ursin, E (1977) 'A multispecies extension to the Beverton and Holt theory of fishing, with accounts of phosphorus circulation and primary production' *Meddelelser Fra Danmarks Fiskeri-og Havundersøgelser* N.S. **7**:319-435

Anon (1994) *Report of the Study Group on the North Sea Plaice Box*, Copenhagen, ICES

Anon (1996) *Manual of Methods of Measuring the Selectivity of Towed Fishing Gears,* Copenhagen, ICES

Danish Presidency (2002) 'Towards a strategy to protect and conserve the marine environment' Danish President's concluding remarks at the Stakeholder Conference *Toward a strategy to protect and conserve the marine environment*, Koge Denmark

Evans, SM, Hunter, JE, Elizal, and Wahju, RI (1994) 'Composition and fate of the catch and bycatch in the Farne Deep (North Sea) Nephrops fishery' *ICES Journal of Marine Science* 51:155-168

FAO [Food and Agriculture Organisation] (1998) *World Review of Fisheries and Aquaculture*, Rome, FAO

Gray, T (2003) Theorising about fisheries governance' Paper given at the Fisheries Governance Workshop, Newcastle University, September 2003

Greenstreet, SPR, Bryant, AD, Broekhuizen, N, Hall, SJ and Heath, MR (1997) 'Seasonal variation in the consumption of food by fish in the North Sea and implications for food web dynamics' *ICES Journal of Marine Science* 54(2):243-266

ICES [International Council for the Exploration of the Sea] (2001) *Report of the ICES Advisory Committee on Fishery Management*, Copenhagen, ICES

ICES [International Council for the Exploration of the Sea] (2002a) *Report of the Workshop on MSVPA in the North Sea*, Copenhagen, ICES

ICES [International Council for the Exploration of the Sea] (2002b) *Report of the ICES Advisory Committee on Fishery Management,* Copenhagen, ICES

ICES [International Council for the Exploration of the Sea] (2003) *Report of the Working Group on the Ecosystem Effects of Fishing Activities,* Copenhagen, ICES

Køge Conference (2002) Conclusions from the conference working session on an ecosystem approach to assessment and management in the Stakeholder Conference *Toward a strategy to protect and conserve the marine environment*, Køge, Denmark: Danish Presidency

Pope, JG (1991) 'The ICES Multispecies Assessment Working Group: evolution, insights, and future problems' *ICES Mar. Sci. Symp.* 193:22-33

Revill, A (1996) *The UK (East Coast) brown shrimp fishing fleet. Inventory of vessels and gear and analysis of effort,* Ijmuiden, RIVO

Rijnsdorp, AD, Piet, GJ and Poos, JJ (2001) 'Effort allocation of the Dutch beam trawl fleet in response to a temporarily closed area in the North Sea' in *ICES Annual Science Conference,* N:01 Oslo, Copenhagen, ICES

Rogers, SI (1996) *A review of closed areas in the United Kingdom Exclusive Economic Zone,* Lowestoft, CEFAS

Shepherd, JG (1993) *Key issues in the conservation of fisheries,* Lowestoft, CEFAS

CHAPTER 15
BRINGING EXPERIENTIAL KNOWLEDGE INTO FISHERIES SCIENCE ADVISORY PROCESSES: LESSONS LEARNED FROM THE CANADIAN EXPERIENCE OF PARTICIPATORY GOVERNANCE

JAKE RICE
Canadian Science Advisory Secretariat, Department of Fisheries and Oceans, 200 Kent Street, Stn. 12S015, Ottawa, Ontario, Canada

Abstract

Canada has made a policy commitment that the science peer review and advisory processes of government departments should be transparent and inclusive of diverse sources of knowledge. During this policy's development, the Canadian Science Advisory Secretariat experimented with many approaches to include fishermen and others with experiential knowledge in the science-based meetings to assess fish stock status and produce harvest advice. Approaches explored included a) "open door", b) institutional representatives, c) invited individuals, d) industry "observers" without full intervention privileges, e) alternating technical meetings of scientists and non-technical meetings with industry. This paper reviews the strengths and weaknesses of each approach.

Among the lessons learned are:

a) Invited individuals with full participation rights has the most strengths and fewest weaknesses.
b) Never designate an individual at a science meeting as a **representative** of an organisation or sector.
c) The presence of media is highly disruptive.
d) Skilled chairs of inclusive meetings are essential (and hard to find)
e) 'Consensus advice' does not mean all participants must agree on a single interpretation of stock status and harvest. It is enough to reach consensus on the risks and the evidence consistent and not consistent with competing interpretations, and let the political process manage the risks.

15.1 Introduction

It is well established that people whose lives are associated with living resources and marine ecosystems acquire substantial knowledge about ecosystem relationships, the status of species in the ecosystem, and the interactions between human activities, such as fishing, and major ecosystem components (Neis and Felt 2000; Murray *et al* this volume; Vodden *et al* this volume). Many other chapters in this book document the potential value of incorporating such information into fisheries management. Such incorporation of knowledge requires not just processes for recording that knowledge, however. It also requires processes for applying that knowledge to the decisions being made and fisheries management plans being developed. This chapter is about the effectiveness of various mechanisms which have been tried by governments to bring that knowledge into the formal scientific advisory processes, as a key step leading to the development of fisheries management strategies and plans.

T. S. Gray (ed.), Participation in Fisheries Governance, 249–268.

The nomenclature used to make reference to this type of knowledge is diverse. Experts make many carefully nuanced distinctions among terms like 'traditional ecological knowledge', 'aboriginal traditional knowledge', 'community knowledge' (Haggan *et al* 2003). I use the collective term 'experiential knowledge' to refer to the broad category of knowledge gained through focused personal experience rather than through designed and controlled experiments or systematic scientific monitoring and data analysis (Stanley and Rice 2003). The term is intended to be broadly inclusive of all types of knowledge gained through experience, and not to differentiate among the heritage of those holding experiential knowledge, or the dynamics of the community in which they live. I do not assume that the experiential knowledge gained is independent of heritage or community context; rather, the processes being discussed for applying that knowledge to decision-making do not differentiate among the various sources of experiential knowledge.

Much of the literature on the use of experiential knowledge in fisheries management focuses on community-based management approaches (Dyer and McGoodwin 1994; Wilson and Delaney this volume; Vodden *et al* this volume). Without judging the relative value of community-based management compared to other approaches, in practice it is currently the exception rather than the rule. Can the benefits of incorporating experiential knowledge in fisheries management be obtained from management systems that are currently more widely used?

The processes used by governmental and international agencies to develop fisheries management strategies and plans are complex and often poorly documented, but typically include both well-structured processes for obtaining advice, consultation, and decision-making; and informal, opaque (not transparent) expressions of politics and power (Parsons 1993; FAO 1997). They can differ substantially according to provisions of national legislation and international treaties. Nonetheless, in essentially the entire developed world, fisheries management and policy bodies receive formal scientific advice from some source (Table 15.1).

These science advisory bodies give science advice a privileged place in government decision-making and policy development. The justification is that science advice is considered to be intrinsically different from most of the other types of input that policy and management experts receive while developing fisheries management plans. Science advice is supposed to be objective, impartial, value-neutral, and empirically-based, whereas 'advice' received from other sources is considered to be to some degree partisan and advocacy-oriented towards the social or economic objectives of the source. The information base for such 'advice' is thought to be selectively filtered by those social and economic objectives, whereas the information base for science advice is considered to be filtered only by professional standards for testing robustness and reliability of results. There is a vigorous debate in the social sciences about the degree to which scientific advice from experts in the natural and physical sciences meets the high standards of objectivity and empiricism (Pickering 1992; Hannigan 1995; Irwin and Michael 2003).

Table 15.1. Examples of the ubiquity of formal science advisory processes associated with various national fisheries management agencies and regional fisheries management organisations (RMFO)

Country / RFMO	*Science advisory body*	*Recipient of advice*
Australia	Fisheries Assessment Group*	Management Advisory Committee
Canada	RAP/ZAP/NAP (Regional, Zonal and National Advisory Processes; coordination by Canadian Science Advisory Secretariat and regional satellite offices*)	Fisheries Resource Conservation Council, Pacific Fisheries Resource Conservation Council, Minister of Fisheries and Oceans
European Union (and member states)	ICES Advisory Committee on Fishery Management and Advisory Committee on Ecosystems	DG Fisheries, Northeast Atlantic Fisheries Commission, International Baltic Sea Fisheries Commission, North Atlantic Salmon Conservation Organisation. (others)
New Zealand	Independent Contractors*	Ministry of Fisheries
United States	Regional Scientific and Statistical Committees, in turn supported by SARC (Northeast Atlantic, Mid-Atlantic), STAR (North Pacific and Pacific), SEDAR (South Atlantic, Gulf of Mexico) committees of the National Marine Fisheries Service	Seven Regional Fisheries Management Councils (Northeast; Mid-Atlantic; South Atlantic; Gulf of Mexico; Caribbean; North Pacific; Pacific
International Commissions for the Conservation of Atlantic Tunas (ICCAT)	Standing Committee on Research and Statistics	ICCAT Commission
North Pacific Anadromous Fish Commission (NPAFC)	Committee on Scientific Research and Statistics	NPAFC Commission
International Pacific Halibut Commission	Scientific Peer Review Committee	Commissioners
Northwest Atlantic Fisheries Organisation (NAFO)	NAFO Scientific Council	NAFO Commission

* Industry Participation permitted

I will not address that debate here, although if science does not meet those standards, then there is little justification for its privileged place in policy formation and management decision-making. For this chapter, however, I will assume that it is possible for science advisors to apply those standards in their work, and advisory processes should strive to meet those standards in their structure, dynamics and outputs. I address the challenge of trying to augment the data, analyses, and information characteristic of contemporary fisheries science with the additional insights and information of holders of experiential knowledge, without losing the goals of objectivity and impartiality that are the basis for the privileged place that science has in decision-making and policy formation. **What can be done to make science advice more inclusive of wider sources of 'knowledge' while still (or finally) deserving its privileged place in governance?** In what follows, I answer this question in the light of my experience of fisheries governance in Canada, where, during the last ten years, five different 'experimental' approaches to the problem of including fishermen's knowledge have been tried.

15.2 The current role of fishermen in science advisory processes

At present, fishermen have very limited roles in the science review and advisory processes used in many regions of the world. Science Advisory Committees of groups such as the Northwest Atlantic Fisheries Organisation (NAFO), the Pacific Salmon Commission, and the International Commissions for the Conservation of Atlantic Tunas (ICCAT) have industry members restricted to observer status. Observers can address the science bodies during their deliberations, but the interventions are scheduled and usually time-limited. The privileges extended to industry observers do not include participating in the give-and-take debate characteristic of challenge-style peer review and advisory meetings, nor in the word-smithing of the consensus advice from the body. The US, Australia, and a number of other counties have Regional Advisory bodies, typically supporting two levels of discussion of stock status and management prior to management decisions. There is, first, a science process (the STAR panels in the US Northeast, SARC Panels in the US Northwest, and SEDAR panels in the US Southeast) where industry observers have very limited intervention privileges. These science advisory processes report to a second-level committee where industry has substantial representation and privileges, and the science advice may or may not form the foundation for their work.

Many of these international and national agencies have begun to acknowledge that fishermen have information and perspectives that can increase the body of information on which science advice can be based, and reduce uncertainties about stock status and trends and impacts of the fishery. Meetings between scientists and industry participants are encouraged prior to the assessment meetings, so industry information can be extracted, consolidated, and used subsequently in the science processes. To my knowledge, the success of these pre-meeting consultations between science and participants in the fishing industry has never been evaluated formally. In fact, it is not even clear what currency would be used to measure ‘success’: the satisfaction of fishermen and scientists with their interactions; the sheer quantity of information which originated with the industry that ended up in the outputs of the science advisory processes; or the reliability of the science advice at the end of the process?

Interestingly, in talking to senior officials of various agencies, one of their key measures of ‘success’ of these two-tiered processes is the degree to which the fishing industry buys into the output of the science advisory process. Their goal is to increase the credibility of the science process, rather than to expand the information input to those processes. If giving industry members an opportunity to input at least narrative information into the science process reduces industry opposition to advice coming out of the science process, then they feel that the efforts have been worthwhile, even if the industry input is not apparent in the science outputs.

The feedback that I have received from both science and industry participants in these two-tiered processes is mixed, however. Uniformly, though, if industry participants do not see the information they contributed to the consultations somehow reflected – or at least acknowledged – in the outputs of the science process, they rapidly become cynical about the separate-and-not-equal role that they are given. They commonly see these two-tiered systems as inherently giving experiential knowledge second-class status. The perception is that where typical fisheries science analyses of surveys, population reconstructions via sequential population analyses (Quinn and Deriso 1999) can be the basis for a parameter value or determination of a trend, information from industry is only referenced when it is consistent with the parameter or trend estimate. If the input from industry does not match the 'science' information, the industry information is either ignored or explained away. This creates stress and distrust between fishermen and scientists.

Moreover, it appears to me that agencies where the role of industry participants in science processes is most tightly constrained, also tend to be agencies where decision-making is most centralised – although again these processes have not been studied formally. These agencies do have consultation processes where the fishing industry is supposed to give input to decision-makers with regard to what management options should be chosen (given the conclusions about stock status and trends). However, when these consultation processes give the industry little real power (and responsibility), then it is a common experience for the industry to try to use their presence in the science process to influence decision-making on management, as well as conclusions about stock status. This increases the tension between the scientists in those meetings and the industry participants, further stressing the entire process.

The issue of fishermen's participation in science advisory processes is particularly divisive in Europe. The International Council for the Exploration of the Sea (ICES) has been discussing with clients of its advice what role, if any, industry members should have at the assessment working group and subsequent advisory committee level. There are deep divisions among member countries, and the positions of agencies which request (and pay for) advice from ICES also differ greatly. Importantly, some countries and some clients have grave reservations about *any* presence of persons associated with the fishing industry in Working Groups or Advisory Committees. The reason given is always that their presence even as silent observers will still politicise the science process, and pressure individual scientists to abandon best science practices in order to avoid angering the industry members who will be following their every word. Some of this debate is summarised in the report of the 5th Dialogue Meeting of ICES from Dublin in 2004 (ICES 2004), where options as artificial as allowing fishing industry members to sit in a different room and watch the science meetings over a video feed were discussed.

Fishermen are not the only stakeholders excluded from a meaningful role in science review and advisory processes. Many environmental groups have well-credentialed scientists who can bring different interpretational ideas to the same data and analyses used in the science review and advisory meetings. These experts may frame different hypotheses, or give credence to different assumptions, even when starting with the same corpus of information. Good science practice embraces many hypotheses as legitimate to test, and should encourage challenging assumptions, so a strong case can be made that these groups should be included fully in the science review and advisory processes. Nonetheless, ENGOs are rarely given full status in the science review and advisory process, again because of concern that they would bring a biased approach to all the discussions – though biased in a different direction than that of the fishing industry.

15.3 Impetus for change in the Canadian science advisory processes

The past decade has seen wider acknowledgement by governments that top-down management does not work as well as stewardship and co-management approaches (Hall-Arber, Hernes *et al*, and Wilson and Delaney, all this volume). In Europe, the Directorate-General for Fisheries (DG Fish) is introducing Regional Advisory Councils (RACs) to give stakeholders a more direct voice in governance, while ensuring the industry input is provided in a structured and transparent manner (EU 2003). In Australia, the Management Advisory Councils, with primarily industry membership, also input directly to the Australian Fisheries Management Authority (Commonwealth of Australia 2003). In Canada, the Fisheries Resource Conservation Council (FRCC) was created in 1993 as the formally designated advisory body to the Minister of Fisheries and Oceans on Atlantic fisheries, composed of a majority of members from the fishing industry, augmented by academics and provincial representatives (FRCC 2004). Two years later, the Pacific Fisheries Resource Conservation Council was created, with a similar makeup and mandate for Pacific salmon fisheries.

In making the management side of governance more inclusive, calls to make the science processes advising the governance systems more inclusive as well have been inescapable. In Canada, a policy of increased inclusiveness was adopted earlier than in most jurisdictions. The collapses of Atlantic cod and other groundfish from the late 1980s to the early 1990s, and widespread closures starting in 1992 with Newfoundland cod (Rice 2002; Rice *et al* 2003) prompted an angry Minister of Fisheries and Oceans to dissolve both the science advisory committee (Canadian Atlantic Fisheries Scientific Advisory Committee - CAFSAC) and the industry advisory board on management (Atlantic Groundfish Advisory Committee - AGAC) (Parsons 1993). AGAC was replaced immediately with the FRCC, whereas the science review and advisory processes were devolved to regional processes, working with general guidelines and a fairly vague mandate (Anon 1994). Assessments were reviewed and conclusions regarding stock status and trends were provided to the FRCC, which was supposed to consult widely with the fishing industry on management options, given the stock status, before formulating management advice to the Minister (behind closed doors). Very quickly, different research centres and regional authorities began to diverge in their approaches to the science review and advisory tasks, and the industry began to use the FRCC consultations to contest the science assessments of stock status, rather than to discuss management options with stock status as a given.

Within a few years, it became clear that greater coordination of the science review and advisory processes was needed to ensure consistency and credibility of the work being presented to the FRCC. The Canadian Stock Assessment Secretariat (CSAS in its earlier form) was created in 1996 with a mandate to coordinate the regionally-based processes. To reverse the trend of using FRCC consultations to contest the science advice, CSAS was also explicitly mandated to make the science processes fully inclusive of academic experts and experiential knowledge, with full participation by persons from the fishing industry as well as environmental organisations.[1]

The mandate to make the science peer review and advisory processes inclusive of more types of knowledge got a boost from a report entitled Science Advice for Government

[1] In 2000, following implementation of Canada's Ocean Act, the mandate of CSAS was expanded to guide and coordinate peer review and provision of advice on all oceans management issues as well as fisheries issues. CSAS then became the Canadian Science Advisory Secretariat

Effectiveness (the 'SAGE Report', CSTA 2000), produced by the Federal Government Council of Science and Technology Advisors. The SAGE Report was the basis for a government-wide policy developed by the Privy Council Office and adopted by the Canadian Cabinet (2000). This policy, which applies to all science advice used in government decision-making, has six fundamental Principles, with associated Guidelines (Table 15.2).

Table 15.2. Extracts of text from Principles and Guidelines from Science Advice for Government Effectiveness (CSTA 2000)

Principle	*Associated Guidelines*
Early Issue Identification: Departments need to anticipate, as early as possible, those issues (representing both challenges and opportunities) for which science advice will be required. A broad base of advice can lead to improvements in the timeliness of issue identification.	• Cast a wide net (consulting internal, external, and international sources); • Support and encourage science and policy staffs to establish linkages with each other and with external and international sources. • Maximise the use of expertise across government departments to identify and address 'horizontal' issues; • [Two other Guidelines].
Inclusiveness: Input should be drawn from a variety of scientific sources and from experts in many disciplines in order to capture the full diversity of scientific schools of thought and opinion so as to enhance the debate and draw in scientific findings, which may not otherwise be considered.	• Science input and advice needs to be sought from a wide range of sources; due weight needs to be given to the 'traditional knowledge' of local peoples; decision makers need to balance the multiple viewpoints received; • While advice from external and international sources needs to be sought regularly, it is especially important to seek such advice in the following situations: o [four listed]; • Decision makers need to be open to both solicited and unsolicited advice from external sources.
Sound Science and Science Advice: The public expects government to employ measures to ensure the quality, integrity, and objectivity of the science and the science advice it utilises, and to ensure that science advice is considered seriously in decision making. Due diligence procedures for assuring quality and reliability, including scientific peer review, need to be built into the science advisory process.	• All advisory processes, including those involving traditional knowledge, need to be subject to due diligence. This should include rigorous internal and external review and assessment of all input, analyses, findings, and recommendations of advisors; • Science advice needs to be supported by research and policy analysis (4 subpoints); • Selection of advisors needs to: o be balanced to reflect the diversity of opinions and to counter potential biases; o include at least some experts from other, not necessarily scientific, disciplines; o [two others]; • Advice providers need to: o clearly distinguish scientific fact and judgement from their personal views; o [two others]. [Further guidelines giving responsibilities of

Principle	*Associated Guidelines*
	Departments and Decision-makers are listed]
Uncertainty and Risk: Science in public policy always contains some uncertainty and often a high degree of uncertainty which must be assessed, communicated, and managed. As such, it is important to consider adopting a risk management approach	[Four guidelines]
Transparency: Democratic governments are expected to employ decision-making processes that are transparent and open to stakeholders. Transparency implies a clear articulation of how decisions are reached, policies are presented in open forums, and the public has access to the findings and advice of scientists as early as possible…decision-makers need to treat the science advisory function as an integral part of the management process	• Departments need to allow scientists freedom to pursue a broad base of inquiry and undertake widespread and thoughtful discussions; • Departments need to publish and disseminate widely all scientific evidence and analysis (other than proprietary information) underlying policy decisions, and show how the science was taken into account in policy formulation; • Decision makers need to explain how the advice they received was used and why the ultimate decision was made; • [Three other guidelines].
Review: 1) subsequent review of science-based decisions to determine whether recent advances in knowledge impact the science and science advice used to inform the decision, and 2) evaluation of the decision making process.	• Departments should capture best practices that emerge from the advisory process and feed these into their guidelines for use of science advice in the future; • [Three other guidelines].

The prominence of Inclusiveness and Transparency as two of the six pillars of science advice in government empowered the CSAS to push aggressively on a programme of assuring the presence of individuals with experiential knowledge in *all* meetings that were to produce science advice to fisheries management or policy.

This institutional mandate to proceed was essential to implementing change, because there was entrenched reluctance in some quarters to bring the fishing industry and environmental groups into the science processes. Reasons for this reluctance were diverse. Parts of the science community were concerned that the presence of fishermen would lower the technical quality of the review, and allow weaker science to be the basis for advice on management. Parts of management and policy sectors were concerned that too much transparency would undermine the effectiveness of the science advice, through revealing how many sources of uncertainty really were present. Throughout all sectors of the department were pockets of suspicion that industry and environmental participants would not respect the rules of objectivity and non-partisan consensus-building, and use the science forum to push their policy agendas. At a practical level, the science review and advisory processes in the four Atlantic Regions (Newfoundland, Scotia-Fundy, Gulf, and Quebec) had evolved in different directions since CAFSAC had been dissolved, and in the Central and Arctic Region and Pacific Region, their processes had never been under the guidance of CAFSAC. These Regional differences meant that a common nation-wide policy on inclusiveness ran into different institutional impediments – some formal and many informal – in different parts of the country, so the path to planning and implementation of consistency was bumpy.

Notwithstanding the reservations in various parts of the department, between 1996 and the early years of this decade, the Department for Oceans (DFO) made its science peer review and Advisory Processes inclusive of people with experiential knowledge in all their advisory tasks at National, Zonal, and Regional scales (NAPs, ZAPs, and RAPs). In all cases, the changes were not just allowing the **presence** of these people at the meeting, but giving them full **membership** in all steps from presenting original material, through challenge and debate of other presentations, to helping to formulate the consensus advice.

In making this fundamental change, we experimented with many different ways to bring those with experiential knowledge into the peer review and advisory processes. In retrospect, these 'experiments' should have been conducted much more formally than they were. When we began, those of us driving the change were unaware of how precedent-breaking our 'experiments' were, and failed to appreciate fully how much more could have been learned from pre-identified performance measures and structured evaluation of the results of each meeting. Rather, we were primarily just trying to make a real-world change successful, and had to take an opportunistic approach to each challenge. We had no control over who from DFO participated in each meeting, and often limited control over who attended from outside the Department. Hence, the degree of shared commitment to make the new processes work was an uncontrolled variable across all our meetings. Moreover, replication and scientific controls of a review and advisory meeting were nearly impossible, so the evaluation methods in which we had been trained were inappropriate.

Despite these short-comings in our 'experimental' approach, over a few years we converged on a number of generalisations from our experience. These have proved a sound basis for codifying the 'rules' of participation in our various types of meetings (Anon 2004). These 'rules' have the flexibility to deal with a variety of types of meetings, while ensuring that all the DFO science review and advisory processes meet the SAGE guidelines for inclusiveness and transparency. The rest of this paper presents the five different models for inclusiveness that we explored, the strengths and weaknesses we identified (particularly weaknesses considered nearly fatal), and what overall lessons we have learned.

15.4 Five approaches explored

In this section, the five different experimental approaches that were tried are explained. They are listed in rank order of the degree of participation by fishermen that they incorporated, from the highest degree to the lowest.

15.4.1 OPEN DOOR

This approach incorporates the highest degree of fishermen's participation, and it may be characterised as 'inclusive participation'. Here, the peer review and advisory meeting is publicised widely through industry and media outlets. Participation of specific individuals from industry, academia, and environmental groups may be encouraged, but all who show up are given full privileges of participation. Anyone can make presentations or challenge other presenters, and all participate in what becomes the consensus advice.

15.4.2 ORGANISATIONAL REPRESENTATIVES

This approach incorporates the next highest degree of fishermen's participation, and it may be characterised as 'representative participation'. Here, for each peer review and advisory meeting, a list of relevant industry organisations, environmental groups, and community associations is assembled. Some or all of these groups are asked to send one or more representatives to the meeting. These representatives have full participation rights in all stages of the meeting; presentation of material, challenge of other presenters, and development of consensus advice. The meeting is closed to those not sent by any organisation which was invited to send one or more representatives. Informal Chatham House rules apply, such that outside the meeting, the substance of the deliberations can be discussed but interventions may not be attributed to individuals or sectors.

15.4.3 INDIVIDUAL INVITEES

This approach incorporates a medium degree of fishermen's participation, and it may be characterised as 'invited participation'. Here, the same list of groups is assembled as in Approach 2, and individuals known to be respected or influential within each group are listed. This list of individuals is often augmented by names of individuals known to be respected among their peers, even if they are not part of formal associations. From this list, CSAS (or the Regional Review and Advisory Offices) picks a slate of individuals who are invited to the meeting. Selection is usually made in consultation with the scientists working on the stock and clients of the advice in Fisheries Management, and often options are discussed with key industry organisations. The selection aims for balance among perspectives (fishermen and environmentalists) and among fleet sectors and harvesters and processors in complex fisheries. All who are invited have full rights of participation, as in the preceding approaches. The meeting is closed to those who have not been invited, and the same informal Chatham House rules are supposed to be followed.

15.4.4 SCIENTISTS MEETING WITH INDUSTRY OBSERVERS

This approach incorporates a lower degree of fishermen's participation, and it may be characterised as 'observer participation'. Here, scientists have a standard assessment meeting with working papers, technical review, and development of consensus advice, in which only scientists are participants. Industry is allowed to be present, usually with selected individuals invited to be observers, or selected associations asked to send an observer. The observers usually are allowed to address the meeting at particular points on the agenda, and sometimes even may be allowed to pose questions of scientists presenting working papers, once the review by the scientists is largely completed. The scientist-to-scientist challenge and response has higher stature in the meeting, and observers have no rights during the development of the scientific advice. This model was never an intended goal of the push to make the meetings inclusive, but for regions sharing trans-boundary stocks with the US, this model was both familiar and strongly promoted by colleagues from the US.

15.4.5 ALTERNATING MEETINGS

This approach incorporates the lowest degree of fishermen's participation, and it may be characterised as 'consultee participation'. Here, non-technical meetings with industry and technical meetings of scientists are held in alternating sequence. This approach was favoured initially by many scientists and departmental administrators, and is similar to the 'sandwich approach' attempted by ICES in 2004. A small group of scientists go to a major fishing centre and have an open meeting with fishermen. The scientists summarise their research results, invite comment from industry participants, and ask the fishermen for their views on the state of the resources. The scientists then go back to their laboratory, conduct analyses, and assemble the general contents of the stock assessment. The scientists then return to the fishing community, discuss their results with the fishermen, and get the fishermen's views on their conclusions. The scientists then revise their assessment as they feel appropriate, hold their peer review and advisory meeting in a scientist-only setting, and prepare a draft of assessment results and conclusions. The results of the assessment are usually discussed one final time at a meeting in the fishing community, with the intent of gaining feedback that will guide improving the clarity of conclusions and draft advice.

15.5 Strengths and weaknesses of each approach

In this section, the five different approaches are evaluated for their success or failure.

15.5.1 OPEN DOOR

15.5.1.1 *Strengths*

This approach ranks as the highest on inclusiveness and transparency. Industry feels highly empowered in this approach, and most like it. They feel it offers them the greatest degree of democratisation of the entire management process, and gives them the fullest opportunity to input to evaluation of stock status and associated scientific advice. In several parts of Canada, the fishing industry is highly fractionated into diverse gear sectors, geographic subdivisions, and linguistic profiles. An 'Open Door' policy means that all industry sectors are present and may participate, no matter how complex the fishery. In this approach, a diversity of perspectives are sometimes presented, so all participants are challenged to defend their contributions to the assessment. With most scientists untrained in conducting rigorous but respectful 'peer review' of experiential knowledge, this approach usually results in different industry sectors cross-examining each others' contributions, while the scientists refrain from appearing to doubt or oppose statements by any single industry sector. However, the sense of ownership of the advisory products by industry is actually highly variable in this approach. It can be very high or very low, depending on the dynamics of the meeting. If the interactions of all the industry sectors, environmentalists, and scientists can be kept constructive, industry confidence in the meeting products can be high.

15.5.1.2 *Weaknesses*

Within one or two assessment cycles, these meetings become unworkably long and large. Very large rooms with layouts not conducive to dialogue are necessary. It becomes nearly impossible to pursue complex topics in depth because dozens of people can be waiting on the speakers' list. Hence there can be long gaps between related interventions, and several different points of discussion can be in play at once. With large numbers of people potentially speaking on each topic, it becomes very difficult to establish the direction in which a real consensus may be found. Individuals can and do orchestrate their interventions to make it appear that a particular point of view has gained far more momentum that it really has. This can be sorted out eventually, when it comes time to establish the point on which consensus has been reached, but only with investment of a very large amount of time. Moreover, notwithstanding full explanations of the 'rules of engagement' for science peer review and advisory meetings, feuds between different sectors of the industry, between industry and environmental groups, or dissatisfactions of anyone with the science components of the assessment or even past departmental actions, all tend to be raised and sometimes played out at the table. Again, this can be managed, but only by very strong and experienced meeting Chairs, who pay a heavy toll for the role they had accepted. Over time it has become very difficult to find individuals adequately knowledgeable of fisheries science who are willing to chair meetings using an 'Open Door' approach.

Furthermore, if the industry is not only divided, but sectors are unequal in size and organisational support, then the format ends up being biased against independent and weakly organised groups. Many fishermen are not comfortable speaking out in large meetings, nor in meetings organised to address science issues, and the combination of these circumstances means that the true amount of original contribution from the industry may be much less than inferred from the number of people present at the meeting.

Also, these meetings can become very expensive if any travel assistance is offered to participants. To be perceived as fair, if support is offered to any participants, all participants should receive the same support. Offering support to no-one immediately biases the 'Open Door' meetings towards the wealthiest sectors and the sectors or interest groups living closest to the meeting venue. This consistently prompts strong protests from those who feel they are being *de facto* disenfranchised from what is supposed to be a science peer review and advisory meeting, not a consultation. Keeping this type of meeting within a budget is very difficult.

Finally, with a highly diverse participation, particularly if there are internal antagonisms among sectors, any achievable consensus is confined to high-level and abstract conclusions. Attempts to move the conclusions and advice to specific points tend to prompt competing sectors to wish to add riders that are unacceptable to other sectors. Uncertainties in science data sources, analyses, and modelling results tend to be emphasised by either industry or environmentalist participants (depending on the direction of the uncertainty) as reasons why no strong conclusions on stock status can be drawn. As a result, the scientific advice from such meetings may be of comparatively little help in supporting hard management decisions, and additional science input is required, usually through informal and sometimes undocumented sources. Finally, once a meeting has been conducted in an 'Open Door' format, it is very difficult to move back to more restrictive formats. Industry feels that their rights are being withdrawn unilaterally, even if they have had the opportunity to participate for a very short time. This approach should only be tried, therefore, if one is ready to live with it for a long time.

15.5.2 ORGANISATIONAL REPRESENTATIVES

15.5.2.1 *Strengths*

This approach also ranks high on inclusiveness and transparency, and where the fishing industry and environmental groups are well organised, the associations feel particularly empowered. 'Organisational Representatives' also facilitates having the experience of all the different sectors presented, and usually someone relatively at ease with meeting formats will contribute the experiential knowledge. Both fishing industry and environmental groups are good at choosing representatives who present their knowledge and perspectives articulately. This results in a good diversity of perspectives being presented, with discussions sometimes becoming quite technical. This trend builds over time, because associations tend to send the same representatives to numerous meetings, so they learn how the meeting dynamics work, and develop the same histories of interactions that have long characterised the traditional science-only assessment review and advisory meetings.

These meetings can be cost-effective logistically and tractable to run, because numbers can be fairly closely controlled. Moreover, if the representatives feel they have been effective in having their experiences and perspectives captured in the science conclusions, there is a fair sense of ownership of the meeting products spread throughout the industry. On the other hand, if one group feels that its interests lost out to those of another group, even if the choice was strictly on objective factual grounds (hypotheses can be refuted, and data sets or analyses shown to be fatally flawed), then the whole sector may reject the legitimacy of the conclusions and advice.

15.5.2.2 *Weaknesses*

Over time, meetings applying this approach usually lose any semblance of pursuing objective, non-partisan science. Just by being named a 'representative', most participants abandon any pretence of objectivity and impartiality. They are there to **represent** the interests of their organisation, and take that role more seriously than helping a science peer review and advisory process achieve its objectives. When individuals are representatives of particular sectors, they commonly come to review and advisory meetings with clear organisational guidance that they cannot agree to any conclusions which are counter to the interests of the organisation. Industry sectors may focus on impeding consensus conclusions which would have detrimental social and economic consequences for their sectors, but representatives of environmental groups can resist just as doggedly any conclusions contrary to policies that their organisation has adopted. In both cases, the factual and analytical evidence for a particular conclusion may be compelling, but their responsibilities as a representative of their organisation commonly take priority. As a result, meetings are frequently characterised by substantial confrontation between competing industry sectors, between industry and environmental groups, and between any of the parties and government scientists, over issues of policy, not science. Strong meeting chairs can reduce this tendency, but in all of DFO there is only a handful of scientists with the technical skills and meeting skills to run such meetings successfully.

15.5.3 INDIVIDUAL INVITEES

15.5.3.1 *Strengths*

This approach allows for good coverage of all perspectives, through care in the selection of the slate of invitees. It thus ranks high on inclusiveness and transparency, again as long as the slate of invitees is balanced and broad. It is possible to get a mix of individuals who understand and respect what a science review and advisory process is, and come prepared to make it succeed in its goals, rather than just to promote sectoral interests. The meetings can be kept at a tractable size and cost-effective, by distributing the range of invited participants carefully across organisations and functional communities according to common interests and experiences. Compared to the two preceding approaches it is also often easier for a meeting Chair to keep the meeting focused on the agenda, and on peer review and integration of all the types of knowledge into science advice.

Individuals who are invited feel an obligation to participate actively, because they understand that there are limited places at the table and they have been selected as having particularly valuable experiential knowledge to contribute, and perspectives to share with other participants. Moreover, as long as they are confident of individual anonymity, industry members can and do make candid interventions which help to establish actual stock status and true activities of the fisheries on the water, even when their interventions are not in the short-term interests of their industry. Invitees from environmentalist perspectives sometimes offer interpretations or perspectives somewhat at variance with the policies of their organisations. It is often possible to reach consensus on science issues of substance and of sufficient specificity to guide management. Individuals acting as individuals often do show common sense, and concede points made or lost on the strength of the evidence (experiential and scientific). Participants often show ownership of the meeting products, to the point where they may explain and defend the advice to their own sector.

15.5.3.2 *Weaknesses*

Meetings in this format are always vulnerable to accusations that the secretariat picked sympathetic external participants, who were known to be predisposed to agree with government experts and policies. Significant effort is, therefore, required to get good and balanced participation. To maintain the credibility of meetings by invitation only, it is necessary to ensure good coverage of even the smaller industry and interest group sectors. This can make it costly to run such meetings if the industry or public interest groups are highly fragmented.

Similarly, if even a few key invitees fail to show up, the credibility of the whole meeting can be placed at risk. In practice this has been more of a problem with invitees from environmental groups than from fishing industry sectors. There has even been speculation that some groups practice this strategically, by accepting invitations and then not showing up at meetings where they expect the evidence will support conclusions that run counter to their policy interests. We also hear reports that over time, individuals from industry who are frequently invited to peer review and advisory meetings because they contribute constructively to the process are pressured by their industry sector to be ‘unavailable’ so alternates who may be more confrontational may have to be invited.

15.5.4 SCIENTISTS MEETING WITH INDUSTRY OBSERVERS

15.5.4.1 *Strengths*

In the Canadian context, many participants from science, industry, and environmental groups were familiar with this format, because of participation in US-run meetings on trans-boundary stocks,. That familiarity usually outweighed concerns of the few people who were uncomfortable with even having industry present in the room for fear that their presence would deter free debate of sensitive issues, such as the quality of catch data. If the industry observers are at least allowed to speak to agenda items or are give a period for asking questions, some degree of inclusvieness is achieved. Transparency is high with this approach, because industry gets to see directly what uses the science meeting made of the information that they have contributed. This feeds back on the scientists to be more candid in any pre-meetings with industry, with regard to lines of reasoning or information sources which industry may support but which the scientist knows will be rejected at the review and advisory meeting. Costs of time (for everyone) and logistics are moderate and controllable, because there is only a single meeting.

15.5.4.2 *Weaknesses*

In practice, this format gives very little chance for meaningful contributions of knowledge from industry. Occasionally industry participants, or science contractors working for the industry, may make polished presentations that look and sound much like the presentations of the scientists themselves. Only in these cases is it likely that the industry 'interventions' will actually carry weight into the fuller peer review process and the formulation of advice. Otherwise, much of the experiential knowledge of the industry (and interpretational hypotheses of environmentalists) is lost by the time the advice is finalised. Moreover, there is a tendency towards grandstanding on both sides during meetings in this format. Compared to science-only meetings, some scientists lean towards more polished presentations, which address the spectators but lack the grist for the rigorous peer review. This frustrates the subsequent process, which has to dig deeply to find the hard-core science for the necessary peer review.

Comparably, if the audience of observers is large, some individuals from industry tend to use their intervention time to gain stature with their own peers, or define 'battle-lines' with the scientists, possibly to position themselves to have greater leverage during the subsequent consultation phase. In the end, again, there is very little sense of ownership by industry of the final product. They may understand the product somewhat better than had they not observed the meeting, but it is still a product of scientists, not a joint product, and there is no assurance that they will see any of their own experiential knowledge in the advice which is produced by the meeting.

15.5.5 ALTERNATING MEETINGS

15.5.5.1 *Strengths*

Scientists are very comfortable with this format. They get to be highly technical with each other, and only a subset who interact well with industry need to attend the industry meetings. Clients of the advice in management and policy are also comfortable with this format. They see industry given ample chance to input to the science process, yet the science process is sheltered from the reality or perception of pressure from partisan directions. Industry gets repeated exposures to the assessment as it develops, and, from meeting to meeting, can pursue its aim of developing support for its own perspective(s). Those who attend all meetings may end up with an in-depth understanding of the assessment.

15.5.5.2 *Weaknesses*
This format is very demanding of time for everyone and costly for industry. If industry participants are actively fishing, or members of environmental groups have other jobs than full-time advocacy on fisheries conservation issues, then every meeting is time away from their source of income. The more that sequenced meetings are used to bring industry along with the assessment, the greater this loss of income and cost for travel and meeting logistics mounts up.

Moreover, the demands of multiple meetings often result in a lack of continuity of industry and environmental group participants from one meeting to the next. This in turn means recovering ground at every meeting, which is perceived as inefficient by the regular participants. Also, the presence or absence of even a few individuals might lead to very different priorities being expressed from meeting to meeting on the industry side, or different importance given to various sources of information on the science side. Consequently, each side may perceive the other as flip-flopping on views and treatment of information, or as unresponsive to past input.

This approach is also weak on real transparency and inclusiveness. Many scientists learned to 'spin' their presentations to appeal to industry, knowing full well that at the science-only review and advisory meeting the scientists would attach different interpretations and weights to the information available. Likewise, industry had no opportunity to see how little or much attention the ultimate science-only meeting gave to the information which they have contributed during the joint meetings, and rarely received justifications when their contributions were not the key determinants of the advice. As a result the fishing industry frequently had little sense of ownership of the final product.

15.6 Lessons learned

The DFO has adopted the third approach of inviting specific individuals as its standard now. This approach has many valuable strengths and more importantly, its key weakness is one about which we can do something. Constant vigilance is necessary to provide a balanced slate of invitees, and to select individuals credible to broad constituencies, not just easy to deal with from within government. This has not proved easy, but it is easier than dealing with the weaknesses of the other options.

We now carefully avoid using the word 'representative' when discussing participation at any science review and advisory meeting. The word itself seems to impel people to take responsibility for protecting the interests of the group they 'represent', and to place that role ahead of any collective interest of objective presentation of the facts. This is anathema to what a science advisory meeting is trying to achieve. However, it has proved hard to avoid this approach completely: in particularly high-profile issues, leaders of industry organisations do lobby the most senior levels of government for invitations to the science review and advisory meetings, and they are rarely placated unless someone very close to a 'representative' ends up with an invitation. Officers of industry unions and executives of environmental organisations are particularly at risk of being a disruptive influence in science meetings. This is far from universal, and there are very responsible individuals in both types of groups – particularly if they feel that they have a chance of winning favourable conclusions on the merit of arguments presented. Nonetheless, as a generalisation, the risk of 'representative advocacy' in a science meeting is higher with officers than with respected individuals from the ranks of either type of organisation. Moreover, even when union officers or environmental group executive members are showing appropriate objectivity in their interventions, their presence alone introduces an undesirable dynamic. The other fishermen usually defer to their union officers, and individual members of environmental groups to their organisations' officers. Once the official has spoken, other individuals from either type of group will rarely offer contrasting experiences or perspectives. This greatly diminishes the potential contribution that experiential knowledge can make to the review and advisory process; presenting only one experience to an assessment group is rarely any better than presenting only one analysis.

The presence of the media in the room is highly disruptive. Many participants are unwilling to speak at all with the media present, while others speak to the 'public' rather than address the agenda item seriously. The press is categorically not allowed in our review and advisory meetings now. However, immediately at the conclusion of any meeting where there is interest from the press, the meeting chair and other individuals nominated by the meeting as a whole will brief the press on the meeting's conclusions. Even if not selected to speak for the meeting, any external participant can stay for the press briefing and speak to the media on the conditions that they make clear they are speaking as individuals, and respect the informal Chatham House rules by not attributing comments to other individuals at the meeting.

Skilled and experienced Chairs for these inclusive meetings are essential but rare. Good chairs need significant technical knowledge, good people skills, and a broad perspective on issues. They also need a clear understanding of what will be done with the products of each meeting, to guide the meeting to produce advice that the clients actually can use in development management and policy, and not just 'advice' that the meeting participants could readily agree on.

The Chairs have to be empowered to not just explain why the meeting needs all participants to be objective and impartial in their interventions, but to enforce those standards. Our practice is to issue one warning to an individual for comments which are either clearly partisan and biased, or disrespectful of other participants or sectors. A second transgression at the same meeting results not just in ejection from the meeting, but has more lasting consequences. External individuals who are ejected once lose the right to any future invitations to any review or advisory meetings. Departmental staff who fail to show respect for external participants – or each other – see the issue taken to the Director of their institute. This power has rarely been used, but even a few instances have conveyed clearly that science review and advisory meetings are serious in maintaining their objectivity and impartiality, while at the same time determined about bringing experiential knowledge into the process.

We have come to make a distinction between 'Facilitators' and 'Chairs'. Facilitators were tried in several types of meeting, but seemed to interpret their job as keeping everyone happy and engaged, and finding a place for everyone's opinion in the meeting conclusions. From all sides, there was agreement that this approach was a major liability in a science challenge-format review, whatever form was used for contributing experiential knowledge. Some ideas and analyses are just plain wrong, and **should** be discarded, and an effective chair has to ensure that happens. As a corollary, scientists need to be educated in how to conduct 'peer review' of experiential knowledge. Standards do exist, and scientists can learn them, but not without some retraining.

It is usually necessary to explain what we are seeking in 'consensus advice'. Consensus does not require universal agreement on one interpretation and one option as superior to all others. Useful consensus is agreement among all participants that:

a. There is sufficient evidence to render some interpretations implausible and some options not viable. These are rejected and the evidence for rejecting such interpretations and options is documented;
b. The available evidence (including experiential knowledge) cannot provide a conclusive *scientific* basis to consider any one interpretation or option 'best'. For each retained interpretation/option we seek agreement on the key evidence consistent with it, and the key evidence that is NOT consistent with it. Industry, environmentalists, and scientists can all agree on that type of consensus, and even often on the weight of evidence.

This information is enough for policy and management to take the next step. The inclusive science review and advisory process has obtained consensus on the descriptions of the risk involved in each option ('probability' from the weight of evidence associated with each option; possible 'consequences' through dialogue). Management and policy then manage the risks, which is their mandate.

The single most important lessons, however, are that it is possible to have inclusive science peer review and advisory meetings on a wide range of issues, and, if done well, that the improved advisory products justify the efforts. Consensus advice from inclusive meetings can be clear, restrictive enough to be useful to managers and policy-setters, and widely supported by diverse participants from the meeting. If the meeting dynamics are constructive, all participants share a sense of ownership in the meeting products, which has many subsequent benefits later in the process of forming policy and management plans. We have found that if the meeting format is correct, discussions in inclusive settings can be objective and non-partisan. Narrative information can be effective from the start, and, over a series of meetings, fishermen readily learn to package their annual experiences in ways that are clear and have impact. Highly technical scientific issues can still be treated professionally in these inclusive formats. It is true that fishermen may not be prepared to participate in every technical debate, but the reality is that in a meeting of a couple of dozen scientists, often only a handful are engaged in debates about some of the more obscure statistical and modelling issues which arise.

15.7 Conclusion

The first attempts at greater inclusiveness in fisheries science advisory meetings are likely to be disappointing to participants from most perspectives. However, professionals in fisheries science tend to forget that the core members of their review and advisory meetings often have been working together for two decades or more. That history contributes to their effectiveness as a group. Even scientifically well-credentialed newcomers to these meetings often are lost for the first meeting or two, and only slowly assimilate into the dynamics of the group. We found that within two or three meetings, invited industry participants became very skilled in presenting their information effectively, and asking questions of the scientists which moved the meeting forward for everyone. The path is not easy, especially if there is a history of antagonism or distrust between industry and government experts, among industry sectors, or between industry and environmentalists. Not every series of meetings has made progress at the same – or even encouraging – rates. However, from the Canadian experience, we conclude that once a commitment is made to make the review and advisory processes inclusive of experiential knowledge, the benefits justify the efforts. If everyone tries to make the meetings work as review and advisory processes, rather than as another setting to argue and lobby, inclusive approaches become the norm for *all* fisheries scientific peer review and advisory meetings, not just the handful of cases when the preconditions for success were mostly met already.

References

Anon (1994) *Proposal for a Regional Assessment Process*, Internal Report for National Science Directors Committee, unpublished

Anon (2004) 'External Participation in the RAP Process' DFO, Government of Canada, www.dfo-mpo.gc.ca

Canadian Cabinet (2000) *A Framework for S&T Advice in Government Decision Making*, Government response to the SAGE report, www.csta-cest.ca

Commonwealth of Australia (2003) *Looking to the Future: A Review of Commonwealth Fisheries Policy*, Department of Agriculture, Fisheries and Forestry, Canberra

CSTA [Council of Science and Technology Advisors] (2000) *Science Advice for Government Effectiveness*, Report of the Council of Science and Technology Advisors, Government of Canada, www.csta-cest.ca

Dyer, CL and McGoodwin, JR (eds) (1994) *Folk Management in the World's Fisheries: Lessons for Modern Fisheries Management*, Boulder, CO, University of Colorado Press

EU [European Union] (2003) 'Fisheries Reform: Commission Proposes Regional Advisory Councils to give Stakeholders a Bigger Say', European Commission Press Release, October

FAO [Food and Agriculture Organisation] (1997) *Fisheries Management*, FAO Technical Guidelines for Responsible Fisheries #4, Rome, FAO

FRCC [Fisheries Resource Conservation Council] (2004) 'What is the FRCC?', www.frcc.ca, Ottawa

Haggan, N, Brignall, C and Wood, L (eds) (2003) *Putting Fishers' Knowledge to* Work, *Fisheries Center Research* Reports 11(1), Vancouver BC, University of British Columbia Press

Hannigan, JA (1995) *Environmental Sociology: a Social Constructionist Perspective*, London, Routledge

ICES [International Council for the Exploration of the Sea] (2004) *Report of the 13th Dialogue Meeting: Advancing Scientific Advice for an Ecosystem Approach to Management*, ICES Cooperative Research Report #267

Irwin, A and Michael, M (2003) *Science, Social Theory and Public Knowledge*, Maidenhead, Open University Press

Neis, B and Felt, L (eds) (2000) *Finding our Sea Legs: Linking Fishery People and their Knowledge with Science and Management*, Memorial University, St. John's, Newfoundland, ISER Books

Parsons, LS (1993) 'Management of Marine Fisheries in Canada' *Canadian Bulletin of Fisheries and Aquatic Sciences* 225, Ottawa, National Research Council of Canada and Department of Fisheries and Oceans

Pickering, A (ed.) (1992) *Science as Practice and Culture*, Chicago, University of Chicago Press

Quinn, TJ and Deriso, RJ (1999) *Quantitative Fish Dynamics*, Oxford, Oxford University Press

Rice, JC (2002) 'Changes to the Large Marine Ecosystem of the Newfoundland – Labrador Shelf' in K Sherman and H-R Skjoldal (eds) *The Changing States of Large Marine Ecosystems of the North Atlantic and Global Environmental Trends*, Oxford, Blackwell Scientific Press

Rice, JC, Shelton, PA, Rivard, D, Chouinard, GA and Fréchet, A (2003) 'Recovering Canadian Atlantic cod stocks: the shape of things to come?' *ICES* CM 2003/U:06

Stanley, RD and Rice, JC (2003) 'Participatory Research in the British Columbia Groundfish Fishery' *Fisheries Center Research Reports* 11(1):44-56, Vancouver BC, University of British Columbia Press

CHAPTER 16
LOCAL ECOLOGICAL KNOWLEDGE, SCIENCE, PARTICIPATION AND FISHERIES GOVERNANCE IN NEWFOUNDLAND AND LABRADOR: A COMPLEX, CONTESTED AND CHANGING RELATIONSHIP

GRANT MURRAY, DEAN BAVINGTON AND BARBARA NEIS
Coasts Under Stress Research Project, Memorial University of Newfoundland, 202 Elizabeth Avenue, St. John's, NL A1E 1L4, Canada

Abstract

Amidst the failures of fisheries across the globe and the perceived failure of scientific fisheries management, some recent scholarship has focused attention on the nature and collection of fishers' knowledge, and on the potential utility of that knowledge to fisheries management. This chapter summarises the results of recent research on fish harvesters' local ecological knowledge (LEK) and its interactions with fisheries science and management in Newfoundland and Labrador, Canada. We treat LEK, science and management as parallel, interacting socio-ecological knowledge systems that are internally complex and dynamic. We begin by characterising the dynamism of LEK in Newfoundland fisheries and then describe the rise of a linked fisheries science and management framework in Canada in the 1970s and 1980s that contributed to the marginalisation of fish harvesters' LEK, particularly that of small boat fishers. We then explore the changing interactions between LEK, governance and science in Newfoundland, associated with a recently shifting international discourse that highlights the need for participation and the devolution of some responsibility and authority for fisheries management from centralised state bureaucracies and government-funded and controlled fisheries science to harvesters and other 'stakeholder' groups. Two case studies, comparing and contrasting the role of harvesters and LEK in the management of Atlantic cod (*Gadus morhua*) and American lobster (*Homarus americanus*) fisheries in Newfoundland and Labrador since 1992, are then used as examples of the interactions between these actors and their knowledge systems in practice. We conclude with a discussion of some of the potential benefits and dangers associated with this emerging contemporary relationship between harvesters and their knowledge, fisheries science, participation and governance in Newfoundland and Labrador.

16.1 Introduction

The last several decades have seen the collapse of fisheries across the globe, and many others are fished to potentially unsustainable levels (McGoodwin 1990; Pauly and Maclean 2003). The collapse and closure of the 'northern' cod (*Gadus morhua*) fishery off Canada's east coast – once one of the largest in the world – is one of the more dramatic examples; an ecological and social catastrophe of sobering dimensions (Figure 16.1).

T. S. Gray (ed.), Participation in Fisheries Governance, 269–290.

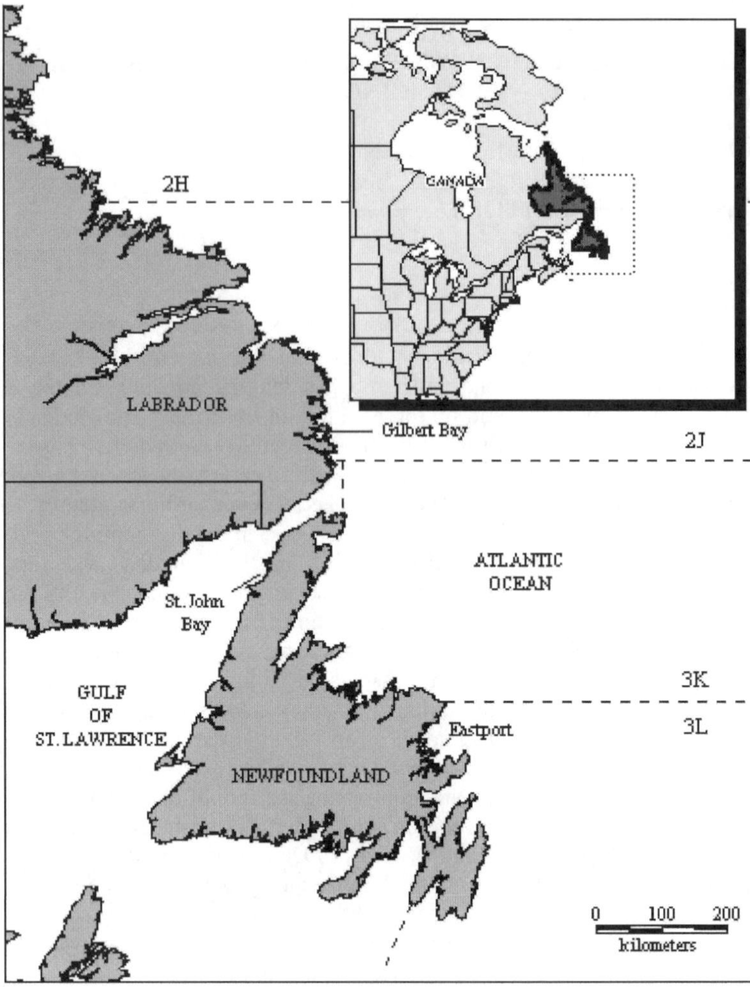

Fig 16.1. Map of the Northwest Atlantic Ocean off Newfoundland and Labrador, Canada illustrating NAFO (North Atlantic Fisheries Organisation) fisheries management divisions 2GHJ 3KL. The Northern cod stocks are generally referred to as those stocks encompassing NAFO divisions 2J3KL

Though differing explanations have been offered about what happened in Newfoundland to precipitate such a crisis, the preponderance of evidence suggests that the collapse stemmed primarily from over-fishing, coupled with inappropriate management measures based on erroneous stock assessments (see Hutchings 1996 for a description of competing hypotheses).

Regardless of the cause of the collapse, there has been a growing crisis in confidence in the scientific, state-controlled fisheries management regimes that were introduced in Canada (and elsewhere) in the 1970s. In Newfoundland, many inshore fishers feel that their claims that stocks were declining in the 1980s were ignored by the Department of Fisheries and Oceans (DFO) (Neis 1992), and some now appear to feel that their claims that cod stocks are rebounding are still being ignored (personal observation). Today, the Newfoundland and Labrador fisheries are dominated (in terms of landed value) by snow crab (*Chionoecetes opilio*), shrimp (*Pandalus spp.),* and, in some communities, lobster (*Homarus americanus*), but many fear that the mistakes of the past are being repeated in these newly expanded fisheries.

As Neis and Felt (2000:12) have pointed out, "(t)here are no panaceas for the current state of the world's fisheries", a statement that certainly holds true for the contemporary fisheries of Newfoundland and Labrador. Amidst a general (in Newfoundland and elsewhere) lack of confidence in centralised, state-based, bureaucratic scientific management, many have called for an increased role for fishers in scientific knowledge production and a movement toward co-management arrangements that involve mixtures of collective, state and, for some advocates, private control over marine resources (Mansfield 2004; Felt *et al* 1997; Pinkerton 1990, 1994; Apostle *et al* 2002; Grafton 1993). Yet the transition to greater user control is far from complete. Some recent scholarship, for example, has suggested that participatory arrangements are best considered as falling somewhere on a spectrum between total government control and total user control, and have identified factors that determine the precise locus of decision-making responsibility and authority (Sen and Nielsen 1996; Jentoft and McCay 1995). Some of this work has suggested that management agency distrust and resistance, as well as a general lack of broadly organised political support, have presented barriers to establishing successful, truly participatory management arrangements (Pinkerton 1999).

Recent scholarship has also focused attention on the nature and collection of fishers' knowledge, and of the potential utility of that knowledge to fisheries science and management (Berkes 1999; Neis and Felt 2000; Neis *et al* 1999). Yet there has been comparatively little research on the way that fishers' knowledge is **actually** incorporated into management decisions, or on the relationship between the inclusion of that knowledge and the participation of fishers in the management process. These questions have become increasingly salient in the wake of recent decisions by the DFO.

Since the collapse of the Northern cod stocks, the Canadian DFO has signalled a desire to increase both the participation of fishers in management and the inclusion of their knowledge in assessing the health of stocks and in setting quotas and designing management regimes. In some areas, it has already begun transferring some formal decision-making powers and responsibilities over to fish harvesters. Over the same time period, however, DFO has undergone significant budgetary cuts that have contributed to downsizing within science and enforcement (particularly of full-time staff), and its mandate has been expanded to include greater responsibility for Oceans, transportation safety, and search and rescue. Moreover, like many other government programmes, DFO has experienced a loss of legitimacy encouraged by influential neo-liberal policy initiatives aimed at transforming

Keynesian welfare bureaucracies, charged with managing common state property, into entrepreneurial, market-oriented agencies focused on establishing self-organising private property regimes (Mansfield 2004; McCarthy and Prudham 2004; Jessop 2002). Under these ideological and material conditions, interactions are changing between the knowledge and participation of fishers, scientists and managers with long-term consequences for the health of fish and fisheries that remain poorly understood.

In this chapter, we address this perceived gap in the research, and are particularly interested in exploring emerging contemporary relationships between fishers' LEK, science, participation and governance in some specific situations in Newfoundland. The discussion is informed by research undertaken as part of the Coasts Under Stress Research Project (CUS), a five-year interdisciplinary project that is examining the dynamic of socio-ecological restructuring on the east and west coasts of Canada, and the implications of this restructuring for the health of people, communities and environments (Dolan *et al* in preparation). Our research starts from the premise that we need to approach fisheries as socio-ecological networks within which different knowledge systems (local knowledge, natural science, governance and social science), and different groups of actors, have interacted at different spatial, temporal and organisational scales to shape the history of fish and fisheries (Perry and Ommer 2003). In much of our work, our intent is to utilise information from different knowledge systems to reconstruct interactions between fisheries, fish, management, industry and communities over several decades. We include results from LEK interviews (with harvester experts), information developed from archival sources, and 'science' data (primarily DFO trawl survey results) (Murray *et al,* forthcoming; Murray and Neis 2004). Generally, we have found that by combining (and contrasting) insights from these different knowledge systems and by looking at processes that have shaped interactions between different groups of actors in these socio-ecological networks, we are able to develop a more nuanced, subtle and effective description/analysis of the history and dynamics of these fisheries (Murray *et al* forthcoming; Bavington *et al* 2004). Related work has explored the feminist political ecology of fishing down marine food webs (Bavington *et al* 2004), and the contested replacement of single species fisheries management with ecosystem-based approaches in Newfoundland and Labrador (Bavington and Kay forthcoming). Research by one of the authors into the multiple meanings of governance in natural resource management has also emphasised the power dynamics involved in shifting the emphasis from managing natural resources to managing the behaviour of human beings and their interactions with natural resource systems (Bavington 2002, and forthcoming). In addition to our own recent work, CUS has helped fund a comparative study of the history of shifting knowledge sources used in management decisions for several different Canadian fisheries (Alcock *et al* 2003); detailed case studies of some Newfoundland lobster fisheries using LEK and science sources (Davis *et al* 2003; Whalen forthcoming); and research on local cod stocks (Gosse 2002; Gosse and Wroblewski 2004).

We draw on this work here using case studies of initiatives related to the management of two species (cod and lobster) in order to illustrate that the interactions between fishers and their knowledge, science, and management in contemporary Canadian fisheries management are fundamentally variable and contingent. Recent shifts in Canadian fisheries management vis-à-vis LEK and fisher participation, for example, must be seen against a contextual backdrop that includes such developments as ecological degradation, technological intensification, changes in fleet characteristics and the numbers of fishers, and radical shifts in effort across marine ecosystems. Broadly stated, our research suggests that the way participation, and the inclusion of fishers' LEK in fisheries science and management plays out depends on the species considered, and the historical, socio-cultural, and geographical contexts surrounding the emergence of the participatory management initiatives. As background for the case studies, the next section discusses our understanding of the dynamic nature of LEK (and issues related to it) in the rapidly changing contexts in Newfoundland and Labrador.

16.2 Local ecological knowledge and participatory fisheries governance in Newfoundland and Labrador

LEK is knowledge derived through experience, or what Franklin (1990) has called 'vernacular knowledge' and others have termed 'tacit knowledge' (Scott 1998). Here, we choose to focus on the inshore sector for three reasons: first, there are far more fishers engaged in the inshore sector; second, in some cases, inshore stocks represent the majority of remaining populations; and third, the LEK of inshore fishers has historically been 'illegible' to managers. It is important to note that by using the term 'LEK' we do not mean to imply an attention only to ecology, or strictly to the bio-physical environment from which humans are too often considered separate. In our research, we are concerned about the physical and biological components of ecosystems (such as the fish, the tides, and water conditions) and collect data on these, but we also ask about fishing and issues related to the larger social and economic context of fishing. Furthermore, we recognise that fishers change – and are changed by – more than bio-physical/ecological conditions. They are also embedded in a complex web of 'social' conditions: management regulations, kinship ties, peer pressure, social support mechanisms, and most importantly perhaps, the global seafood market. LEK is therefore a **socio-ecological product**, reflecting social and ecological times and places as well as culture and other institutions, and is mediated by labour processes, technologies, modes of management, economic, and ecological conditions.[1]

However, while local, and related (in this case) to the harvesting of fish in particular times and places with particular kinds of gear, LEK should not necessarily be considered fragmented or simply an instrumental strategy for achieving specific goals (Neis and Felt 2000). When viewed as only an instrumental strategy to achieve specific goals, the way can be paved for co-optation and the mining of LEK to serve the goals of fisheries managers and others with the power to expropriate knowledge and use it as a means to serve their ends rather than those of fishers, as is emphasised by post-colonial theorists (Banerjee and Linstead 2004), critical management scholars (Parker 2003; Willmott 1993) and labour process theorists (Taylor 2002; Braverman 1974). Rather, as Berkes (1993:4-5) has pointed out, LEK should be seen as "an integrated system of knowledge, practice and beliefs."

[1] The same is true of fisheries science, as explorations of its early development and broader social studies of science make clear (Smith 1994; Holm 1996; 2001; Latour 1987, 2004; Bavington, forthcoming).

One of the reasons why we have used a large sample in our own research, stems from the complexity of LEK. LEK is fundamentally dynamic, subject to changes that are driven by economic, socio-cultural, managerial and ecological factors. Fishers' observations are acquired during the harvesting of fish, but are mediated by (and shift with changes in) a variety of factors, including knowledge transmitted orally from previous generations (which is subject to memory); the specific areas harvesting activities occur in; when fishing takes place; what gear and specific techniques are used; the species and sizes that are targeted; changing ecological and bio-physical conditions; volatile markets for their fish; shifting management regulations and scientific information; and the technologies fishers deploy (Fischer 2000). LEK is also influenced by gender (Power 2000), generational, technological and other divisions of labour, whereby elements of the knowledge system of a particular group may be dispersed between individuals.

Change in Newfoundland's fisheries has been particularly evident in the post World War II period, when the rapid adoption of technological innovations dramatically transformed social-economic relations and the way that fish were harvested, particularly in the offshore sector.[2] Neis and Kean (2003) explain the dynamics of 'fishing up' in Newfoundland, focusing on the inshore sector. They describe inter-related processes of spatial, temporal and ecological intensification and spatial, temporal and ecological expansion. In Coasts Under Stress, we have found dramatic increases in effort in the inshore fishery across such axes as vessel size and materials, engine size, the use of electronics, and the amount and type of gear used. Associated with this technological intensification there have been spatial shifts at varying scales as fishers, often equipped with bigger, longer-ranging vessels with more powerful hauling gear, have been able to access both deeper and more distant waters. Indeed, many of these vessels have nearly obliterated the distinction between the inshore and off-shore sectors. In combination, this technological intensification and spatial expansion has led to 'ecological intensification', which involves increased pressure on existing stocks through changes in mesh sizes and harvesting locations, as well as ecological expansion through a shift to new stocks and new species as others are depleted. As cod stocks have collapsed, for example, fishers have moved rapidly to different species – a trend that is illustrated in Figure 16.2.[3]

---

[2] The late 1950s saw the arrival of an international fleet of highly efficient factory freezer otter trawlers that were able to harvest fish at unprecedented levels. Overall landings increased dramatically during this time, reaching a peak in approximately 1968, after which landings declined precipitously reflecting the dominance in the cod fishery of the offshore, mobile gear large-vessel sector (as compared to the fixed gear/inshore sector). It should also be noted that before 1977 and the declaration of the Canadian 200 mile Exclusive Economic Zone (EEZ), a large proportion of these landings were from foreign vessels. After 1977, there was a relative shift in overall productivity to fixed gear (a rough proxy for the inshore sector). This period also saw a dramatic shift towards Canadian vessels.

[3] Groundfish here includes cod, plaice, turbot, haddock and redfish. Pelagic includes herring and capelin. Crustaceans includes snowcrab and shrimp. 'Other' includes all other species.

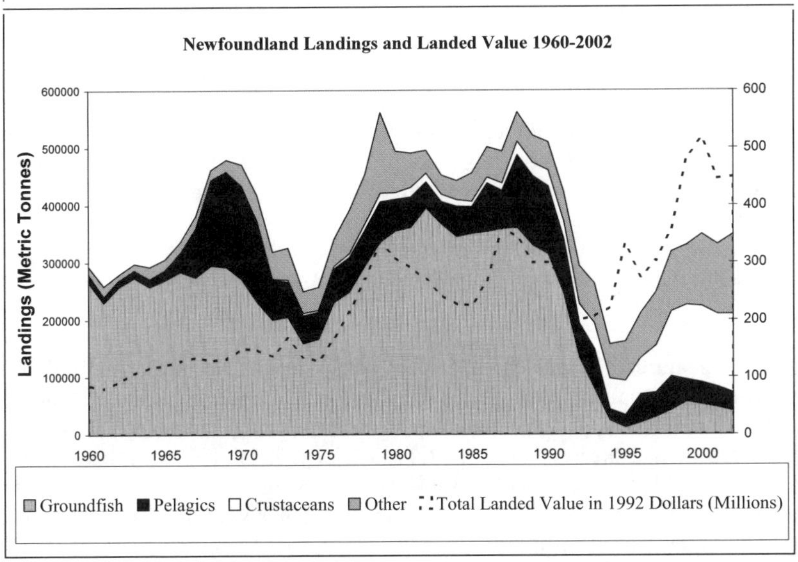

Fig. 16.2. Newfoundland Landings and Landed Value 1960-2002

Of course, these patterns are not the same across all fisheries. Though not readily apparent in Figure 16.2, for example, we have found that lobster landings have also increased in economic importance for many small-boat, coastal harvesters who continue to fish from smaller boats and stay closer to home.[4] It is important to realise, however, that lobsters are far from evenly distributed in Newfoundland, which is situated at the edge of their range. Lobsters are not found in Labrador at all, for example. Furthermore, a combination of state regulations and local customs regulates where any fisher can fish for lobster – generally these areas have been fishers' traditional grounds and are located close to their homes. Access to lobster stocks, unlike to mobile populations of cod, is therefore limited by geography and a combination of state and local rules. One of the benefits of LEK is that it can help us understand local rules and the way their effectiveness and enforcement interacts with environmental change, fisheries policies and practices and with fisheries management over time.

As the socio-ecological system in which these fishers are embedded has changed, so too has the knowledge and orientation of these fishers to each other, the fish and to their work. Also changed is the very process of learning (Johnson *et al* 2004; Murray and Neis 2004). For example, in the 1950s, fishers describe fishing for cod with cod traps very close to shore in hand-built vessels, using no more than landmarks, a compass, and knowledge of the sea to navigate. Most describe learning to fish as simply a part of growing up on the water – by

[4] Indeed, by presenting aggregate data as in Figure 16.2 (which shows total landings by Newfoundland fishermen) we are masking a great deal of spatial, sectoral, and individual variability in target species. Fishers we interviewed show a great variety of adaptive strategies for coping with shifting socio-ecological conditions.

being immersed in a fishing culture and learning from family members by way of doing. By contrast, today some fishers describe ranging up to 180 miles off-shore for deep-water crab (and/or shrimp) stocks, in 45-65 foot fibre-glassed and steel-hulled vessels that can cost up to a million dollars. Navigation and communications now requires a sophisticated understanding of GPS, computer technology and radios. Consequently, many successful skippers and crew members now take additional training and learn new techniques in institutional settings in fishery professionalisation colleges such as those located in the capital city of St. John's. Likewise, fishers' interactions with markets have shifted from a reliance on selling salt-fish (cod) to local merchants at the end of the season for pre-set prices, to becoming enmeshed in global seafood markets for fresh products with volatile prices.

LEK therefore features a high degree of complexity and is *not* **standardised** in terms of temporal scale, territorial coverage, technology, effort and expertise. While LEK is 'held' by groups of resource users, it is unevenly distributed among them, and is as diverse and dynamic as are their fisheries (Neis *et al* 1999; Felt 1994). The complexity, unevenness, opacity, and embeddedness of LEK present some challenges when seeking to incorporate it into a management framework. For this reason, the fisher who stands up at a consultative meeting and claims, based on his or her experience, that there are no more fish (as many inshore fishers in Newfoundland did in the 1980s), is not often given the same authoritative weight as a scientist, who can present a graph showing steady abundances based on years of 'objective' evidence. This is not to suggest that scientists are 'right' and fishers are 'wrong' – indeed, the experiences of the 1980s suggests that at times the converse was true. As indicated by the failure of the Canadian government to reduce total allowable catches (TACs) for northern cod to levels recommended by scientists in the late 1980s (Finlayson 1994), it is also not to suggest that management is necessarily dictated by the results of scientific research. On the other hand, we do not mean to suggest that fishers are always 'right', or that what they say is necessarily based on the knowledge they have acquired through fishing (as opposed to politics, emotion, or impressions). We suggest, therefore, that the nature of the observational processes and the conclusions drawn from those observations must be carefully understood for both science and LEK. The other point we wish to make is that science is more easily translatable and moveable than is LEK – as part of a written and professional culture, it is designed to produce charts, graphs, quantified summaries and abstract generalisations. This gives it power, especially in the context of management deployed from above and afar.

Social scientists and others have sometimes sought to collect and translate LEK into a form that is more legible, mobile and more easily used in fisheries management activities, or for other reasons, like historical reconstructions, as in our research (Murray *et al* forthcoming). It is important to note, however, that these researchers are often located in "centers of calculation" far from the places where LEK is created (Holm 2003). The translation process necessary to bring knowledge to centres of calculation is frequently located in unequal fields of power (biophysical, social, cultural, political and economic in nature) that fundamentally influence how LEK and the fishers who create it can participate in fisheries management.

Furthermore, LEK is an enormously complex knowledge system, and no researcher can hope to capture it in its entirety (Mailhot 1993); nor, for that matter can any one fisher hope to 'know it all'.

Table 16.1. Types of information from different knowledge systems

LEK	*Archival Sources*	*Science (Trawl Data)*
Local Taxonomies Usage patterns for non-commercial species Fishing areas (location and depth) Fish Behaviour and Biology (including migrations, spawning areas, ecosystem interactions, observed size, local stock structure) Vessel Characteristics (including size, engine size, range, materials, electronics usage, hauling equipment) Gear (type and amount) Species targeted Markets utilised Crew size and composition (e.g. kinship ties) Training Amount and type of participation in management and interactions with fisheries managers)	Licensing and participation Management (approaches, policies and regulations) Fleet characteristics Landings and landed value Processing facilities	Abundance Distribution (including depth)

One of the basic realisations of our work has been (with science and with LEK) that the outcome of our translation process depends directly on the research questions we asked. We have also realised that LEK changes during the research process in ways that depend on the research approach, protocol and process. Furthermore, the way LEK translations are received and used, is also an extremely important factor in determining the ultimate results of integrating LEK into fisheries management and science. The translation process, in other words, depends on what is being translated, who is doing the translation, how that translation is done and how it is accepted and used in science and management. In our own research, for example, our focus has been on how a variety of parameters have changed during our study period, 1950-present. Collectively, our research has involved a large sample (over 150) of semi-structured interviews with expert fishers in different parts of Newfoundland, during which we have collected information on a variety of topics that are summarised in Table 16.1 (see Murray *et al* forthcoming for a description of sampling and methodology).

We have sought to translate this work using Nud*ist, Excel and MapInfo software and have suggested ways in which LEK could be ethically collected to be used in fisheries science and management (Neis *et al* 1999).

16.3 Trends in Canadian fisheries science and management: From single species populations and top-down statecraft to complex ecosystems and participatory governance

The history of fisheries science and governance shows constant change. Fisheries have not always been managed in Newfoundland and Labrador, and how they are presently being managed is in the process of fundamental reform (Bavington forthcoming). While numerous community-based forms of regulation founded on local norms, incorporating, for example, bans on fishing technologies and control over access to fisheries resources, existed before the development of scientific fisheries management (Cadigan 1999; Matthews 1993), fisheries resources were generally not understood as being in need of, or amenable to, management in the sense of controlling and carefully using the fishery resources, until well into the twentieth century (Smith 1994). In this section, we describe some general trends in Canadian fisheries management. We draw heavily on the case of cod (both because it is well-studied, and because it has been historically dominant), but we acknowledge that management approaches in other fisheries have had somewhat different trajectories.

Historians of fisheries science and management argue that scientific fisheries management began to emerge in the last half of the nineteenth century in response to demands by diverse fishing interests who needed to discover what caused fluctuations in landings in order to guarantee a safe environment for capital investment in fishing economies that were shifting from peasant-mercantile to industrial-market forms (Smith 1994, Ommer 2002). After much work, the biological concept of single species fish populations (that could be mapped and their dynamics predicted) emerged as the focus of fisheries management regimes around the world (Smith 1994). With the development of bio-economic models in the 1950s that modelled humans fishing for single species fish populations as rational economic actors, a manageable model of both the fish and fishers was constructed (Holm 1996). States and international fishing organisations began mapping single species populations and setting TACs that would produce maximum economic yields. These TACs, however, were extremely hard to enforce. Ocean resources were considered open access – fish were the property of no one until they were captured and pulled onboard a fishing boat. Nations proposed to end this open access regime and create the conditions for scientific fisheries management, by eventually extending economic jurisdiction to 200 miles from their shorelines (to enclose most of the productive continental shelf fisheries), and by introducing licensing and other systems within their fisheries. This process of enclosure created domestic fish populations owned by nation states. Canada declared its 200 mile Exclusive Economic Zone (EEZ) on January 1st 1977, and created the DFO to rebuild and scientifically manage the newly nationalised fish populations.

Claims that such zones would reduce the likelihood of over-fishing and improve the overall benefit of fish resources for humanity, however, were based on a series of problematic techno-utopian assumptions (Finlayson 1994). To manage this new national territory and the fish that were contained within it, DFO focused primarily on offshore areas and fisheries. With respect to cod, migrating offshore populations were assumed to be the most numerous and productive and therefore were deemed the most economically important. Data for fisheries management models were obtained by DFO in randomised annual survey trawls and from catch-per-unit-effort (CPUE) data kept by the offshore dragger fleet. Since there were fewer offshore vessels than in the inshore fleet, and the offshore fleet captured a large portion of the catch, caught over a massive area with relatively standard gear (otter trawls), data from the offshore dragger fleet were assumed to provide consistent statistical information for single species fisheries management models (Finlayson 1994). Relatively little research focused on the inshore sector and on fish in the coastal bays, though there were far more vessels and fishers engaged in this sector. The inshore fishery and the knowledge its participants produced was illegible to the DFO. Data, when it was collected from this sector, was not used in fisheries management modelling or in the setting of the TAC after the extension of the EEZ. By the mid-1980s, fisheries management models for the Northern cod seemed to show a growth of the stock since 1977, despite persistent claims from inshore fishers that TACs were being set too high, and their observation that fish were scarce and becoming smaller-at-age inshore (Neis 1992). At that time, Canadian cod fisheries management was perceived to be one of the best examples of successful scientific management in the world.

The story of the collapse of Newfoundland and Labrador's northern cod stocks is well known and widely studied (Hutchings and Meyers 1994, 1995; Neis *et al* 1999; Hutchings *et al* 1997; Steele *et al* 1992). The closure of the fishery resulted in massive social upheaval and a significant de-legitimisation of DFO's population dynamics models, leading to a plethora of critiques of fisheries management and a reassessment of the assumptions in the models used to manage the fishery. The collapse of the northern cod fishery also coincided with a shifting of international scientific and managerial discourses. The single species population models that had served as the foundation of fisheries management since its inception, began to be challenged by fisheries ecologists who argued that these single species, statistically defined populations were simplified fictions, good for enabling centralised state-led management perhaps, but reflecting an impoverished view of reality (Busch *et al* 2003; FAO 2003a, b; Garcia *et al* 2003; Caddy and Cochrane 2001; Caddy 1999). They proposed thinking about fish in relation to their biological and physical contexts, resulting in a switch from population to ecosystem thinking in fisheries science. In addition to requiring an increased variety and amount of knowledge, ecosystem thinking led advocates to emphasise that fisheries management is more about managing human behaviour and interactions with ecosystems than the fish themselves (Bavington 2002; Maguire *et al* 1995). This shift from population to ecosystem thinking was joined by additional innovations in managerial thought that influenced the context within which the DFO had to attempt to implement ecosystem-based fisheries management. These new managerial innovations emphasised participation in networks of fisheries governance that included expert state-based managers, but also representatives from the fishing industry, academia, environmentalists and the broader community, as emphasised in Canada's new Ocean's Act and Strategy (DFO 2002).

Moreover, with the rise of neo-liberal ideology globally, central planning and top-down state-led bureaucratic management have fallen into disregard. In their place, a focus on downsizing the public sector, privatisation of government services and downloading responsibilities for management and regulation onto individual resource users and the corporate sector has taken hold. State-led bureaucratic governance, with its emphasis on top-down expertise, rational legal frameworks and inflexible proceduralism, has been replaced by a new emphasis on flexible, entrepreneurial governance, emphasising bottom-up participation by economically interested stakeholders, innovation, public-private partnerships and the creation of responsible self-managing individual and corporate 'citizens.'[5] Under neo-liberalism, state agencies no longer are seen as being capable or responsible for delivering management services such as fisheries management alone. Rather, agencies like DFO must coordinate and enable industry to carry out self-regulation and monitoring, increasingly through new technologies (such as Global Positioning Systems (GPS) tracking 'black boxes') and innovative financial incentives disciplined through the market. As McCarthy and Prudham (2004:276) observe, in this neo-liberal climate of market managerialism, "collaboration and partnership become the new mantras of regulatory relations between capital and citizen (underpinned by the discursive rebirth of capital *as* citizen), less and less mediated by formal, state institutions."

Neo-liberal market governance impacted DFO in the decade following the moratorium. The science and enforcement branches of the agency are now asked to do substantially more with significantly less in the way of human and financial resources. Faced with limited resources (including research days-at-sea on fully equipped research vessels and enforcement and policy staff) and dealing with cod stock remnants in poorly understood coastal areas, the participation of fishers in the collection of scientific data and the management of ecosystems has become a necessity. Through license buybacks, professionalisation, deregulation and privatisation of the seafood processing sector and fish quotas, the Department is seeking to create a new, self-managing fishing industry. While not complete or uncontested, this policy focus has created opportunities for forms of community-based management for some species (such as lobster) and enrolled a select group of fishers in new monitoring and management programs associated with cod. Alcock *et al* (2003) found that managers in the crab and herring fisheries also followed a similar pattern vis-à-vis the sources of information DFO claims to use in making management decisions: an early dependence on LEK, followed by a dependence on economic sources; then by science; and, finally, the reappearance of LEK. What is not clear, however, is precisely **how** LEK is utilised in making management decisions.

The broad changes in science and governance outlined above – when combined with the specific biophysical characteristics of the species and the rapid contextual changes and changing nature of LEK itself outlined earlier – provide an opportunity to explore specifically how fishers and their LEK are actually participating in contemporary fisheries science and governance regimes in Newfoundland and Labrador, as well as some of the consequences of this new participation, for fish, fishers and other actors in the fisheries governance network. The following section looks at two examples of new approaches to science and governance in the context of Newfoundland and Labrador fisheries for cod and lobster.

---

[5] As McCarthy and Prudham (2004:276) argue, "neoliberal notions of citizenship and social action are discursively repackaged in the image of homo-economicus, the ideal, entrepreneurial, self-made individual."

16.4 Case studies

16.4.1 FISHERS AND SCIENTISTS AND THE SENTINEL COD FISHERIES PROGRAMME

As noted above, before the moratoria of the 1990s, the knowledge and interests of inshore fishers played little role in cod fisheries management in Newfoundland and Labrador. Single species management models focused on assessing and allocating cod stocks and relied primarily on data from offshore scientific surveys and commercial catch rates. Scientific data were reviewed by scientists behind closed doors and then in closed door industry meetings. However, since the moratoria were announced, this has changed somewhat, and inshore fishers have been presented with a number of participatory management structures that they can now become involved with. For example, the Fisheries Resource Conservation Council (FRCC) was created in 1993 to form a partnership between scientific and academic expertise and all sectors of the fishing industry, though the Minister of Fisheries and Oceans maintains ultimate decision making authority. The council holds open public meetings to gather information from fishery stakeholders and delivers public recommendations to the Minister of Fisheries and Oceans on issues such as TACs, scientific research and assessment priorities and other conservation measures in the Atlantic fishery. It remains clear, however, that this arrangement does not represent active citizen control of the fisheries examined by the FRCC (Bavington forthcoming; Gray 2002; Arstein 1969).

In part to address these gaps and to make inshore cod populations legible for management purposes, DFO also began enrolling a select number of cod fishers in 'Sentinel' fisheries programmes. In the 1990s, it became apparent that offshore cod stocks had been over-harvested to the point of endangerment. In some areas, the only cod populations that showed any sign of relative health lived primarily in the coastal bays (Wroblewski, 2000; DFO 2003). In addition, many of the large, commercial groundfish trawlers (which had been a primary source of data for stock assessments) were sold off to other countries and the closure or dramatic reduction in fisheries reduced the catch rate data available to the stock assessment process. Whereas information on inshore stocks and from small-scale fishers had been marginalised within DFO since the 1970s, after 1992 the Department had to develop new tools to assess and manage the remaining cod populations.

The Sentinel fishery was organised collectively by DFO and the Fish, Food and Allied Workers' Union, which represents Newfoundland and Labrador fish harvesters (FRCC 2002). The Sentinel fishery was designed to make inshore populations legible to fishery managers by prescribing fishing practices that would yield statistically significant annual results. Fishers were to fish in specific locations using standardised gear. They also participated in tagging programs, fish measurement and the harvesting of otoliths for use by scientific technicians in aging the fish harvested. Due to the need for statistically significant information to feed into single species population models, fishers deploying gill nets were favoured over other forms of fishing gear, such as cod traps or baited hand lines, for which it was more difficult to accurately and consistently quantify catch-per-unit effort (FRCC 2003).

The Sentinel fishery programme was also designed to reduce the gaps between scientists and harvesters, though to close this gap requires focused research that has not been carried out. However, it can be said that through the Sentinel fishery programme, the inshore fishery and coastal cod were translated into 'manageable objects'. Fishers who were selected to participate in the program were paid to fish for DFO. This represented the enrolment of fishers as technicians, and solicited a very narrow and transformed slice of inshore LEK without seeking to substantially change the relative decision-making power of fishers, scientists and fisheries managers. The approach built on the use of logbook programs that had been common in some other fisheries, such as the capelin trap fishery, for several years (Neis and Morris 2002) but added new components like protocols for the amount of gear, gear design, gear location as well as scientific sampling to these programmes.

Perhaps not surprisingly, Sentinel fisheries for cod are contested fisheries. Some harvesters have sought to influence the design of the programme, and disagreements about the accuracy of the catch rate information and interpretations of that information have erupted between Sentinel and non-Sentinel harvesters and between harvesters and scientists. Some fishers argue that their **experiences** on the water do not confirm the **experimental** results of the Sentinel survey (FRCC 2003). This may reflect the fact that stock assessment science relies heavily on catch rate information developed under controlled, repeated (and therefore statistically valid) circumstances. Commercial fishers, on the other hand, are able to sustain high catch rates as stocks decline by varying effort across time and space (Harris 1990). The protocol involved in fishing for science, in other words, is quite deliberately different from commercial fishing practice. That said, greater familiarity with scientific methods appears to have increased rather than reduced scepticism among some harvesters, contrary to intentions. Utilising short time series, limited knowledge about fish behaviour and random, and sparse sampling of a mobile and scarce population can, as fishers' criticisms suggest, mean high levels of uncertainty about stock abundance. The Sentinel fishery reminds us that science and management initiatives, as well as changes in fisheries, can change LEK and, in LEK as in science, high uncertainty tends to be associated with "interpretive flexibility" (Finlayson 1994).

Elsewhere, relationships between fishers (and their LEK), science and governance in the cod fishery have taken a form that is somewhat different from that in the Sentinel fishery. Gilbert Bay, Labrador, provides one such example, and a partial counterpoint to how this relationship is structured in the context of the Sentinel fishery. Due in part to the new Canadian Oceans Act and new legislation related to Marine Protected Areas, there are now several areas in Newfoundland that have been designated as special areas of interest. These areas of interest may become protected areas. Gilbert Bay is a candidate for protected area status because of the scientifically documented presence of a genetically distinct bay stock of cod. Gilbert Bay contrasts with the Sentinel cod fishery in the comprehensiveness of the research program that has been carried out on the local cod and in the involvement of local stakeholders, including non-fishers, in management discussions related to this area (Wroblewski 2000; Gosse and Wroblewski 2004). More research is needed, however, to explore the dynamics of the evolving relationship between harvesters, LEK and science and governance in this Bay.

16.4.2 SCIENTISTS AND FISHERS AND THE EASTPORT LOBSTER FISHERY

There has been a commercial lobster fishery in many parts of Newfoundland since the late nineteenth century, although in most areas this fishery was secondary to the cod fishery until the 1990s. Over-fishing of lobster in the early twentieth century resulted in a multi-year closure in the 1920s (Rogers 2002). Since then, the state has assumed increasing control over the lobster fishery, and management now includes limited entry, restrictions on the capture of egg-bearing females, minimum size restrictions, open seasons, trap limits, and the creation of lobster fishing areas. Size limits were introduced in Newfoundland in 1939, and by 1945, fishers were restricted to certain fishing districts (Parsons 1993) which, coupled with strong local norms regulating lobster fishing grounds, limited the mobility of fishers. Limited entry was first introduced in Atlantic Canada in 1968 (in Nova Scotia), and was extended to Newfoundland in 1975, and, by the mid 1970s, many part-time and recreational fishers were excluded, so that 'legitimate' fishers could increase their returns. Trap limits were established by 1975. Some limited attempts to include lobster fishers in management were also initiated. For example, local advisory committees were put in place in 1975 for each lobster fishing district. This structure was modified by the early 1980s with the creation of Lobster Fishing Area Advisory Committees (LFAACs), which included elected fishers and processors, the DFO, economists and others. Management plans regulating lobster fishing in each Lobster Fishing Area (LFA) were to be modified only after discussion with local committees, though ultimate decision-making authority rests in the hands of the DFO.

Despite these conservation initiatives, and partly because of the decline in cod landings in the 1980s, lobster stocks in Newfoundland came under heavy fishing pressure in the 1980s and 1990s, as harvesters with lobster licenses shifted effort to this fishery. Harvesters in the Eastport Peninsula area of northeast coast Newfoundland responded to this shifting effort and to concerns about the encroachment of outsiders onto their lobster grounds in the 1990s, by establishing the Eastport Peninsula Lobster Protection Committee (EPLPC) (Rowe and Feltham 2000). This Committee undertook to take control over management of their local lobster fishery and was eventually able to negotiate with DFO and neighbouring harvesters the introduction of an exclusive fishing zone for harvesters from their area and an adjoining buffer area. This zone represents a portion of the LFA that covers the larger area. They then introduced a series of other conservation measures including v-notching, closed areas and self-policing for poaching. University and DFO scientists have actively supported this initiative, and a substantial amount of science carried out in the area with harvester involvement has been an integral part of the initiative. In contrast to many other areas of Newfoundland, lobster catch rates appear to be more resilient in this area, rather than continuing to decline, and there is a relatively high level of support for the initiative among local harvesters (Davis *et al* 2003).

Overall, the Eastport co-management initiative has been quite effective at bridging gaps between harvesters and science; drawing on LEK to inform science and management and to 'test' the effectiveness of particular management initiatives; and protecting the lobster resource in this area. Relative to other lobster fishing areas, catch rates are doing better in the Eastport area, local community territories have been largely sustained, and v-notching and other conservation initiatives appear to be well-supported by local harvesters (Davis *et al* 2003; Rowe and Feltham 2000). Attempts are underway to extend the Eastport approach to other lobster fishing areas in Newfoundland where it has been coupled with a new requirement for harvesters to produce an integrated management plan before fisheries can proceed.

A striking feature of the Eastport (and Gilbert Bay) initiatives, however, is the extent to which scientific and other resources have been concentrated in these areas relative to others. For example, while significant public resources have been dedicated to helping the EPLPC get established and to sustaining this initiative, this has not been the pattern throughout Newfoundland. The lobster fishers in St. John Bay on Newfoundland's west coast, for example, where there is a much more intensive lobster fishery and where lobster is much more important to most harvesters' incomes, have received very little attention and support from DFO and other public agencies, such as the Memorial University. This is despite the fact that mismanagement of licensing policy by DFO in the 1980s contributed to the emerging crisis in the St. John Bay lobster fishery. For example, the number of fishers in the St. John Bay LFA was dramatically increased when displaced fishers from another LFA were moved (Whalen forthcoming).

What appears to be happening as DFO moves from a centralised to a decentralised mode of governance, is the abandonment of a formal, if ultimately problematic, approach that concentrated on studying and managing all the stocks and commercial species in a particular area. While the claim to 'universality' in that earlier approach has been legitimately questioned by those who have looked at the science and management from the perspective of inshore harvesters (Neis 1992; Finlayson 1994), the shift to 'targeted' programs where resources are directed to particular groups and situations, while others are neglected because of cuts in the science program at DFO, is also very problematic from the point of view of the disadvantaged harvesters and relevant species/populations.

16.5 Conclusion

We have suggested that fisheries are best approached as socio-ecological networks where various actors and different knowledge systems (local knowledge, natural science, governance and social science) have interacted at different scales to shape the history of fish and fisheries (Perry and Ommer 2003). These knowledge systems are fundamentally dynamic, and have shifted internally, and in relation to each other during our study period. One of the basic findings of our work has been that there is, therefore, no single relationship between fishers and their LEK, scientists and managers in the fisheries of Newfoundland and Labrador. Generally, we find that these relationships, although always complex, have become less uniform as a result of deregulation and the transfer of greater responsibility for management and enforcement and infrastructure costs for fisheries onto harvesters. The precise nature of the relationship between scientists, managers, fishers (and their knowledge), and the way that the rhetoric of DFO will play out in practice, is contingent on a variety of factors.

As the lobster case study shows, the ways in which neo-liberal inspired participation is playing itself out in fisheries governance vary, depending on the species, the fishers, and the fishing communities involved, with access to resources and support dictated by local organisational resources, and, to some degree, personal connections. In Eastport, lobster fishers were self-organised and motivated; part of a fishery with an established tradition of exclusive, local usufruct rights; an integral part of scientific research in the area; and significantly supported by the DFO. This interaction resulted in several positive outcomes for the fishers and the fish. Yet these outcomes have been far from universal, even within the lobster fishery – St. John Bay is a case in point.

On a more general level, we would suggest several additional factors that might influence the way that fishers and their knowledge are incorporated into fisheries management, including: the capabilities and aspirations of user groups; whether there are top-down or bottom-up approaches; the relative difficulty of decisions; the specific management tasks; the stage in the management process at which user groups are involved; and the presence or absence of non-governmental organisations. These factors strongly influence where a particular arrangement might fall along the 'co-management spectrum' between total government control and total user control (Sen and Nielsen 1996; Jentoft and McCay 1995). This reflects the fundamental linkage between the nature of the participatory arrangement, the way that fishers LEK is incorporated in management decisions, and resultant outcomes.

We have also suggested that the knowledge produced by LEK research reflects what is being translated (the research questions), who is doing the translating, how that translation is done, and how the translations are received and put to use by natural scientists and fisheries managers. We point out that fishers' knowledge is subject to co-optation due to the inherent power imbalances between researchers, managers and fishers. As a legacy of single species approaches that were so dominant earlier, managers often depend on statistically valid, 'objective' indicators like the trawl survey and commercial CPUE data that DFO used to rely almost exclusively on in conducting research. There is a real danger, therefore, that despite a rhetoric that supports the inclusion of fishers' knowledge in management decision-making, the practices of translating this information can serve to co-opt fishers into scenarios where they simply serve as low cost technicians for producing model-ready data. In this sense, these new approaches to management may use participation as a controlling tool (means) rather than to increase freedom or serve democratic ends. Certainly, the inclusion of fishers' LEK (or modified versions thereof) in management decisions that are divorced from the active and equitable participation of the fishers themselves can lead to situations like that of the contested Sentinel cod fishery. We therefore suggest that the translation process should be critically examined, and that effective, ethical, and equitable participation of fishers, scientists and fisheries managers in these research avenues is essential. To test the validity of our translation process in our own research, for example, we have used feedback sessions with fishers to evaluate the soundness of our research and check our results (Murray *et al* forthcoming). Other desirable options might include more participatory forms of research, involving fishers more directly in all stages of the research process itself.

At the same time, our research has highlighted that LEK itself is a socio-ecological creation, reflecting time, space and a range of changing socio-ecological conditions. LEK is therefore fundamentally dynamic in a way that demands further attention. As noted, the LEK that is held by the fishers of today is radically different from that held by their mothers and fathers. We suggest that exploring how LEK itself has changed in the context of rapidly changing technologies, shifting markets, changing management, and ecological change, will be a critical avenue of research. This is true for both older fisheries, and for more recently established fisheries such as those for snow crab and northern shrimp. Furthermore, we would suggest that the relationship of fishers to the fish (and to each other) may have changed, suggesting additional questions. Can we assume, for example, that fishers participating in increasingly corporate, technologically sophisticated, debt-driven, market-oriented fisheries have the same relationship to the fish as their parents? Are they more or less likely to have conservation in mind? If so, why? Just as we have learned to approach scientific knowledge with a critical eye, we must also be careful not to unduly lionise fishers (or their LEK) and assume that the uncritical inclusion of LEK in management will benefit the fish, or the fishers and fishing communities over the long term. Critical areas of research therefore involve not only the dynamic nature of LEK itself, but also how LEK might interact with science and management against this dynamic contextual backdrop and the dangers associated with self-regulation and dwindling, often piece-meal resources for independent and effective fisheries science. The potential involvement of LEK in participatory fisheries management points to the need for LEK researchers who can draw attention to the dangers, while also seeking out the potential benefits, associated with particular initiatives. We therefore stress the importance of contextual, contingent and empirical evaluations of the relationship of LEK to fisheries management that can help us move towards improved conservation and towards the enhancement of equality, freedom, and solidarity on the water and in these communities (Bavington *et al* 2004).

The recent signalling by the DFO of a desire both to incorporate LEK into stock assessment, and to include fishers in the decision-making process, can be seen as a means of maintaining legitimacy in the wake of the collapse of groundfish stocks in Atlantic Canada (and fisheries collapses world-wide), and of questioning previous top-down approaches based on scientific single species stock assessments. This helps to explain why, despite its failure in administering the cod fishery and subsequent cuts to its budget, DFO has maintained its dominance within fisheries science and management, and is leading the way on new forms of oceans management, including the ecosystem-based approach, MPAs, sustainability and other policy goals contained in Canada's Ocean's Strategy. We agree that managing with scarcity and managing more effectively at the level of individual populations requires a different approach to science and governance than was typical of Canadian fisheries in the 1980s. There are real reasons to increase the potential for effective involvement of harvesters in science and governance, but such initiatives are unlikely to enhance the capacity for stock recovery or the longer term benefits to harvesters overall if the partnerships are poorly developed and designed, and if financial and other supports are limited to short term, philanthropic interventions in selected areas. While centralised, bureaucratic science and governance have been shown to have serious shortcomings from the point of view of conservation and from the point of view of many fishing people (Hutchings *et al* 1997; Finlayson 1994), the deinstitutionalisation of science and the related shift from a universal and comprehensive strategy for the protection of fish stocks to industry self-policing, deregulation, and targeted programs for 'deserving' communities and fishers poses new threats to fish, fishers and fishery-dependent regions.

Acknowledgements

The authors wish to thank the Social Sciences and Humanities Research Council of Canada (SSHRC), and the Natural Sciences and Engineering Research Council of Canada (NSERC) who have provided the major funds for the 'Coasts Under Stress' Project through the SSHRC Major Collaborative Research Initiative (MCRI) program. Funding was also provided by the host universities: Memorial University of Newfoundland and the University of Victoria.

References

Alcock, E, Ings, D and Schneider, D (2003) 'From local knowledge to science and back: evolving use of local knowledge in fisheries science' in JJ Heymans (ed) *Ecosystem Models of Newfoundland and Labrador: Additional Information and Analyses for 'Back To the Future',* Vancouver BC, University of British Columbia Press

Apostle, R, McCay, B and Mikalsen, K (2002) *Enclosing the Commons: Individual Transferable Quotas in the Nova Scotia Fishery,* St. John's Newfoundland, ISER (Institute for Social and Economic Research) Books

Arnstein, S (1969) 'A ladder of citizen participation in the USA' *Journal of American Institute of Planners* 35:216-224

Banerjee, S and Linstead, S (2004) 'Masking subversion: neocolonial embeddedness in anthropological accounts of indigenous management' *Human Relations* 57(2):221-247

Bavington, D (2001) 'From jigging to farming' *Alternatives* 27(4):16-21

Bavington, D (2002) 'Managerial ecology and its discontents: exploring the complexities of control, careful use and coping in resource and environmental management' *Environments* 30(3):3-21

Bavington, D (forthcoming) *Handling Fish and People: Managerial Ecology in Newfoundland and Labrador Cod Fisheries,* Unpublished Dissertation, Waterloo Ontario, Wilfrid Laurier University, Dept. of Geography & Environmental Studies

Bavington, D, Grzetic, B and Neis, B (2004) 'The feminist political ecology of fishing down: reflections from Newfoundland and Labrador' *Studies in Political Economy* 73:159-182

Bavington, D and Kay, J (forthcoming) 'Ecosystem-based insights on Northwest Atlantic fisheries systems in an age of globalization' in M Schechter, W Taylor and L Wolfson (eds) *Globalization: Effects on Fisheries Resources,* Cambridge, Cambridge University Press

Berkes, F (1999) *Sacred Ecology: Traditional Ecological Knowledge and Resource Management,* Philadelphia, Taylor & Francis

Berkes, F (1993) 'Traditional ecological knowledge in perspective' in J Inglis (ed) *Traditional Ecological Knowledge: Concepts and Cases,* Ottawa, Canadian Museum of Nature, International Program on Traditional Ecological Knowledge

Braverman, H (1974) *Labor and Monopoly Capital,* New York, Monthly Review Press

Busch, W, Brown, B and Mayer, G (eds) (2003) *Strategic Guidance for Implementing an Ecosystem-based Approach to Fisheries Management,* Silver Springs, Maryland, United States Department of Commerce, National Oceanic and Atmospheric Administration, NMSF

Caddy, J (1999) 'Fisheries management in the twenty-first century: will new paradigms apply?' *Rev. Fish Biol. Fish.* 9:1-43

Caddy, J and Cochrane, K (2001) 'A review of fisheries management past, present and some future perspectives for the third millennium' *Oceans and Coastal Management* 44:653-682

Cadigan, S (1999) 'The moral economy of the commons: ecology and equity in the Newfoundland cod fishery, 1815-1855' *Labour/Le Travail* 43(Spring):9-42

Cooke, B and Kothari, U (eds) (2002) *Participation: The New Tyranny?*, London, Zed Books

Davis, R, Whalen, J and Neis, B (2003) *The Establishment and Maintenance of Locally Managed Lobster*

Conservation Programs on the Eastport Peninsula, Unpublished Report, Coasts Under Stress, St. John's, Newfoundland, Memorial University

DFO [Department of Fisheries and Oceans] (2002) *Canada's Ocean Strategy: Policy and Operational Framework for Integrated Management of Estuarine, Coastal and Marine Environments in Canada*, Ottawa, Fisheries and Oceans Canada

DFO [Department of Fisheries and Oceans] (2003) 'Closure of the cod fisheries and action plan to assist affected individuals and communities' 24 April, DFO Website

Dolan, H, Taylor, M, Neis, B, Eyles, J, Montevecchi, B and Ommer, R (in preparation) 'Restructuring and health in Canada's coastal communities: a socio-ecological model of restructuring and health' Submitted to *EcoHealth*, September 2004

FAO [Food and Agriculture Organisation] (2003a) 'The ecosystem approach to marine capture fisheries' *FAO Technical Guidelines for Responsible Fisheries*, No. 4(Suppl.2)

FAO [Food and Agriculture Organisation] (2003b) 'Report of the expert consultation on ecosystem-based fisheries management', Reykjavik, Iceland, 16-19 September, *FAO Fisheries Report* No. 690, Rome

Felt, LF (1994) 'Two tales of a fish: the social construction of indigenous knowledge among Atlantic Canadian salmon fishers' in CL Dyer and JR McGoodwin (eds) *Folk Management in the World's Fisheries: Lessons for Modern Fisheries Management*, Niwot, University Press of Colorado

Felt, LF, Neis, B and McCay, B (1997) 'Co-management' in J Boreman, BS Nakashima, JA Wilson and RL Kendall (eds) *Northwest Atlantic Groundfish: Perspectives on a Fishery Collapse,* Bethesda Maryland, American Fisheries Society

Finlayson, AC (1994) *Fishing for Truth: A Sociological Analysis of Northern Cod Stock Assessments from 1977-1990*, St. John's Newfoundland, ISER Books

Fischer, J (2000) 'Participatory research in ecological fieldwork: a Nicaraguan study' in B Neis and L Felt (eds) *Finding our Sea Legs: Linking Fishery People and Their Knowledge with Science and Management,* St. John's Newfoundland, ISER Books

Franklin, U (1990) *The Real World of Technology,* CBC Massey Lectures, Toronto, Canadian Broadcasting Corporation

FRCC [Fisheries Resource Conservation Council] (2002) Public Consultation on Northern Cod Stock Status in 3Ps, 8 December, Sunnyside Newfoundland

FRCC [Fisheries Resource Conservation Council] (2003) Public Consultation on Northern Cod Stock Status in 2J3KL, 4 March, St. John's Newfoundland

Garcia, SM, Zerbi, A, Aliaume, C, Do Chi, T and Lasserre, G (2003) 'The ecosystem approach to fisheries: issues, terminology, principles, institutional foundations, implementation and outlook' *FAO Fisheries Technical Paper* 443, Rome, FAO

Gosse, KR (2002) *Brown Cod and Bay Stocks: Science and Fish Harvester's Knowledge of Colouration in Populations of Atlantic Cod (Gadus morhua) in Newfoundland and Labrador*, St John's Newfoundland, Memorial University, Environmental Science Programme

Gosse, KR and Wroblewski, JS (2004) 'Variant colourations of Atlantic cod (*Gadus morhua*) in Newfoundland and Labrador nearshore waters' *ICES Journal of Marine Science* 61:752-759

Grafton, RQ (1993) *Individual Transferable Quotas and the Groundfish Fisheries of Atlantic Canada,* Ottawa, Department of Fisheries and Oceans

Gray, T (2002) 'Fisheries science and fishers' knowledge' *ENSUS 2002*, Marine Science and Technology for Environmental Sustainability Conference, University of Newcastle Upon Tyne, UK, December

Harris, L (1990) *Independent Review of the State of the Northern Cod Stock*, Prepared for the Canadian Department of Fisheries and Oceans, May 15

Holm, P (1996) 'Fisheries management and the domestication of nature' *Sociologia Ruralis* 36(2):177-188

Holm, P (2001) *The Invisible Revolution: The Construction of Institutional Change in the Fisheries*, Unpublished Doctoral Dissertation, Norway, Tromso University, Norwegian College of Fisheries Science

Holm, P (2003) 'Crossing the border: on the relationship between science and fishermen's knowledge in a resource management context' *MAST* 2(1):5-33

Hutchings, J (1996) 'Spatial and temporal variation in the density of Northern cod and a review of hypotheses for the stock's collapse' *Canadian Journal of Fisheries and Aquatic* Sciences 53: 943-962

Hutchings, J and Ferguson, M (2000) 'Temporal changes in harvesting dynamics of Canadian inshore fisheries for Northern Atlantic cod (Gadus morhua)' *Canadian Journal of Fisheries and Aquatic Sciences* 57:805-814

Hutchings, J and Meyers, R (1994) 'What can be learned from the collapse of a renewable resource? Atlantic cod (Gadus morhua), of Newfoundland and Labrador' *Canadian Journal of Fisheries and Aquatic Sciences* 51: 2126-2146

Hutchings, J and Meyers, R (1995) 'The biological collapse of Atlantic cod off Newfoundland and Labrador: an exploration of historical changes in exploitation, harvesting technology, and management' in R Arnason and L Felt (eds) *The North Atlantic Fisheries: Successes, Failures, and Challenges*, Charlottetown, University of Prince Edward Island, Institute of Island Studies

Hutchings, J, Neis, B and Ripley, P (2002) 'The "nature" of cod, *Gadus morhua'* in R Ommer (ed) *The Resilient Outport: Ecology, Economy, and Society in Rural Newfoundland*, St John's Newfoundland, ISER Books

Hutchings, J, Walters, C and Haedrich, RL (1997) 'Is scientific inquiry incompatible with government information control?' *Canadian Journal of Fisheries and Aquatic Sciences* 51:2363-2378

Huxley, TH (1883) *Inaugural Address to the Fisheries Exhibition,* London

Jentoft, S and McCay, B (1995) 'User participation in fisheries management: lessons drawn from international experience' *Marine Policy* 19(3):227-246

Jessop, (2002) *The Future of the Capitalist State*, Oxford, Polity

Johnson, JP, Murray, G and Neis, B (2004) 'How knowledge changes in the fisheries' presented at the XI World Congress of Rural Sociology on *Globalisation, Risks and Resistance in Rural Economies and Societies*, Trondheim Norway, 25-30 July

Latour, B (1987) *Science in Action*, Cambridge Massachusetts, Harvard University Press

Latour, B (2004) *Politics of Nature: How to Bring the Sciences into Democracy*, Boston, Harvard University Press

MacGarvin, M (2002) 'Fisheries: taking stock' in *Late Lessons From Early Warnings: The Precautionary Principle 1896-2000*, Environmental Issues Report 22, European Environmental Agency, Luxembourg, Office for Official Publications of the European Communities

Maguire, JJ, Neis, B and Sinclair, PR (1995) 'What are we managing anyway? The need for an interdisciplinary approach to managing fisheries ecosystems' *The Dalhousie Law Journal,* Special edition on the Atlantic Canadian Fishery:141-153

Mailhot, J (1993) *Traditional Ecological Knowledge: The Diversity of Knowledge Systems and Their Study*, Montreal, The Great Whale Public Review Support Office

Mansfield, B (2004) 'Neoliberalism in the oceans: "rationalization", property rights, and the commons question' *Geoforum* 35:313-326

Matthews, DR (1993) *Controlling Common Property: Regulating Canada's East Coast Fishery*, Toronto, University of Toronto Press

McCarthy, J and Prudham, S (2004) 'Neoliberal nature and the nature of neoliberalism' *Geoforum* 35:275-283

McGoodwin, JR (1990) *The Crisis in the World's Fisheries*, Stanford, Stanford University Press

Murray, GD, Neis, B, Ings, D, Schneider, D, Alcock, E, Bennett, E and Chaffey, H (forthcoming) 'Local ecological knowledge in the historical reconstruction of marine socio-environmental systems: methods, procedures and challenges' in B Neis and J Lutz (eds) *How Knowledge Moves*, Montreal, McGill-Queens University Press

Murray, GD and Neis, B (2004) 'Lessons learned from reconstructing interactions between local ecological knowledge, fisheries science and fisheries management in the commercial fisheries of Newfoundland and Labrador, Canada', presented at *The Commons in an Age of Global Transition: Challenges, Risks and Opportunities*, The Tenth Biennial Conference of the International Association for the Study of Common Property (IASCP), Oaxaca Mexico, 9-13 August

Neis, B 1992) 'Fisher's ecological knowledge and stock assessment in Newfoundland' *Newfoundland Studies* 8(2):155-178

Neis, B and Felt, L (eds) (2000) *Finding our Sea Legs: Linking Fishery People and Their Knowledge with Science and Management*, St. John's Newfoundland, ISER Books

Neis, B and Morris, M (2002) 'Fishers' ecological knowledge and stock assessment: understanding the capelin (Maillotus villosus) and capelin fisheries in the Bonavista region of Newfoundland' in R Ommer (ed) *The Resilient Outport: Ecology, Economy and Society in Rural Newfoundland*, St. John's Newfoundland, ISER Books

Neis, B, Schneider, DC, Felt, LF, Haedrich, RL, Hutchings, JA and Fischer, J (1999) 'Northern cod stock assessment: what can be learned from interviewing resource users?' *Canadian Journal of Fisheries and Aquatic Sciences*, 56:1944-63

Ommer, R (ed) (2002) The Resilient Outport: Ecology, Economy, and Society in Rural Newfoundland, St. John's Newfoundland, ISER Books

Parker, M (2003) *Against Management*, Malden, Maryland, Blackwell Publishers

Parsons, LS (1993) 'Management of Marine Fisheries in Canada' *Canadian Bulletin of Fisheries and Aquatic Sciences* 225, Ottawa, National Research Council of Canada and Department of Fisheries and Oceans

Pauly, D and Maclean, J (2003) *In a Perfect Ocean: the State of Fisheries and Ecosystems in the North Atlantic Ocean,* Washington, Island Press

Perry, RI and Ommer, RE (2003) 'Scale issues in marine ecosystems and human interactions' *Fisheries Oceanography* 12(4/5):513-522

Pinkerton, E (1990) *Co-operative Management of Local Fisheries,* Vancouver, University of British Columbia Press

Pinkerton, E (1994) 'Cooperative management as a strategy for the sustainable management of fisheries' *Canadian Journal of Fisheries and Aquatic Sciences* 51:2363-78

Pinkerton, E (1999) 'Factors in Overcoming Barriers to Implementing Co-management in British Columbia Salmon Fisheries' *Conservation Ecology* 3(2) [online]

Power, N (2000) 'Woman processing workers as knowledgeable resource users: connecting gender, local lnowledge, and development in the Newfoundland Fishery' in B Neis and L Felt, (eds) *Finding our Sea Legs: Linking Fishery People and Their Knowledge with Science and Management,* St. John's Newfoundland, ISER Books

Rogers, K (2002) 'Micro-level historical reconstruction of Newfoundland and Labrador fisheries between 1891-2000: findings and issues, back to the future, methods and results' in *Symposium on the Restoration of Past Ecosystems as Policy Goals for Fisheries*, Vancouver BC, University of British Columbia Fisheries Centre, February

Rowe, S and Feltham, G (2000) 'Eastport Peninsula lobster conservation: integrating harvester's local knowledge and fisheries science for resource co-management' in B Neis and L Felt (eds) (2000) *Finding our Sea Legs: Linking Fishery People and Their Knowledge with Science and Management,* St. John's Newfoundland, ISER Books

Scott, J (1998) *Seeing Like a State*, London, Yale University Press

Sen, S, and Nielsen, JR (1996) 'Fisheries co-management: a comparative analysis' *Marine Policy* 20(5):405-418

Smith, T (1994) *Scaling Fisheries: the Science of Measuring the Effects of Fishing, 1855-1955*, Cambridge, Cambridge University Press

Steele, DH, Andersen, R and Green, JM (1992) 'The managed commercial annihilation of northern cod' *Newfoundland Studies* 8:34-68

Taylor, H (2002) 'Insights into participation from critical management and labour process perspectives' in B Cooke and U Kothari (eds) *Participation: The New Tyranny?*, London, Zed Books

Whalen, J (in preparation) *Using Harvesters' Knowledge to Develop an Individual Based Computer Simulation Model of the St. John Bay, Newfoundland Lobster (Homarus americanus) Fishery,* Unpublished Masters dissertation, St Johns Newfoundland, Memorial University, Department of Geography.

Willmott, H (1993) 'Strength is ignorance; slavery is freedom: managing culture in modern organizations' *Journal of Management Studies* 30(4):515–52

Wroblewski, J (2000) 'The colour of cod: fishers and scientists identify a local cod stock in Gilbert Bay, Southern Labrador' in B Neis and L Felt (eds) *Finding our Sea Legs: Linking Fishery People and Their Knowledge with Science and Management*, St. John's Newfoundland, ISER Books

CHAPTER 17
A COMPARATIVE ANALYSIS OF THREE MODES OF COLLABORATIVE LEARNING IN FISHERIES GOVERNANCE: HIERARCHY, NETWORKS AND COMMUNITY

KELLY VODDEN, ROSEMARY OMMER AND DAVID SCHNEIDER
Coasts Under Stress Project, Memorial University of Newfoundland, 202, Elizabeth Avenue, St John's, NL A1E 1L4, Canada

Abstract

This chapter discusses three different ways of using collaborative learning in fisheries governance, all of which have been applied in the Coasts Under Stress (CUS) project in Canada. The three modes are: hierarchy; networks; and community. The hierarchical mode entails top-down computer modelling techniques, in which the experiential knowledge that is gathered from fishers' haul data is integrated with scientists' survey data into management plans. The networks mode entails developing an understanding of complex marine ecosystems by sharing knowledge between individuals and groups interacting in discussions about ecosystem structures and recovery strategies. The community mode entails the involvement of local communities in knowledge sharing. Our finding is that, in whatever mode it occurs, collaborative learning is of inestimable value in improving fisheries governance, especially by removing mutual misunderstandings. But techniques of collaborative learning cost time and money, and governments must be willing to devote the necessary resources to make them work.

17.1 Introduction

This chapter begins with an assumption that we have a duty to govern our interactions with social-ecological systems in a responsible fashion, but that we have failed to carry out this duty in recent years, partly because we have been using inadequate modes of governance. In particular, we have not fully recognised the complexity of the marine ecosystem, or the vital contribution to our understanding of this complexity that can be made by experiential knowledge. Indeed, the complexity of the ecosystem requires a correspondingly complex governance response. Governance of social-ecological systems today involves networks of interdependent public, private and non-government interests planning and making decisions at scales ranging from the local to the global in an "emerging multi-level governance regime" (Environment Canada 1999; Phillips and Orsini 2002).

Such improved governance for sustainable development requires improved systems for **managing knowledge** in the face of complexity, uncertainty and serious knowledge gaps that exist even in the midst of our 'information age' (Bouder 2002). Gaps and challenges in knowledge management are particularly apparent in the marine and coastal environment (Wolfe 2000). The Coasts Under Stress (CUS) project drew together over seventy investigators to generate knowledge of coastal social-ecological systems. Specifically, the research was designed to identify the complex non-linear interactions

T. S. Gray (ed.), Participation in Fisheries Governance, 291–306.

between social and environmental restructuring, as that operated across scales to affect the health and resilience of social-ecological systems (Murray *et al* this volume). CUS, itself an experiment in knowledge governance, was organised not around a central hypothesis with Cartesian partitioning into discrete subcomponents, but around one meta-concern (healthy social-ecological systems) in which various aspects of system well-being were first examined relatively distinctly and then integrated as the pathways between them were identified. Using the logo of a seastar, which mirrors the approach, one 'arm', for example, sought to understand how different kinds of **knowledge** (local and formally scientific) about ecosystem dynamics help to influence decision-making, which in turn affects human and environmental health. In this chapter, we discuss the creation and dissemination of knowledge relative to the dimensions of space and time using three CUS case study examples, one conceptual and two applied, to understand how cross-scale knowledge movements contribute to evolving forms of collaborative, adaptive multi-scale governance, necessary to the creation and maintenance of resilient socio-ecological systems.

17.2 The importance of scale

An important feature of improved fisheries governance lies in the scale of what we might call our 'home place' or 'community', to which we feel a sense of belonging and relatedness. In CUS we have found through the examination of wetlands stewardship programmes, for example, that there are very different interpretations of the notion of 'stewardship' across interests and scales (McLaren *et al* forthcoming). Stewardship ties in with the geographer's concept of *genre de vie* or **lifeworld**. Environmental aspects of the lifeworld include "sense of place, social space, time-space rhythms, and the lived dialectic between home and horizon" (Buttimer in Seamon 2004:1), and the term **landscape** used by cultural geographers when referring to relations between the natural environment and human society (Rose 1993). Much more complex than a space that can be indicated by a boundary on a map, landscape is both a home and a site of struggle, both "embedded in place and constructed and reconstructed by forces larger than itself", a complex, unstable material, and at the same time an ideological entity (Mitchell 2001:271; cf Lippard 1997). Thus the notion of clearly defining boundaries of 'community' in either time or space for purposes of governance is problematic. This is particularly true for First Nations, where the concept of boundary gives way to that of a functioning interlinked social-ecological cultural-spatial system whose limits cannot therefore easily translate to a boundary line drawn on a map.

Mitchell (2001) speaks of the **politics** of scale, using the example of how relation to place is packaged and sold to meet the needs of industries, such as tourism, which compete for access to space and resources with traditional activities (such as logging or fishing) that are very much a part of this original sense of identity. We found many of these ideological and material struggles over locality in our CUS research areas on both the east and west coasts of Canada. Other struggles over scale included debates about whether fisheries decision-making should take place at the local or regional, provincial, national or even international scale. For example, should cod fisheries be managed by individual bay, province, nation and/or internationally? Mitchell (2001) argues that any space is at once both local and global. CUS research (Vodden 2004; Ommer *et al* forthcoming), as well as that of resilience researchers such as Yorque *et al* (2002) suggest that while governance must increasingly involve multiple nested scales, regional scales, such as watersheds or catchment areas, warrant special attention, being an appropriate level at which to consider issues of sustainability since, at that scale, people and ecosystems maintain a close connection. This suggestion moves governance thinking closer to that of First Nations and their traditional territories, since ecology and society are clearly interdependent in such a view.

How, then, can we transfer knowledge of social-ecological systems across various scales? Since we need to work across physical and temporal scales if we are to grasp the complex relationships involved between people (individual, household, community…), their location (village, coastline, region…) and their environment (land, sea, fauna, biota and atmosphere), considered over time, multi-scale analysis is obviously required, as is cross-disciplinary work. One needs a range of space, time, organisational and conceptual levels of analysis to understand the structure and dynamics of social-ecological systems. In doing so, many different tools can be employed. The challenge is to investigate the utility of these tools in cross-scale analysis and knowledge transfer involving participants outside of the realm or discipline from which they evolved. The three case discussed below focus on three different modes of fisheries governance: 1) hierarchical space-time modelling; 2) network ecological modelling; and 3) community research activity.

17.3 Case studies

17.3.1 HIERARCHICAL SPACE-TIME MODELLING

This mode of fisheries governance is a top-down approach, making use of computer modelling to integrate the information gathered by fishers with the statistical data generated by scientific surveys. It is an example of interdisciplinary collaboration between fishers and scientists, though neither participated directly in management decisions. Figure 17.1 shows how this knowledge gathering activity is carried out.

Figure 17.1 shows one of the principal information gathering activities in fisheries science: the stratified random survey to estimate stock size. The unit of information is the biomass of fish in a single haul of the net (fixed duration of 10 minutes). A single survey of a large area (such as NAFO zones 2J and 3KL east of Newfoundland) might consist of 900 single hauls placed randomly in a larger area, the frame (F) of all possible hauls in 2J3KL. The area measured is $\sum$H, which takes about a month. The number of fish in the entire area (the frame) is then inferred from the sample ($\sum H \rightarrow F$). This measurement at one point in time is then used to represent the status of the stock for the year. The length of the horizontal arrow shows the burden placed on statistical inference from a sample to the frame of all possible sample locations.

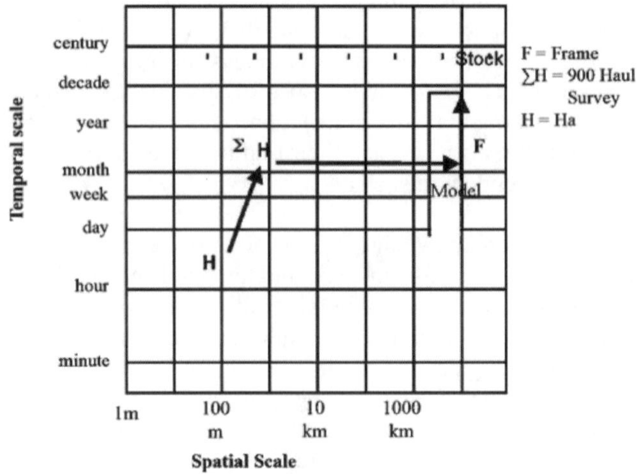

Fig. 17.1. The space and time scales of knowledge assembly by stratified random survey and computational model in fisheries research

Figure 17.2 contrasts the information gathering activity in fishery science with the information gained through the experience of an individual fisher and a fishing community. The information gathered during the course of a single trip occurs over a small area. Experience from each trip accumulates during the course of a career (and even multiple generations and careers), usually over a restricted area. The experience of the individuals in a community will cover a larger area.

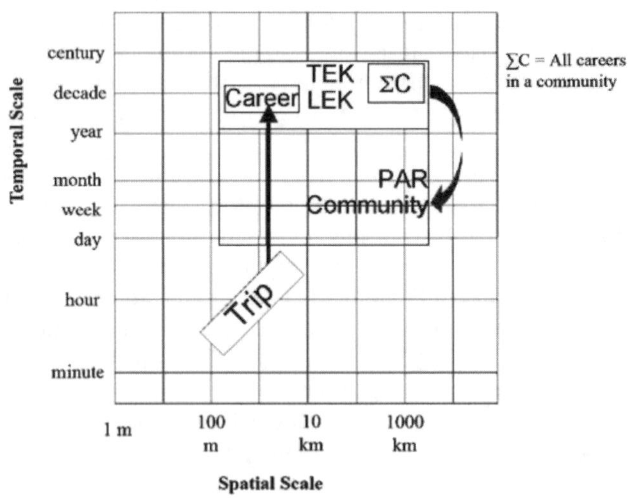

Fig. 17.2. The space and time scales of knowledge assembly by participatory action research (PAR) including the traditional or local ecological knowledge (TEK, LEK) of a community of people engaged in fishing.

A comparison of these diagrams illustrates the tendency of local knowledge to provide more spatially specific but temporally extensive knowledge, while conventional 'western' science tends toward larger scale models over limited time frames (illustrated further by the examples provided below). Both knowledge systems are required to acquire a more complete understanding at multiple scales. Faced with two different dilemmas: how to reduce the burden of inference in a stratified random survey and the burden of scaling up, Figure 17.1 shows the first step in solving the problem. Knowledge gathered on the hauls is fed into a model, which includes estimates of mortality rates and other knowledge, which is combined in computational form, to allow a generalising of haul data at point F. Figure 17.2 shows a similar kind of process for the fishing experience of individual people, in which recurrent patterns are identified through analysis of data banks of individual trips and careers. Finally, the information and analysis from both research methodologies is returned to the participants (be they government scientists or community people) and, ideally, these two groups work jointly to pool information and insights, and thus come to a shared management plan. This process represents cross-scale knowledge-sharing but also shared decision-making or fisheries co-management, a relatively well-researched form of multi-level governance (Vodden 1999; Pinkerton and Weinstein 1995; Berkes, 1996).

In future iterations one might compare diagrams for different generations of scientists and communities over time, or for different areas, to examine issues of comparative scale, such as how management plans may need to be modified over time or space. Further development of the model might seek ways to integrate such functions into Figure 17.2 to reveal how, for example, observations may vary based on occupation or function within a community (whether it be communities of place, interest or both). Discussion of the diagrams as suggested is likely to illustrate very different interpretations of the various kinds of scale and their interactions, as well as the ways that such a tool can be used and improved. Discussion about values and objectives embedded into the process of designing and using the tool may also result.

The example of this hierarchical model is but one illustration of how interdisciplinary and collaborative explorations of topics critical to sustainable coastal development can evolve and lead to innovations in coastal social-ecological systems knowledge. CUS researchers developed and examined several such fisheries-related modelling tools and techniques. One was created, for example, to identify candidate areas for marine protected area (MPA) designation in Atlantic Canada, taking into account ecological parameters such as species richness and species-at-risk protection as well as social considerations such as likelihood of acceptance and conflict with economic interests (Baker 2003). Neither the development of the model, nor consideration of its outcomes, however, involved participants outside of the academic community. Stakeholder involvement is considered an important next step for checking data and assumptions made in the model and for implementing any decisions about MPA designation and implementation that may come about as a result (Ommer *et al* forthcoming).

17.3.2 NETWORK ECOLOGICAL MODELLING

A second modelling procedure developed to provide direction for the restoration of depleted marine ecosystems made use of networks of interested parties to generate the necessary information. This mode of governance is exemplified in the Back to the Future (BTF) project undertaken by CUS. BTF 'reconstructs' systems as they might have been before they were depleted by modern industrial fishing, and then quantifies the socio-economic and ecological tradeoffs inherent in restoration. Using a whole ecosystem approach, restoration goals (an optimal restorable biomass) are set and strategies developed, including simulated optimal patterns of fishing, to rebuild the ecosystem and recapture lost productivity potential (Ommer *et al* forthcoming). The temporal dimension of scale is critical to this process, which considers present, past and future ecosystem states. Social, economic and ecological costs and benefits of varying restoration goals are evaluated using an intergenerational discounting technique that allows for division of the benefit stream between current and future generations (Ainsworth and Sumaila 2003). Methods are being sought for incorporating cultural and ecosystem service values as well.

To date, the BTF model has been applied in British Columbia and Newfoundland, Canada and in several other locations internationally (Ainsworth *et al* 2002; Heymans and Pitcher 2002; Ainsworth and Pitcher 2004; UBC 2004). BTF takes advantage of advances in ecosystem, spatial and bio-economic modelling to integrate both quantitative and qualitative information. It also adds a participative element missing in the MPA modelling discussed above. Community interviews and workshops, national and international conferences and publications are utilised to generate discussion on model development, outcomes and restoration strategies with scientists, managers, First Nations and other resource users. New model iterations have been produced as a result of these interactions. The process is innovative in its ability to integrate information from a wide range of sources. Respect and recognition of all knowledge systems is emphasised.

Despite international recognition and application of the modelling techniques it employs, BTF has not yet achieved the widespread sharing of knowledge and facilitation of multi-level dialogue among Canadian academic, community, fishing and government sectors that it warrants. Available resources for engagement of non-academic partners are limited and there has been as yet no 'take-up' by those who are exploring and implementing ecosystem-based management (Haggan 2004). Reflections on these difficulties identify the challenge of generating genuine commitment to knowledge sharing in the polarised climate of today's fishery where knowledge may be hoarded as a bargaining chip or 'edge' over other actors (Haggan 2000). As an 'honest broker' of information, the university can play an important role in encouraging knowledge sharing. Further, such iterative processes may build intellectual and social capital over time that will build a collective understanding and improve modelling techniques and outcomes, particularly as additional parties provide input and data (Ommer *et al* forthcoming; UBC 2004). This is an essential precondition for agreement on reinvestment in natural capital at the necessary scale.

Two additional (and linked) challenges encountered in BTF include the problem of matching often micro-spatial scale local and traditional knowledge (LK/TK) with macro-spatial models, along with problems associated with the diverse and changing nature of LK/TK and methods of capturing this knowledge in an appropriate and respectful manner. BTF work draws upon global scale examinations of fishing down marine webs (Pauly *et al* 1998) while seeking to address the shortcomings of fisheries landings data at the provincial-scale with the knowledge of harvesters, First Nations and non-government organisations. Local knowledge has been used, for example, to provide the micro-spatial and temporal information required to reconstruct cod migration patterns and suggest relationships between these patterns and states of recovery. It has also demonstrated spatial, temporal and ecological expansion and intensification in fisheries. The gathering and translation of such knowledge has not come without problems, however, including community tensions and disagreements due to deep conflicts between small boat and large boat fish harvesters (UBC 2004). Such tensions illustrate the complexity and diversity of LK (Murray *et al* this volume).

Collaboration on the reconstruction of past abundance and imagination of possible futures may well build greater mutual understanding of ecosystem structure and function and of recommended recovery strategies, but the question remains whether the will or resources are sufficient to continue this work and further refine these innovative tools. Further, can policy-makers who currently dominate fisheries governance, such as the federal Department of Fisheries and Oceans (DFO), be brought fully into the dialogue so that this new and integrated knowledge is translated into improved decision-making?

Individuals and groups involved in discussion about tools such as time-scale diagrams and ecological models will change over time, as will analytical tools, fish stocks and ecosystems themselves. The collection of these individuals, groups and resources can be viewed as a network. Clark (2000) describes networks as two or more people communicating back and forth. Actor networks are described by Ommer and Sinclair (forthcoming:7) as "constantly emergent linkages among people, wildlife and materials" (all actors, or agents producing effects), though the capacity for responsibility for actions is not the same for all participants. The network concept "allows for uncertainty, chaotic outcomes, positive or network transforming feedbacks, and system complexity". Actor-network theory assumes that social structure, like notions of scale discussed above, "is not free-standing…but a site of struggle, a relational effect that recursively generates and reproduces itself…" (Law 1992:5). Actors within a network carry varying degrees of power and influence but have some common purpose for communicating and/or working together (Creech 2001). Knowledge then is a product of networks, networks that include an increasingly broad range of actors and are subject to shifting power dynamics. The recognition of local actors and social scientists beyond the field of fisheries economics as valid participants in the fisheries management knowledge network demonstrates such a shift. Under-utilisation of BTF techniques indicates, however, that rigidities remain in the system. In the case of BTF, it appears that, despite the development of a promising analytical tool, the potential for knowledge sharing and transfer between the academic community and other actors in the Canadian fisheries policy network has not been fully realised.

17.3.3 COMMUNITY RESEARCH ACTIVITY

The third mode of fisheries governance using experiential knowledge is community-based. This mode signifies another change in Canada's fisheries networks – towards the creation of community-based research institutions to facilitate local participation in knowledge creation and dissemination with respect to fisheries and other interrelated aspects of social-ecological systems. This aspect of CUS research involved the examination of such local institutions, their interactions with other institutional actors, and their impacts on knowledge flow and decision-making (Vodden and Bannister forthcoming; Gibson *et al* forthcoming). The example of the Indian Bay Ecosystem Corporation (IBEC) demonstrates the use of modelling as a participative tool leading to fisheries policy change at scales well beyond the local. It further suggests the importance of local institutions as critical nodes in actor networks that can facilitate local input into broader-scale processes, while also bringing those knowledge generation and decision-making processes 'closer to home' and to the ecosystems and resources they impact.

Indian Bay is a fishing and former logging community on Northwest Arm at the head of Bonavista Bay, on the northeast coast of Newfoundland. A major forest fire put an end to logging as the area's primary industry in 1966. The Indian Bay watershed, an extensive freshwater system made up of more than seventeen lakes[1] feeding into the Indian Bay River and ultimately to Bonavista Bay, was renowned nationally for exceptional angling, and especially for trophy size brook trout, *Salvelinus fontinalis* (Power 1997). The system is highly productive in terms of both fish diversity and growth rates. While in operation, the forest company restricted access to the lakes. When the company pulled out of the area, and road access was provided, coupled with technological developments such as the snowmobile, the system soon became overexploited by both locals and visitors. Anglers could harvest twenty-four fish per day (or ten lbs plus one fish) legally, yet even these generous restrictions were not enforced and often ignored. Trout populations hit what is thought to be an all-time low in the mid-1980s (Gibson *et al* forthcoming). Indian Bay and neighbouring communities decided something needed to be done to protect what was an economically, socially, culturally and ecologically valuable resource. Two local development associations teamed up to form the Indian Bay/Cape Freels Ecosystem Committee, which later evolved into IBEC.

More than 5000 bags of garbage were removed from the system in the early years, along with heavy equipment left behind by the logging industry. Habitat improvements included the removal of debris, pulpwood jams, logging dams, old beaver dams and culverts; the reconstruction of collapsed bridges; and riverbank erosion control. The group conducted a public awareness campaign and a user survey that confirmed concern about the declining stocks and suggested local values related to the fish resources (Gibson *et al* forthcoming). Some government support had been obtained at this point, particularly in the form of funding from federal human resources and environment departments. However, the committee realised that to go the next step in fisheries recovery and management, partnerships with universities and provincial and federal fisheries managers would be essential.

Unlike other Canadian provinces, the Province of Newfoundland has not accepted full responsibility for the management of freshwater fisheries, because it claims that it does not have the resources to devote to scientific study related to trout management. However, although the DFO "does not currently do any science on trout" (DFO manager 2004) it did devise a Trout Management Plan (expired in 2003 and now under review). The Province got involved in trout research, in cooperation with IBEC, DFO and others,

[1] Small and medium sized freshwater bodies are called ponds in Newfoundland and Labrador.

during a federal/provincial agreement in the early 1990s and has continued a limited program since that time. Brook trout are the most commonly harvested freshwater fish in Newfoundland. Variability in life history, distribution and productivity across a wide range of habitats in Newfoundland and Labrador make management of this species "highly data dependent" (van Zyll de Jong *et al* 2002:267). Current management regulations are, however, largely determined in the absence of data and, before those in Indian Bay, few studies had been done on brook trout dynamics or on simulation exercises to examine the effectiveness of management strategies. By the late 1980s, concerns regarding declines in trout stocks were present well beyond Indian Bay. IBEC was formed in the absence of strong, defensible management (van Zyll de Jong *et al* 2002).

It was the IBEC's original intention to build a hatchery for the purpose of restocking brook trout. But this recovery strategy was not implemented because of scientific advice given by the Salmonid Research Group at Memorial University regarding potential deterioration in genetic diversity and declines in numbers and size of wild fish (Gibson *et al* forthcoming). Instead, a multi-year research and trout assessment project was launched by IBEC in partnership with Memorial University, and provincial and federal fisheries departments (including not only the Province of Newfoundland but also the Ontario Ministry of Natural Resources) to: a) provide the scientific data needed to evaluation fish population status; b) provide a model for managing exploitation in brook trout fisheries; and c) offer advice for management decisions (van Zyll de Jong 2002). Together the research partners showed that the water quality of the system was good, and that growth rates of trout in Indian Bay remained the same as before, and were greater than stocks for the Avalon Peninsula (southeast Newfoundland). They derived models to show that stocks would recover if special regulations for the system were made.

While provincial and academic partners were crucial to the development of the model itself, the research questions were established by community concerns (knowledge of local problems). Local staff and students were employed to collect data, and community members were brought in through the IBEC Board of Directors and open community meetings to discuss results and recommend regulatory changes. Special experimental regulations were put in place for some lakes, and two were closed for fishing as a result of this research and public support. IBEC recommended to the federal DFO that special regulations be implemented. DFO agreed, and seasons were shortened, and a bag limit of six fish or two lbs plus one fish was applied in 1994, from which time angling began improving (increased fish size). Legislative and policy measures in Indian Bay are now accompanied by monitoring and enforcement activities undertaken by local staff, again in cooperation with federal and provincial departments. Brook trout sizes have increased as a result, though modelling results suggest that the fishery is presently operating near full capacity.

Present regulations are not effective when effort is high, and size restrictions as well as species-based regulations are needed to contain increased effort while sustaining high quality fishing. IBEC had been in discussion with DFO about such regulatory changes for nearly ten years, and at IBEC's 2004 annual meeting with government and industry partners, a senior DFO fisheries manager announced that new regulations had finally been passed to allow for such management measures. IBEC was asked to bring a proposed management plan for the watershed to their upcoming AGM and, if approved by the community and consistent with the regulatory changes, DFO would also endorse the plan. Similar changes were to be rolled out across the Province throughout 2004 and 2005. The IBEC Manager now chairs the Province's Trout Advisory Working Group, yet another indicator of the local organisation's province-wide influence. Further, local students have completed their training and graduate research within the watershed and then gone on to work in their careers as provincial fisheries managers.

The decline of the trout resource and its recovery in Indian Bay is an excellent example of where negligence, overexploitation and lack of knowledge led to the collapse of a resource, but where local initiative and the 'bottom-up' approach, coupled with knowledge and decision-making partnerships with 'top-down' scientists and policy-makers, at least in part, has helped to restore the system. Their collaborative research efforts continue to inform and improve fisheries management at multiple scales from the immediate region to the province as a whole, and even nationally and internationally. Corporate and individual decisions within the watershed have been influenced by IBEC, while requests for information and interest in applying this approach to freshwater fisheries management have come from as far afield as Africa, the UK and US (IBEC 2003). IBEC then has a potential application in addressing challenges in freshwater and recreational fisheries worldwide.

IBEC also offers an opportunity for cross-fertilisation of lessons from these fisheries to other resources and ecosystems (for example, marine fisheries). It has broadened both its spatial and functional scale of discretionary reach over time, and now includes an ecosystem-based approach to research and planning rather than single species or even fisheries-focused efforts. Studies have been undertaken relating not just to fisheries but also to forestry practices, invertebrates, plants, anthropology and rural development. Research on logging impacts in the watershed led to some areas being set aside for protection as pristine study areas under agreement with the logging companies. An IBEC-developed land use plan resulted in still others being excluded from cabin development. There are plans underway to build more laboratory facilities, accommodation and an interpretation centre, and to create a nationally important field research centre (the Indian Bay Centre for Cooperative Ecological Studies). With these developments has come a broader focus on the ecosystem as a whole, including attention to the coastal area and linkages between freshwater and marine systems (Vodden 2004). However, this ecosystem focus would not have been possible initially. It has taken time to build capacity, knowledge and awareness about the linkages between the health of freshwater fisheries and the social-ecological systems of which they are a part.

However, all has not been smooth sailing within the IBEC. As in the BTF, attempts to engage fisheries stakeholders; community conflict; and differing opinions portrayed as knowledge, within a community have added complexity to the process. For instance, because of conflicting objectives, misinformation, lack of communication between local, academic and government partners, and inaction by those who could have intervened, a controversy emerged in the community about the impacts of fisheries research (particularly mortalities due to fyke netting). This brought the research to a temporary end, but, as Gibson *et al* (forthcoming) point out, the efforts of a few key people and their commitment to local involvement in managing the watershed and its fishery, have managed to re-establish and even expand research and monitoring activities, demonstrating the resilience of the community-based system. The presence of a local institution to facilitate discussion of these conflicts led to their resolution and a greater understanding among misinformed parties on both extremes of the debate, a learning process that will improve future research efforts and collaboration.

Nevertheless, despite the apparent resilience of the Indian Bay model, now nearly twenty years old, reliance on key individuals introduces an element of vulnerability, as does the need for ongoing, yet fragile, public, government and academic assistance. Reliance on strained public financial resources for community engagement, and for all of their ongoing work, is a challenge in the IBEC case, as it was in the BTF case. Volunteer and leadership energy has limitations, and the organisation is struggling to develop a more entrepreneurial institutional model under pressures of government downsizing and cost-cutting associated with neo-liberalism and the post-industrial era. While the organisation has been able to build understanding and negotiate compromises and win-win solutions that satisfy the very different values of local users, academic and government partners, their ability to continue to do so requires ongoing effort supported by organisational infrastructure, human resources and relationships.

In the case of IBEC, the watershed has proven a workable scale for the application of stewardship and collaborative governance. Aside from their logic as a biophysical scale with relatively clear ecological boundaries, watersheds are scales to which local communities can increasingly relate, particularly as watershed-level institutions are established and environmental awareness increases. People tend to feel a direct connection with the watersheds in which they live and rely upon for their drinking water, recreation, subsistence and/or culture. For senior governments, and academic researchers, the watershed is a scale of significant size to warrant their attention. This ability to relate to the watershed scale is an important consideration in stewardship and in knowledge generation and sharing. People are more apt to care for something they understand and feel some direct connection to. Awareness of the watershed scale (what it is and how it functions, both in terms of the larger catchment area as well as the subsystems within it) can be fostered through education and information sharing as well as through direct participation in watershed activities. Public or citizen knowledge, then, is an important knowledge type. IBEC emphasises the importance of simultaneously reaching out both above and below, engaging the grassroots along with external partners (at higher levels in the spatial hierarchy).

17.4 Discussion

These examples highlight the role of knowledge sharing, communication and social learning among coastal interests, including the application of this learning to the development of new institutions and new analytical tools. They also raise the question of whether such processes of collaborative enquiry will lead to more sustainable fisheries management decisions and, ultimately, a lighter social-ecological footprint on a globally-imperilled resource. This, of course, means thinking about how people act, their values and ethics, the networks they can create, their learning systems and the role of communications across scales in developing knowledge systems that are not only adaptive and evolving but also collaborative. It means opening up the knowledge process to input and critique from a wide range of diverse actors and avoiding the kind of 'closed shop' government science that has been widely criticised for its failure in the past (Hutchings *et al* 1997). We have inherited a legacy of barriers in understanding between actors from diverse disciplines, cultures and locales which strongly suggests that we are dealing with unproductive rigidities in our current knowledge systems, such as the unsustainable fisheries practices that led, and still lead, to ecosystem collapse. New models of collaboration – such as CUS itself and the knowledge partnership case studies it examined – show the breaking down of *some* of these barriers. Many remain. Social learning theory suggests how this change (breaking down barriers to knowledge creation and exchange) might occur. An extension of an organisational learning theory, it suggests that 'change' is a fundamental feature of modern life and hence imperative to develop social systems that are able to learn and adapt through "reflection-in-action" (Schön 1983; Argyris and Schön 1974, 1978). Networks of actors and informational 'feedback loops' that operate throughout these networks, are fundamental to the development and operation of such learning systems (Smith 2001).

Not all learning, however, is beneficial (Huber 1991). One can learn, for example, how to cheat, steal or pollute. What is required for social learning to produce positive sustainability outcomes is a re-evaluation not only of technique but also of purpose and values, with the hope that predominant values will align themselves with those compatible with ecological approaches to sustainable development. Each of our case studies and the tools and models they employed involved an examination of values and comparison of management outcomes when different values are emphasised (conservation versus economics or recreation). In each case, determining measures of effectiveness and productivity was dependent on objectives and values (weighting to social, ecological or economic outcomes). Values are a lens through which we interpret knowledge and decide what the important questions are to examine (McLaren *et al* forthcoming). Stenmark (2002) notes how all environmental policies are based on attitudes and judgements, more often implicitly rather than explicitly stated, about what is valuable, what should be encouraged and about whose interests should be satisfied. Values and goals determine where we want to go. Scientific information helps us get there, but knowledge obtained through lived experience can also tell us where we should be going, and what values should guide us along the way. Values then can be learned. The key is to recognise the importance of values, articulate and make them explicit wherever possible and openly discuss them. Stewardship, although it requires knowledge, is also a values exercise (McLaren *et al* forthcoming). Argyris and Schön (1978) emphasise the importance of learning strategies that involve a re-evaluation of governing goals, policies, values and "mental maps". Others have described this as second-order, large-scale or 'deep structure' change (Bartunek and Moch 1987; Gersick 1991).

Our research suggests that inclusion of local actors in knowledge networks helps open up these difficult dialogues about values, a discussion often avoided by scientists who aim to be 'entirely objective and value-free', despite the now demonstrated incapacity of anyone to be so. It is necessary to hold such discussions in order to develop mutual understandings and meanings and decisions that are as widely acceptable as possible. Such a process becomes particularly useful when conclusive scientific evidence is not available, as is often the cases in such complex problems as fisheries management (Bouder 2002). Workshops and conferences involving a wide range of actors were utilised in both BTF and IBEC as a tool for facilitating this dialogue.

However, our examination of knowledge partnerships in CUS offers lessons of caution as well as optimism. Collective learning is costly. Learning systems must be designed, managed and supported. They require investment in developing group policies, protocols and norms, as well as activities and organisational infrastructure. Nodes within the network, at multiple scales and within various communities of interest, must be maintained, their role being to facilitate knowledge exchange. This requires a policy framework and ongoing support for knowledge exchange and investment by private, public and civil society sectors. Networks tend to be more successful, requiring less investment of time and resources, where their component actors and nodes are more cohesive, and their resources are more equitably distributed. Yet this is rarely the case on Canada's coastlines, where controversy rages over issues of aquaculture, fisheries, forestry, tourism and offshore oil and gas development. The need for collaborative knowledge systems is great, but the capacity at present is fledgling and under-resourced. Strong leadership support and capacity for collaboration at all levels is also required but often lacking.

What of the implications of this collaborative learning process for management decision-making structures? Collaborative learning processes will, inevitably, precipitate changes in management structures: both detailed management changes in a fishery within a region based on a new, shared understanding of stock size; and strategic value-driven changes to wetlands stewardship programmes or discounting techniques. Such changes will have to be based on learning, facilitated by innovations ranging from old analytical tools revised for modern purposes (for instance, the time-scale diagram), to new technologies in computer modelling, or conferences and research projects that lead to deep structure debates. These examples illustrate the potential for extending communications networks to knowledge networks to multi-level coastal governance arrangements where decisions are made in a more equitable, collective manner.

However, given what we know about rigidity in the present fisheries governance system, is it realistic to believe that the potential of collaborative learning systems will ever be realised? Learning systems require both monitoring and an ability/willingness to respond to feedback. Will an improved time-scale diagram, better modelling techniques or more effective local institutions matter? Without a deeper value and culture change, we think not. As Bouder (2002) points out, we need more than new institutions, we need existing institutions to better integrate economic, environmental and social objectives within their mandates. Fortunately, institutions can change significantly in terms of the rules and norms that guide their behaviour, not just their form and structure (Lowndes and Wilson 2001). Tools such as those discussed in this chapter can provide not only better information (feedback) but also can help to facilitate discussion about differing perspectives and ,over time, they can yield shared understandings.

Our experience in CUS has demonstrated some ability and willingness in Canada's federal institutions to respond to feedback about the kind of deep, value-driven change required to adapt to large-scale changes in social-ecological systems. Support and willingness to collaborate with strong local institutions has been demonstrated through IBEC and Murray *et al* (this volume). On a broader scale, Canada's Oceans Act and its Strategy and Action Plan call for collaborative decision-making processes that involve multiple stakeholders and are based on principles such as ecosystem management and the precautionary approach. This new policy framework explicitly recognises a role for collaborative knowledge generation. Also promising is Canada's investment in knowledge networks, including those specifically designed to inform coastal and oceans policy such as the Ocean Management Research Network and Canadian Policy Research Network. The varied ways in which the Oceans policy is (and is not) being implemented across Canada, demonstrates that actors at scales below the national level are, indeed, important in policy outcomes.

Although significant barriers to knowledge flow continue to exist, even among members within these various networks, CUS innovations in interdisciplinary work demonstrate that boundaries can be crossed, and they provide hope that rigidities can be overcome, utilising nested multi-scale structures linked vertically and horizontally through extensive and ongoing communication, together with the development of new tools and techniques employed within knowledge networks. Common to these multi-level structures is a recognition of the importance of the local and regional scale, but also an acknowledgement that local is not enough. National and even international actors must be designed into the multi-level network, with different tools employed for effective and ongoing communication at all levels. Such efforts are difficult and costly but necessary if complex realities are to be addressed.

A 2004 International Conference on New Approaches to Rural Policy called for a "new rural governance, based on consultation, negotiation and partnerships among government, businesses and communities". The facilitation of "knowledge pooling" across and within levels of government and between public and private sectors was described as one of three key areas for investment in such a new governance system (FRB *et al* 2004). Such investments could start with the education of individuals, including school children and the public-at-large, to facilitate their engagement in sustainability debates and increase their discretionary reach. Resourcing and capacity-building must then extend to scales beyond the individual to the support of various nodes within knowledge networks that help facilitate participation and knowledge flow. We have provided examples above of nodes at both functional and territorial/spatial scales. Continued research and reporting on measures to increase the success of knowledge networks can play an important role in their continued development.

17.5 Conclusion

Our conclusion is that collaborative learning is a pre-condition of a successful management response to the complexities of marine ecosystems, and that the more extensive the net is spread to include the largest number of contributors to the collective learning experience, the better. Establishing and maintaining networks of actors across territorial and functional scales, each with varied and shifting boundaries and interpretations, is a massive challenge. But our three examples offer promise, drawn from the perspective of coastal knowledge management as evidenced in the recent experience of the CUS project, a major interdisciplinary, collaborative research effort. Visions for future policy-making create a picture of a new multi-level, multi-interest governance regime where government is not the only govern"er" yet remains a significant supporter of and participant in knowledge and policy networks capable of addressing complex realities. Integration of shared knowledge represents a starting point for developing sustainable governance systems, a much-needed shift in human systems in the face of large-scale human-induced ecological changes that threaten the very survival of marine ecosystems.

References

Ainsworth, C, Heymans, J, Pitcher, T and Vasconcellos, M (2002) 'Ecosystem models of northern British Columbia for the time periods 2000, 1950, 1900 and 1750' *Fisheries Center Research Reports* 10(4):41pp, Vancouver BC, University of British Columbia Press

Ainsworth, C and Pitcher, T (2004) 'Back to the future in Northern BC: Evaluating historic goals for restoration', presented at the 4th World Fisheries Congress, American Fisheries Society, 2-6 May

Ainsworth, CH and Sumaila, UR (2003) 'Intergenerational discounting and the conservation of fisheries resources: a case study in Canadian Atlantic cod' *Fisheries Center Research Reports* 11(3):26-33, Vancouver BC, University of British Columbia Press

Argyris, C and Schön, DA (1974) *Theory in Practice: Increasing Professional Effectiveness,* San Francisco, Jossey-Bass

Argyris, C and Schön, DA (1978) *Organizational Learning: A Theory of Action Perspective*, Reading Massachusetts, Addison-Wesley

Baker, K (2003) *Potential Sites for Marine Protected Areas in Canada's North Atlantic*, MES Thesis, Department of Environmental Science, Memorial University of Newfoundland and Labrador, St. John's, Newfoundland and Labrador

Bartunek, J and Moch, M (1987) 'First-order, second-order, and third-order change and organizational development interventions: a cognitive approach' *Journal of Applied Behavioral Science* 23:483-500

Berkes, F (1996) 'Social systems, ecological systems, and property rights' in S Hanna, C Folke, and KG Maler (eds) *Rights to Nature*, Washington DC, Island Press

Bouder, F (2002) *Improving Policy Coherence and Integration for Sustainable Development: A Checklist*, Paris, OECD Policy Brief

Clark, H (2000) *Formal Knowledge Networks: A Study of Canadian Experiences*, Winnipeg, International Institute for Sustainable Development

Creech, H (2001) *Form Follows Function: Management and Governance of a Knowledge Network*, Winnipeg, International Institute for Sustainable Development.

DFO manager (2004) Department of Fisheries and Oceans Canada, personal communication

Environment Canada (1999) 'The New World of environment policy making: theory and practice' presented at the ESAC Annual Meeting, Sherbrooke, Quebec

FRB [Federal Reserve Bank of Kansas City], OECD [Organisation for Economic Cooperation and Development], Rural Policy Research Institute and The Countryside Agency (2004) *International Conference on New Approaches to Rural Policy: Lessons from Around the World*, Washington DC, 25-26 March

Gersick, CJG (1991) 'Revolutionary change theories: a multilevel exploration of the punctuated equilibrium paradigm' *Academy of Management Review* 16:10-36

Gibson, J, Haedrich, R, Kennedy, J, Vodden, K and Wernerheim, M (forthcoming) 'Barriers and diversions to knowledge flow' in J Lutz and B Neis (eds) *Making and Moving Knowledge*, Montreal, McGill-Queen's Press

Haggan, N (2000) 'Back to the future and creative justice: recalling and restoring forgotten abundance in Canada's marine ecosystems' in H Coward, R Ommer and T Pitcher (eds) *Just Fish: Ethics in the Canadian Coastal Fisheries*, St. Johns Newfoundland, ISER Books

Haggan, N (2004) personal communication

Heymans, JJ and Pitcher, TJ (2002) 'A model of the marine ecosystem of Newfoundland and Southern Labrador (2J3KLNO) in the time periods 1985-1987 and 1995-1997' *Fisheries Center Research Reports*, 10(5):5-43, Vancouver BC, University of British Columbia Press

Huber, GP (1991) 'Organizational learning: the contributing process and the literatures' *Organizational Science* 2:88-115

Hutchings, J, Walters, C and Haedrich, R (1997) 'Is scientific inquiry incompatible with government information control?' *Can. J. Fish. Aquat. Sci.* 54:1998-1210

IBEC [Indian Bay Ecosystem Corporation] (2003) *Phase I Feasibility Study, Indian Bay, Newfoundland*, Indian Bay, NL Centre for Terrestrial and Aquatic Field Studies

Law, J (1992) *Notes on the Theory of the Actor Network: Ordering, Strategy and Heterogeneity*, Lancaster University Centre for Science Studies, Cartmel College

Lippard, L (1997) *The Lure of the Local: Sense of Place in a Multi-centred Society*, New York, The New York Press

Lowndes V and Wilson, D (2001) 'Social capital and local governance: exploring the institutional design variable' *Political Studies* 49:629-647

McLaren, B, Hollis, T, Roach, C, Blanchard, K, Chaurette E and Bavington, D (forthcoming) 'Knowledge flows, conservation values and municipal wetlands stewardship', in J Lutz and B Neis (eds) *Making and Moving Knowledge*, Montreal, McGill-Queen's Press

Mitchell, D (2001) 'The lure of the local: landscape studies at the end of a troubled century' *Progress in Human Geography* 25(2):269-281

Ommer, R, Coward, H and Parrish, C (forthcoming) 'Knowledge, uncertainty and wisdom' in J Lutz and B Neis (eds) *Making and Moving Knowledge*, Montreal, McGill-Queen's Press

Ommer, R and Sinclair, P (forthcoming) 'Introduction' in R Ommer and P Sinclair (eds) *Power, Agency and Nature: Shaping Coastal Society and Environment*, St. John's Newfoundland, ISER Books

Pauly, D, Christensen, V, Dalsgaard, J, Froese, R and Torres, FJR (1998) 'Fishing down marine food webs' *Science* 279:860-863

Phillips, S and Orsini, M (2002) *Mapping the Links: Citizen Involvement in Policy Processes*, Ottawa, Canadian Policy Research Network Discussion Paper No. F21

Pinkerton, E and Weinstein, M (1995) *Fisheries That Work: Sustainability Through Community-based Management*, a Report to the David Suzuki Foundation, Vancouver BC, David Suzuki Foundation

Power, W (1997) 'Rebuilding trouters' heaven' *Newfoundland Sportsman* 7(4 -July/August):26-31

Rose, G (1993) *Feminism and Geography: the Limits of Geographical Knowledge*, Minneapolis, University of Minnesota Press

Schön, D (1983) *The Reflective Practitioner: How Professionals Think in Action*, London, Temple Smith

Seamon, D (2004) 'Grasping the dynamism of urban place: contributions from the work of Christopher Alexander, Bill Hillier, and Daniel Kemmis' in T Mels (ed) *Reanimating Places: a Geography of Rhythms*, London, Ashgate

Smith, MK (2001) 'Donald Schön: learning, reflection and change' *Encyclopedia of Informal Education*, [online]

Stenmark, M (2002) *Environmental Ethics and Policy Making*, Aldershot, Ashgate

UBC Fisheries Center (2004) *Back to the Future* webpage www.fisheries.ubc.ca/projects/btf/

Van Zyll de Jong, M, Lester, N, Korver, R, Norris, W, and Wicks, B (2002) 'Managing the exploitation of brook trout, *Salvelinus fontinalis* (Mitchill), populations in Newfoundland lakes' in IG Cowx (ed) *Ecology and Management of Lake and Reservoir Fisheries*, Oxford, Fishing News Books

Vodden, K and Bannister, K (forthcoming) 'Circularising knowledge flows: institutional structures, policies and practices for community-university partnerships', in J Lutz and B Neis (eds) *Making and Moving Knowledge*, Montreal, McGill-Queen's Press

Vodden, K (2004) 'Watershed management in Canada: lessons and implications for institutional design', paper presented to the *United Nations Environment Programme Hilltops to Oceans Partnership Conference*, Cairns, Australia, May

Vodden, K (1999) *Nanwakola: Co-management and Sustainable Community Economic Development in a British Columbia Fishing Village*, Master's Thesis, Department of Geography, Simon Fraser University, Burnaby, British Columbia

Wolfe, L (2000) *Patterns of Equivocality in Decision Making for Marine and Terrestrial Ecosystems: Towards the Design of Equivocality-Reducing Management Systems*, PhD Dissertation, Simon Fraser University School of Resource and Environmental Management, Burnaby, British Columbia

Yorque, R, Walker, B, Holling, CS, Gunderson, LH, Folke, C, Carpenter, S and Brock, W (2002) 'Toward an integrative synthesis' in LH Gunderson and CS Holling (eds) *Panarchy: Understanding Transformation in Human and Natural Systems*, Washington, Island Press

CHAPTER 18
GETTING THE SCALE(S) RIGHT IN OCEAN FISHERIES MANAGEMENT: AN ARGUMENT FOR DECENTRALISED, PARTICIPATORY GOVERNANCE

JAMES WILSON
School of Marine Sciences, University of Maine, Orono, Maine, 04469, USA

Abstract

The focus of this chapter is on the problem of scale in fisheries governance. This is the problem of what is the appropriate scale of the marine ecosystem for fisheries management purposes. Current fisheries management regimes largely bypass this problem by focusing their attention on scale-less, single species populations. But such an approach rests on an inadequate mental model that ignores the complexity of the marine ecosystem. By contrast, the ecosystem-based approach offers an alternative mental model that deals with this complexity, not by bypassing it, but by scaling down to local ecosystem levels, which are best managed by decentralised, co-management governance arrangements that make full use of resource users' knowledge and also ensure accountability.

18.1 Introduction

Management of ocean fisheries is usually carried out at a broad geographical scale, often at the level of the nation state or some broader scale international political entity. There are numerous reasons one might cite for this choice of scale – for example, the absence of finer scale political boundaries; the costliness of ocean observing; the difficulty of conceiving and managing an ecology that is poorly known and understood; the need to match the scale of science with the scale of political authority; and the scientific belief that fish stocks are generally mobile and range over large areas (Degnbol 2001). Whatever the reason, most developed nations are caught up in institutional arrangements that require that they act as if a broad-scale, single-species approach is appropriate for the management of fisheries.

As our understanding of ocean ecosystems expands, there is growing reason to be sceptical about the scale of these institutional arrangements and their derivative scientific perspective. The current turmoil in fisheries science is telling evidence of the breadth of this scepticism. There are basically two reasons to be sceptical: an empirical reason and a theoretical reason. The empirical reason is the very poor results – the major failures – that generally obtain in ocean fisheries management (Pitcher and Pauly 1998). The theoretical reason is the serious difference between the holistic concept of the ocean held by ecologists, and the discrete, single species fish population model used as the conceptual basis for most management (Hutchings 2000). These empirical and theoretical reasons challenge some of the basic assumptions implicit in the design of our management institutions, and in the policy instruments such as quotas and individual transferable quotas (ITQs) which are generally favoured among managers and economists.

T. S. Gray (ed.), Participation in Fisheries Governance, 307–318.

Our conception of the biological structure and processes of the ocean is critical to our interpretation of the way human activity impacts the ocean, and to our sense of what we need to learn and do to manage those impacts for sustainability. The conventional scientific view is one that attempts to find a workable solution given political realities; the costly measurement and observation problems encountered when working in the ocean; and our fundamental computational and conceptual limitations. As a result, almost by default, conventional fisheries management science has simplified the complexity of the ocean into a series of scale-less (or single scale), independent, single-species models driven by assumptions of density-dependent, equilibrating processes. Every scientist ('every' may be too strong a generalisation) realises that this is a gross simplification of the ocean. The operative question, however, is whether it is an adequate simplification – that is, one that captures the essence of fish population dynamics and provides us with guidance about human behaviour that is appropriate for sustainability.

This question of operability is not easily resolved. For example, the manifest failures of fisheries management cannot necessarily be explained as the result of the inadequacies of this science. It is a commonly held view among scientists and others, especially those who have been the architects of conventional management, that the science is basically correct and that the failures we observe are simply the result of a lack of political will: that is, politicians and managers are unwilling to do what scientists deem necessary (Rosenberg *et al* 1993; Ludwig *et al* 1993). This proposition is not easily subject to scientific proof, nor is any proof generally thought to be necessary.

An alternative explanation for the failures of fisheries management is rooted, not in the idea that there is insufficient political will, but rather in the idea that both our science and governance processes are designed around a conceptual simplification that seriously mischaracterises ocean ecosystems. This proposition is also not easily proved but it is certainly worth exploring, not least because it forms a critical part of the argument for co-management. In this chapter, I explore this proposition, first, by presenting an alternative mental model of the important processes in ocean ecosystems, focusing especially on its relevance for the governance problem; second, by providing an interpretation of how current fishing regimes affect the ocean based on that model; and, third, by suggesting how we might reorganise our governance institutions and re-direct our science – really our overall approach to learning – so that we are better able to adapt to a spatially and temporally complex biological system.

18.2 An alternative mental model of the system

Conventional theories of fisheries management rely upon a mathematically elegant conception of ocean processes. The ocean system is viewed as a collection of independent, scale-less populations driven by density-dependent processes that create strong equilibrium tendencies. Because of those tendencies, the impacts of human interventions in the system are assumed to have predictable outcomes. This predictability implies control and the ability to manage each population for sustainability. In this chapter, I challenge these assumptions, and outline an alternative mental model of the structure and dynamics of an ocean system. This alternative model emphasises questions of scale and complexity, and is slanted towards those aspects of the system that appear to be particularly important to the design of governing institutions. Creating a mental model is an important exercise because it defines what we think the world is like, the collective learning problem we believe we face and, from that, the nature of the restraints we have to place on our own activities if we are interested in sustaining the resource. Probably just as important, it allows us to understand one another's perspective and makes a constructive dialogue more likely.

It is generally accepted that evolution has been the principal shaper of the structure and dynamics of living systems, including the ocean. A casual, or intense, reading of the literature on marine ecology strongly reinforces that view. It is, however, a view that is, by and large, absent from most of the science that is used for fisheries management. The reason for this, it seems, is that fisheries scientists tend to operate in the belief that the processes of concern to them – that is, the annual, or short-term, fluctuations in the abundance of fish populations – take place at a temporal scale that is essentially irrelevant to evolutionary processes. Occasionally, there is mention of fishing exerting a selection pressure that leads to earlier maturity among fished populations, but generally there seems to be a sense that the evolved structure of the system remains more or less constant. But the evidence certainly seems to contradict that basic assumption. For this reason, an important question that is not often asked is how fishing affects that structure – not just the genetic structure of individual populations, but also the cultural or learned behavioural attributes of fish populations and the interactions of those populations with one another, with the abiotic world and with humans. A large part of the scepticism about conventional fisheries management derives from a concern that the way we manage (or do not manage) fishing has led to a significant erosion, or deconstruction, of evolved population and ecosystem structure (Pitcher and Pauly 1998; Myers and Worm 2003; Jackson *et al* 2001).

Consider a physical environment with a diverse geology and topography, widely varying currents and tides, different salinities, temperatures and chemical circumstances – all subject to continuous perturbations by storms, climate and anthropogenic disturbances. In short, consider a physical environment that is diverse in space and time. For any living organism, finding the right place in that diversity is critical to survival. A cobbly, rather than a sandy bottom, for example, is important to a young lobster simply because cobble, compared to most other environments, provides better protection from predators, and adequate food and other resources that enhance survival (Walters 2000). The abundance of lobsters in areas that have a lot of cobble is the result of a long evolutionary process in which those lobsters that have led the early part of their life in cobble have had a higher rate of survival than lobsters that settle in areas that are dominated by, say, sandy bottom. Lobsters have evolved in close association with a large number of other organisms, some of which tend to eat lobsters, some of which lobsters like to eat and some with which lobsters compete for food and shelter. Given the particular physical and behavioural traits of young lobsters, cobble happens to be a place that lends itself well to the survival of lobsters. Something similar is true of every other organism in the ocean at any time in its life. There are places and times where the physical and biological circumstances are favourable to its growth and survival, and places and times where just the opposite is true. As a result, the co-evolved physical and behavioural traits of each organism combine with the varied abiotic characteristics of the system to create a spatially diverse and dynamic environment (Levin 1999).

Storms, seasonal changes, climate change and geological processes all alter the physical environment of the system at a variety of temporal and spatial scales. For each species, the impact of these physical perturbations is compounded by the adaptive responses to those same physical changes by the many species with which it interacts. A local storm might flush nutrients and fresh water from the land into the coastal zone, which might lead to an earlier phytoplankton bloom, which might lead to a whole cascade of biological responses, all of which take time to work their way through the system. For some species, these changes might provide an immediate new opportunity for increased growth and survival. For others, these changes might prove detrimental to growth and survival. If one species tends to be strongly favoured by a particular change, it may become very abundant in the near term, but that change simply leads to an opportunity for its predators in the longer term, and vice versa.

The interactions of species through these co-evolved, adaptive mechanisms constrain the proliferation (and the demise) of each population and, thereby, give order to a system that might otherwise lack any structure or persistence (Kaufman 1995). But that order is not the kind of order that is easy to summarise in a single mathematical equation or system of equations, nor is it reasonable to describe it as one with strong equilibrium tendencies, at least at the temporal and spatial scale of interest to fisheries management. There tend to be lags of various lengths that limit the constraining effect of other populations, and many species tend to be functionally very similar (Steneck 2001), with the result that very small changes in the circumstances surrounding, say, recruitment processes, will lead often to large changes in the fortunes of one rather than another species. The difficulty of prediction is made even greater because fish tend to move around and follow usually complex life histories. Populations that range over a broad area may have relatively discrete components that have adapted (genetically or behaviourally) to particular spawning grounds but then spend other parts of their life histories mixed up with the remainder of the larger meta-population. Some species at certain stages of their life may range very widely; others may be relatively sedentary; some may be both, and exhibit a variety of intermediate behaviours (Robichaud and Rose 2004; Kritzer and Sale 2002). Some may have eggs and larvae that drift widely; others may choose spawning areas and adopt behavioural responses that keep or entrain their eggs and larvae within very local areas. Local ecological regions may act as relatively coherent, almost closed systems for any short period of time but for longer periods they tend to become more and more open.

Generally, interactions with both the physical and biological environment tend to be characterised by thresholds, exponential growth and a variety of other non-linear events (Holland 1998; Levin 1999; Ulanowicz 1997; Pahl-Wostl 1995). The result of all this complexity is a patchy, diverse, dynamic and difficult-to-predict environment, especially if one focuses on particular species. But even at the level of functionally similar species and at the level of the ecosystem, stability is often hard to find. Marine systems seem to be prone to system flips, or alternative system states, that compound the problems of predictability (Gunderson *et al* 2002).

From an economist's perspective, the aspects of this kind of system that are especially important are: 1) the difficulty of ever knowing what populations are actually being fished, given the apparent localisations, mixing and overlapping of different stocks; 2) the absence of species-specific predictability and the resulting inability to engineer particular biological outcomes; 3) the very high information costs associated with monitoring the spatial and temporal complexity of biological diversity (Wilson 2002); and 4) the long time lags and other complex temporal dynamics, such as abrupt systems 'flips' or alternative states, that make it very difficult to learn about the dynamics of the system and, especially, the effects of fishing.

18.3 Changing our view of what is needed for sustainability

Conventionally, we have tried to cope with the complexity of the ocean by adopting an analytical perspective that asks the question 'how will the abundance of these populations change if all other things in the system remain stable?' This is a reasonable question if what is meant by stability is an environment that is not perfectly stable but one which exhibits statistical regularities that make the assumption of stability a good bet. It is also a reasonable question if the spatial structure of the system, and of the stocks therein, are simple and discrete, so we can match effort to particular stocks.

If, however, natural systems do not exhibit relatively stable dynamics and spatial simplicity, or if fishing disrupts or further destabilises a relatively unstable natural system – in other words, if the system conforms to the mental model outlined above – then the logic of this approach to simplification is questionable. The implication is that we have to search for a different way to simplify the complexity of the ocean: that is, a different way to fish that maintains the co-evolved relationships that are the ecological structure of the system. Generally, this means finding ways to establish rules about place, time and technologically specific ways to fish, ways that are sensitive to the particular spatial and behavioural adaptations of fish (Pitcher and Pauly 1998; Hutchings 2000). Importantly, it also means a psychological shift away from the idea that we can control individual populations (in the sense of producing particular statistically reliable outcomes) and towards the idea that, at a minimum, we have to find ways to manage and maintain ecological structure and processes.

This is a very different objective from the myopic concern with optimal fishing mortality that is the conceptual foundation of so much of conventional management activity. From a social and economic perspective, this changed objective is important, because it leads to a very different idea of the kinds of information and knowledge that we need to manage the system. In a single species approach, the information requirements relate principally to the changing abundance of each species. Collecting and analysing this data is not an easy task by any means, but it is a task that can be accomplished so that major trends, at least, are apparent over a period of a few years. However, if our scientific conception shifts so that it emphasises system-wide implications of species' adaptations, the information problem increases drastically. The broad-scale factors of conventional concern – the numerical abundance of each population – remain part of the equation, but the finer scale temporal and spatial attributes of each population, and the structure of the system as a whole, must also be considered. For example, for each stock it is important to know where it spawns and when. Its nursery grounds and habitat become important considerations, as does its range and the important interactions it has over its life cycle with other elements of the system. Behavioural, or cultural, patterns such as pre-spawning courtship activity, along with the spatial temporal patterns of these and many other aspects of the stock, also need to be understood. At the system level, there has to be an understanding of the link between oceanographic features and the organisation of biological activity – for instance, the particular local and wider adaptations of each species and the pattern of trophic linkages. And there has to be some coherent sense of how the spatial/temporal adaptations of the various species to one another and to the physical/oceanographic environment, add up to a constrained, orderly system. All of this amounts to a very large and, potentially very costly, information problem. It may be that over time we can sort these requirements down to a critical few that are important to sustainability. Most marine ecologists already have a good theoretical idea of what those requirements are likely to be, but we still have to learn what those requirements are for particular times and places. In other words, even if eventually we can parse our management requirements into a small set, until we get to that point we will have to incur large learning costs.

18.4 How to reorganise fisheries governance

When viewed 'in the raw', that is, without the filter of any prior theory, ocean systems appear immeasurably complex. If we are to adapt our behaviour to this complexity it is clear that we have to find ways to simplify that apparent complexity. This is what fisheries scientists have been trying to do for the last half-century or longer. However, the problem has always been addressed as if it was a classical problem in physics – searching for a way to condense the essence of the system into a few well-chosen equations, or lately, simulations. There has been little or no formal analytical recognition of the scientific limitations we have had to impose upon ourselves because of the way we have organised our scientific and fisheries management enterprises. In particular, broad scale management makes it very costly to monitor at an intensity that is meaningful for an ecosystem-based approach to management. Consequently, for all practical purposes we have closed off that scientific option. By re-designing the management enterprise we can act to reduce the costs of monitoring ocean systems, relax many of the limitations on our science (and more broadly our collective learning problem) and, consequently, delve a little more deeply and practically into the complexity of ocean systems.

The appropriate, efficient way to re-organise fisheries management depends almost entirely on the organisation of the ocean ecosystem (Simon 1996). What I mean by ecosystem organisation is the spatial pattern of coherent interactions, that is, of systems and sub-systems. That organisation is not simple but it does show regularities in time and space. Much of the system's physical (non-living) oceanographic attributes – its topography, currents, chemical make-up and pattern of seasonal change – are the most regular elements of the system. They are configured in ways that are strongly place-based and multi-scalar; recognisable, regular patterns occur at, say, the scale of the North Atlantic as well as at the scale of a small embayment. In fact, one can divide the North Atlantic, or any other large system, into a nested hierarchy of spatially defined, somewhat independent components, ranging in size from very small estuaries, to the North Atlantic as a whole, with each component displaying regularities that strongly reflect its unique oceanographic circumstances.

Without these regularities, learning and adaptation are not possible. The behaviours of fish, fishermen and the whole system reflect these place-specific, abiotic regularities. But populations of living organisms (and usually fishermen) are not confined, generally, to particular places. Depending upon their particular adaptation, they may move from one part of the ocean to another very quickly or very slowly – tunas vs. tunicates. For any temporal scale that one might choose, one's perception of the connectedness – of the systematic coherence – of places changes, as does one's perception of the relevant range of individual stocks of fish (O'Neill *et al* 1986). Over a very short period, say a matter of days, everything is more or less stationary, and any particular small place might be viewed as relatively independent from others. There is, at this short temporal scale, a local connectedness among organisms, but that connectedness dissipates quickly with distance. Over a period of a season or a year, the mobility of organisms increases and the extent of connectedness enlarges. Over a decade, there is still broader connectedness. For individual species this might result, depending upon their characteristics, in broad-scale but patchy populations, discrete localised populations or meta-populations whose internal dynamics occur at both fine and broad scales. Consequently, the perceived spatial organisation of the system depends upon the temporal scale of interest to the observer – that is, the observer's focal scale (O'Neill *et al* 1986). For periods of short duration, small and generally quickly changing sub-systems are the appropriate scale; for longer periods, larger and slowly changing sub-systems are relevant.

But, if one thinks of the observer as the collective action that we call 'management', no single scale is appropriate; all scales have to be addressed in ecosystem-based management. The critical role of organisation – specifically some sort of decentralised, multi-scale decision-making process – is that it gives us the collective ability to simultaneously address both fine and broad-scale aspects of the system. By matching (or approximating) the temporal and spatial scales of our organisation with those of the ocean, we can more easily partition the overall problem of learning about the system into sets of smaller, more tractable, place-based problems. Good boundaries, that is, ones that capture the internal coherence of sub-systems, create tighter feedback and make it easier to learn about each sub-system (Levin 1999). This organisational approach is not a conceptual simplification of the system such as a scientist might strive for, but it is a simplification that makes it easier to understand the system and solve the problem of human adaptation, and that, after all, is exactly what science is trying to do.

However, the current centralised, hierarchical mode of fisheries governance does not facilitate such an ecosystem-based management approach, and there are many reasons why we might expect re-organisation to make our management problem easier. A clear advantage of multi-scale organisation is the information/communication costs economies it offers (Arrow 1974; Williamson 1985). Centralised organisations operating in complex environments have to pass an enormous amount of information up, down and across the organisation. At a minimum, information is gathered locally, passed up the chain, coordinated, analysed, decided upon and then passed down the chain. The costs of transmitting and coordinating that information; the possibilities of distortion and misunderstandings; and problems arising from untimely responses – collectively transactions costs – can all be very large. However, if these costs are not incurred, the foregone information can seriously impair the effectiveness of the organisation. Nevertheless, for budget and other, usually political, reasons, centralised organisations frequently economise by adopting policies that try to dispense with the need for much of the information about the complexity of the system. These attempts are generally guided by a theory – scientific or managerial – that describes the essential, simplified information that captures the essence of the system and stays within the confines of what is deemed to be economical by the managerial system. Because theory defines the flow of scientific information (such as what needs to be monitored and what can be ignored; what needs to be analysed and what can be ignored), if it is inadequate or overly constrained by the need to economise, it will blind the organisation to the 'true' nature of its environment, and seriously impair the resultant policies it develops and its assessment of the outcomes of those policies. This is essentially the argument stated earlier about why we depend so heavily on single species management and, consequently, why our current institutions and policies are poorly adapted for ecosystem-based management.

The alternative to 'economising-by-ignoring' is to decentralise the organisation. But it should be realised that decentralisation is feasible and efficient only when the environment can be partitioned into relatively coherent, or self-contained sub-systems (Simon 1996). Thus, in fisheries there is a critical link between the organisation of the natural system and the organisation of management. Given relatively coherent sub-systems, local decision makers can be given authority over certain events, and the impact of decisions made under that authority are likely to be principally local. Consequently, provided their incentives are aligned with the goals of the organisation as a whole, and that their authority is limited to events with a local impact, local decision makers can be trusted to make decisions consistent with those goals. As a result, the overall organisation can avoid many of the transactions costs of centralised administration. The extent of efficiencies that can be achieved in this way depends critically upon two things: the extent of local coherence in the biological sub-system (that is, the degree to which the results of local decision making are actually retained within the locality), and the incentives of local decision makers (that is, the degree to which their self-interest is aligned with the broader goals of the organisation).

The transactions cost savings of decentralisation, while important, are not its most important attribute. Most important is the way decentralisation allows us to 'fit' our collective activities to the environment (Ostrom 1991). It is the equivalent of giving an organism new sensory capabilities. Put differently, it enhances our ability to learn about the environment in which we are operating. Learning is the essence of adaptive management (Walters 1986; Vodden *et al*, this volume). In a complex environment that means the ability to observe, learn and react at multiple scales. Events at a local scale are often incomprehensible without some understanding of broader scale phenomena. For instance, when we observe local system erosion, often we are not able to perceive the way such local events accumulate to broader phenomena. Conversely, events at a broad scale are often incomprehensible without knowledge of what is happening at a finer scale (O'Neill *et al* 1986). Major shifts in system structure, for example, are events we tend to notice and understand only after the fact because we tend not to see the fine scale erosion of the system. The problem at both fine and broader scales is that observers – individuals and scientists and organisations – are generally confined to a single scale, and communication across scales is sparse, biased and uncoordinated. As a result, the extent to which we can collectively assemble observations, learn about, and respond to significant events in the system is highly impaired. Put differently, in the absence of coordinated multi-scale organisation, the quality of information available for management is very poor, and management results can be expected to reflect that poor quality ('garbage in, garbage out', as they used to say).

Decentralisation, in these kinds of complex environments, also has the great advantage of lowering the cost and risk of experimentation and, consequently, of creating a greater likelihood that experiments and learning will be undertaken. By contrast, experimentation at a broad-scale in a heterogeneous environment is difficult, because it is hard to find broad-scale – that is, uniform – policies that are appropriate to the different circumstances of different areas. Because uniform policies almost always disadvantage some parties more than others, experiments that proceed under the assumption of uniformity tend to breed strong pockets of political opposition. The result usually is inflexibility, little or no experimentation and a retarded ability to learn by doing. It is much easier to adapt new policies – experiments – to relatively homogeneous local conditions than to a broad heterogeneous area (Wilson 2002). This means that political problems of experimentation are less, and the probability of learning about the system and finding new, effective policies is much higher.

It should go without saying that these adaptive advantages of decentralisation are not likely to be realised if local decision makers do not have the incentive to act in a way that is consistent with the broad social goal of conservation. This is a point that economists and other social scientists have been making for years, with some effect. Most of the material in this volume is devoted to the argument that the institutional arrangements generally known as participative governance are most likely to lead to incentive alignment – what economic theory defines as a necessary condition for an efficient solution to management problems. Consideration of complexity in ecosystems adds only one point to this literature, but it is an important point, that rights to fish have to be designed to match the circumstances in which they are embedded. The conventional arguments that lead fisheries economists to argue so fervently for ITQs, for example, are based on single-species, scale-less theories of population dynamics, and the idea that it is possible to generate statistically reliable biological outcomes through restraints on a single variable – fishing mortality. In this kind of simple biological situation, the establishment of rights to fishing mortality (such as ITQs) leads to both efficiency and incentive alignment, principally because in these assumed circumstances such rights allow a meaningful control over future biological states (NRC 1999).

In a spatially complex biological environment, however, where fish stocks mix and overlap; where predictable outcomes from human interventions are difficult to predict; and where ecological interactions are important, the fundamental premise of this approach is questionable. First, it may not (except in special circumstances) be possible to match fishing effort with particular stocks, and, as a result, quotas applied in a way that does not discriminate among multiple stocks may simply encourage a kind of pulse fishing. For each 'local' stock, conditions of open access obtain, and the regulatory environment induces a race to locate and fish down localised stocks (Wilson *et al* 2000; Frank and Brinkman 2001). The results can be expected to be similar, except perhaps in the scale of their incidence, to the devastation of the distant water fleets of the nineteen sixties. Second, even if it were possible to match effort to particular stocks, the complexity of the ocean is not likely to yield a statistically reliable control over biological outcomes, and, consequently, can provide little rational basis for long term, self-interested restraint that is consistent with sustainability. Species-specific mortality control, and access rights based on that control, do not address the common pool, ecological externalities that need to be internalised for sustainability. In short, they do not generate the conditions required for socially efficient property rights. Consequently, in a complex fishery governed by a regime of species-specific fishing rights, individual incentives to maintain ecosystem structure are basically non-existent, and do nothing to resolve the fundamental incentives of open access. Even if fishermen understand the importance for stock sustainability of fish habitat, behaviour and other ecological factors, any steps that they might take to conserve those aspects of the environment generate costs that they incur but that their less scrupulous competitors do not. They cannot capture the benefits of their own restraint and, consequently, are not likely to behave in a way that is consistent with sustainability of the resource. As a result, one would expect these kinds of rights regimes to contribute to the long-term erosion of ecological structure and processes.

Finding rights that will avoid the erosion of ecological function is likely to be the most difficult and important part of ecosystem-based management. More than anything else, the complexity of these systems requires rights systems that facilitate collective learning – both individual and scientific. The argument here suggests particular criteria that should be met by individual rights arrangements: Fishing rights should:

1. Extend to all the species in the regime (to the extent possible) in order to internalise the externalities that would arise with more narrowly defined rights, and for the purpose of generating systemic rather than simply species-specific knowledge.
2. Be place-based, multi-scalar and associated with a distinct oceanographic regime in order to increase the capture of feedback about the impact of human and natural activity, and contribute, thereby, to learning and accountability.
3. Be embedded in a decentralised governance regime in order to mobilise and coordinate through collective forums operating at multiple scales, the knowledge of individual users (and scientists), thereby improving the quality of information available to decision-makers, and promoting learning and accountability.

Ecosystem-based co-management, conducted in a regime of individual owner operators, old-fashioned yeomen if one prefers, is likely to be the only organisational form able to successfully meet these criteria. The reason for this is perhaps best explained by a comparison of individual incentives under corporate and under owner-operator regimes. In a relatively simple system in which the principal controller of sustainability is fishing mortality, the decision-making discretion of individual fishermen is limited. In these circumstances, corporate owners of fishing rights can effectively monitor the results of employees' decisions and assure the alignment of individual incentives with corporate objectives. Ownership and decision-making authority can be safely and economically separated (Rosen 1993). In a complex environment, on the other hand, where there are multiple drivers of sustainability the discretion of individual fishermen decision-makers, especially with regard to fine-scale phenomenon such as habitat, is greatly increased, as are the costs of monitoring their behaviour. The likelihood that corrupt incentives will arise increases, as do, consequently, the dangers of separating ownership and decision-making authority. For example, a corporate employee skipper has little or no incentive to avoid towing in habitat critical to the system if he can't be monitored effectively. His catch and remuneration go up and there is no loss to him because he has no stake in the future of the system. For all practical purposes he has the incentives of a roving bandit and lacks the fundamental accountability imposed by the market (Olson 2000). Owner-operators on the other hand, can have a stake in the health of the system if their rights are systemic, if their ability to cash out of the system is constrained so that they don't become financial roving bandits and if they have some way to reach collective agreement on mutual restraint. In this sense, the agenda of co-management is the creation of basic accountability in a complex environment.

18.5 Conclusion

In this chapter, I have noted that, from an economist's perspective, the appropriate organisation of common pool institutions and the definition of individual access rights for the management of renewable resources depend critically upon the nature of the biological regime being managed. In ocean fisheries, that regime has until recently been characterised as a relatively simple collection of independent, scale-less, single species populations. Current institutions and access rights are designed with that conception of ocean ecosystems in mind. However, over the last decade and more, there has been growing reason to be sceptical about the efficacy of these arrangements. The generally very poor results of conventional management, together with the scientific evidence and theories that emphasise the spatial and temporal complexity of ocean ecosystems, strongly suggest the need to manage in a way that addresses this complexity. But recognition of complexity also requires recognition of the fact that our current institutions and rights systems are designed with a particular learning and control problem in mind, one that derives from a mental model of very simple, ecologically isolated, single-species populations. These institutions constrain the kind of science that can be done, and lead to rights systems that fail to internalise the drivers of system sustainability. In order to adapt human activity to the complexity of the ocean, that is, to introduce ecosystem-based management, it will be necessary to reform the organisation of management and of individual access rights so that it is possible to deal efficiently with the large information and learning requirements generated by these systems. Broadly, this means some form of place-based, decentralised organisation that is congruent with the multi-scale spatial structure of the ecosystem. It also means rights that are place-based; are specified in terms of access to a local sub-system (not species specific); are lodged in individual, owner-operators; and are embedded in a co-management governance arrangement that is able to assure individual accountability and mobilise the knowledge of individual users and science for the purpose of management.

References

Arrow, K (1974) *The Limits of Organization*, New York, WW Norton

Degnbol, P (2001) 'Science and the user perspective – the scale gap and the need for shared indicators' presented at *People and the Sea*, Inaugural Conference of the Centre for Maritime Research (MARE), Amsterdam

Frank, KT and Brinkman, D (2001) 'Contemporary management issues confronting fisheries science' *Journal of Sea Research* 45:173-187

Gunderson, LH, Holling, CS, Pritchard, L and Peterson, G (2002) 'Resilience of large-scale resource systems' in LH Gunderson and L Pritchard (eds) *Resilience and the Behavior of Large-scale Systems*, Washington DC, Island Press

Holland, J (1998) *Emergence*, Cambridge, Perseus Books

Hutchings, JA (2000) 'Numerical assessment in the front seat, ecology and evolution in the back seat: time to change drivers in fisheries and aquatic sciences?' *Marine Ecology Progress Series* 208:299-303

Jackson, JBC, Kirby, MX, Berger, WH, Bjorndal, KA, Botsford, LW, Bourque, BJ, Bradbury, RH, Cooke, R, Erlandson, J, Estes, JA, Hughes, TP, Kidwell, S, Lange, CB, Lenihan, HS, Pandolfi, JM, Peterson, CH, Steneck, RS, Tegner, MJ and Warner, RR (2001) 'Historical overfishing and the recent collapse of coastal ecosystems' *Science* 293:629-638

Kaufman, S (1995) *At Home in the Universe: the Search for Laws of Self-organization and Complexity*, Oxford and New York, Oxford University Press

Kritzer, JP and Sale, PF (2002) 'Metapopulation ecology in the sea' *Fish and Fisheries* 4:1-10

Levin, S (1999) *Fragile Dominion,* Cambridge, Perseus Books

Ludwig, D, Hilborn, R and Walters, CJ (1993) 'Uncertainty, resource exploitation, and conservation: lessons from history' *Science* 260:17-36

Myers, RA and Worm, B (2003) 'Rapid worldwide depletion of predatory fish communities' *Nature* 423:280-283

NRC (National Research Council) (1999) *Sharing the Fish; Toward a National Policy on Individual Fishing Quotas*, Committee to Review Individual Fishing Quotas, Commission on Geosciences, Environment, and Resources, Washington DC, National Academy Press

Olson, M (2000) *Power and Prosperity: Outgrowing Communist and Capitalist Dictatorships*, New York, Basic Books

O'Neill, RV, DeAngelis, DL, Waide, JB and Allen, TFH (1986) *A Hierarchical Concept of Ecosystems*, Princeton, Princeton University Press

Ostrom, V (1991) 'Polycentricity: the structural basis of self-governing systems' in V Ostrom (ed.) *The Meaning of American Federalism: Constituting a Self-Governing Society*, San Francisco, ICS Press

Pahl-Wostl, C (1995) *The Dynamic Nature of Ecosystems: Chaos and Order Entwined*, Chichester, John Wiley

Pitcher, T and Pauly, D (1998) 'Rebuilding ecosystems, not sustainability, as the proper goal of fishery management' in TJ Pitcher, PLB Hart and D Pauly (eds) *Reinventing Fisheries Management*, Dordrecht, Kluwer Academic Publishers

Robichaud, D and Rose, GA (2004) 'Migratory behaviour and range in Atlantic cod: inference from a century of tagging' *Fish and Fisheries* 5:185-214

Rosen, S (1993) 'Transactions costs and internal labor markets' in OE Williamson and SG Winter (eds) *The Nature of the Firm: Origins, Evolution, and Development*, New York, Oxford University Press

Rosenberg AA, Fogarty, MJ, Sissenwine, MP, Beddington, JR and Shepherd JG (1993) 'Achieving sustainable use of renewable resources' *Science* 262:828-829

Simon, H (1996) *The Sciences of the Artificial* (3rd Edition), Cambridge MA, MIT Press

Steneck, RS (2001) 'Functional groups' in S Levin (ed.) *Encyclopedia of Biodiversity* Vol 1, New York, Academic Press

Ulanowicz, R (1997) *Ecology: The Ascendant Perspective*, New York, Columbia University Press

Walters, CJ (1986) *Adaptive Management of Renewable Resources*, New York, McGraw-Hill

Walters, CJ (2000) 'Natural selection for predation avoidance tactics: implications for marine population and community dynamics' *Marine Ecology Progress Series* 208:309-313

Williamson, O (1985) *The Economic Institutions of Capitalism*, New York, Free Press

Wilson, J, Ostrom, E, Low, B and Costanza, R (2000) 'Scale misperceptions and the spatial dynamics of a social-ecological system' *Ecological Economics* 31:243-57

Wilson, J (2002) 'Scientific uncertainty, complex systems and the design of common pool institutions', in P Stern, E Ostrom, T Dietz and N Dolsak (eds) *The Drama of the Commons*, Washington DC, National Research Council, Committee on Human Dimensions of Global Climate Change

CHAPTER 19
SCIENTIFIC KNOWLEDGE AND PARTICIPATION IN THE GOVERNANCE OF FISHERIES IN THE NORTH SEA

DOUGLAS CLYDE WILSON and ALYNE ELIZABETH DELANEY
Institute for Fisheries Management and Coastal Community Development
P. O .Box 104, DK-9850 Hirtshals, Denmark

Abstract

The participatory mode of fisheries governance is based on effective communications that are able to bring together the viewpoints of many stakeholders so that management decisions can be generated. This chapter offers a discussion of the relationship between stakeholder participation as it is taking place on a European scale and the generation of formal scientific knowledge for the management of fish stocks under the Common Fisheries Policy (CFP). It examines the demersal stocks in the North Sea in particular. Stakeholder participation has been an important factor leading to demands for changes on the ways in which formal scientific advice is generated and communicated. The impacts on scientific deliberations of three such demands are examined: a) a demand that advice shift from the fish stock to the fishery as its basic unit of reference; b) a demand that advice not be open to different interpretation by the various stakeholders; and c) a demand that the results of existing technical fisheries management measures be examined when preparing advice. The chapter concludes that a flatter decision making hierarchy could make possible both a richer knowledge base and greater public support for management decisions.

19.1 Introduction

The participatory mode in fisheries governance begins with a shared understanding of what is going on in the sea. Sharing such an understanding implies an approach to developing the scientific basis of management decisions that has itself, in some sense, been participatory. This idea should raise some eyebrows. Our common sense understanding of science, for good reasons, does not include the idea of 'participation'. Science is supposed to yield objective knowledge, not participatory compromises. The role of science in fisheries management is precisely to provide objective information about the situation that then can be used to make participatory decisions about responses to the situation. Participation, after all, is a polite word for politics, and science is supposed to be shielded from politics.

Jasanoff (2002) is one of a growing number of voices expressing an alternative approach to science and policy. While the West has spent the past 30 years developing institutions that are supposed to protect policy-relevant science from politics, she argues, this vision has never been achieved. Within every policy arena where science is relevant it has been continually re-entangled in politics. However, the world is changing and this entanglement is taking on different kinds of meanings. In the past, when totalitarianism and nuclear weapons were the defining images of science and technology, the danger

T. S. Gray (ed.), Participation in Fisheries Governance, 319–341.

was a "monolithic alliance of science and technology with the state" (Jasanoff 2002:367). Now, however,

> ...it is the turn of civil societies to insist that the production of policy relevant knowledge should be made available for public scrutiny and input. To politicise science in *this* way – that is, by making it publicly transparent and accountable – is not the same as allowing science to be captured by the special interests of state and industry. Public accountability, carefully institutionalised, can only promote the interests of democracy. (Jasanoff 2002:368)

A clear case can be made for the application of Jasanoff's reasoning to fisheries. The early post-war years were indeed often characterised by an overly close cooperation between the fishing industry and the agencies responsible for the assessment and monitoring of fish stocks (McEvoy 1986), so a 'monolithic alliance' justifying itself with biased science was a real danger. It is hardly a danger today. Fishers face regulatory agencies staffed by scientists who have strongly embraced the precautionary principle (Wilson *et al* 2002) and any openings for fishers' participation in the scientific aspects of management could only conceivably exist within a civil society context in which marine conservationists would also have standing. Checks and balances would be in place that would allow Jasanoff's (2002) public transparency and accountability to make a positive contribution to the accuracy and legitimacy of the science being used by management decisions. Hence, we do not see a problem stemming from the basic idea of civil society participation in fisheries science.

The question is how such participation should be achieved. We believe that one important key lies in understanding the relationship between the physical (and social) scale of what is being managed and institutions doing the management. There are numerous examples of a knowledge base for fisheries management being produced through cooperation between scientists and fishers on **small** scales that are perceived by most stakeholders as useful and legitimate (Wilson 1999). However, collaborative programmes dealing with **large-scale** fisheries have been much more focused on involving fishers in particular roles, often as data gatherers or reviewers of completed science, without achieving participation in a broad sense (Bernstein and Iuddicello 2000). In fact, in our assessment, social scientists do not know very much about how to 'carefully institutionalise' large-scale institutions to allow participation in science to 'promote the interests of democracy'.

We examine aspects of the production of scientific knowledge for fisheries under the EU Common Fisheries Policy (CFP) in general and in relation to North Sea cod in particular. As the data presented in this paper will demonstrate, the issue is not in any important sense a problem of unresponsive bureaucrats or a lack of political will. It is a problem of institutional coordination; it is about the possibilities and constraints in how institutions make and communicate decisions. In our close observation of examples of these processes we actually found a good deal of **accountability** traceable to the concerns of the fishing industry, as well as extensive and honest attempts by many fisheries scientists to be **transparent** about how scientific decisions are arrived at. Yet this accountability and transparency has in no way led to a knowledge base for fisheries management perceived by stakeholders as useful and legitimate. Indeed, the scientific structures of the CFP are so reviled that some groups of fishers openly and actively resist providing it with any data at all.

Our intent is to provide a systematic analysis of some of the impacts that the current forms of participation have on the science structures of the CFP. This is not meant to be a 'big picture' analysis of policy making in the CFP. We recognise that there are many other factors influencing policy beyond the kinds of participation with which we are concerned. In fact, we cannot assess its relative importance as a driver of policy decisions, but we do know that it exists and that lessons can therefore be drawn from it about how to improve governance institutions. The participation that is happening now is a long way from the ideal of participation. It takes place mainly through the intervention of politicians, lobbying by European and national organisations representing the fishing industry (and conservationists), and at times through active political resistance. This participation is strongly influenced by rivalry between and among fisheries and member states. Nevertheless, it is a form of participation that generates forms of accountability, demands transparency, and has real impacts, both positive and negative, on the way science is done. Indeed, what is happening with fisheries science in the CFP is, we believe, quite representative of stakeholder participation as it is actually carried out on large scales. Our hope is that an empirical analysis of the impacts of current large-scale, participatory practices, however flawed, on the generation of scientific advice will be a greater contribution to improving those practices than if we were to write an essay on how it might be done.

19.2 Theory and methods

We believe that thirty years of both case studies and comparative research has established that participatory approaches do increase the legitimacy of and cooperation with environmental policy. The interesting research question is now how to make such approaches work well at larger rather than local scales. To begin such an investigation there are many theoretical traditions to choose from. Within anthropology and sociology most of these approaches are based on the examination of small-scale processes and do not provide tools to examine scale. This weakness is not shared by political science and empirical work in that discipline has indeed made important contributions to our understanding of both scale and participatory institutions (Ostrom 1990). They have achieved this, however, by using game theory or other approaches based on an atomistic theory of motivation grounded in instrumental rationality. This works very well as long as it is applied to institutional contexts where the 'game' metaphor is a good fit with the way people actually approach situations.

This very strength, however, makes it less useful for understanding how to improve participatory processes in other institutional contexts where assuming that the process is a game is tantamount to assuming that the process will, at least to some degree, fail. To the extent that people interact as tactical opponents in a process they weaken those aspects of the outcomes that are the central aim of participation: legitimacy and cooperation. We cannot ignore the fact that people do interact as tactical opponents, but we will achieve little progress in improving participatory governance within large-scale institutions if we assume, either as a simplification or through an empirically uninformed understanding of society, that it is the only way that people interact.

It is these considerations that point us toward Habermas (1984, 1987) whose *Theory of Communicative Action* uses a dual theory of motivation in which actors are oriented both to instrumentally rational goals and towards achieving a mutual understanding based on what he calls 'communicative rationality' – the rationality that allows people to make sense of what they are saying to one another. This is a unique and important contribution, even if a somewhat incomplete one. The coordination of social action requires the existence of mutual understandings. Habermas's investigation of the underlying logic of communicative rationality, which he presents in tandem with a theory of social systems, provides a starting point for analysing the relationship between communications and scale. For fisheries management, this has been developed further (Wilson and McCay 1998; Wilson and Jentoft 1999; Wilson 2003) in the direction of examining how the rationalities of both instrumental competition and communication to achieve coordinated actions, coexist within social conflicts and help shape institutions. While there is no space for a full description, the main points of the theory relevant to the present paper are that:

- Institutions are shaped by both overt and tacit bargaining between groups in pursuit of diverse and/or conflicting goals;
- Institutions need communicative mechanisms to coordinate social action;
- Such mechanisms have strengths and weaknesses that are scale-dependent;
- The mechanisms that work well on small scales allow institutions to have greater sensitivity to social values and factual truth;
- The mechanisms that are effective at coordinating behaviour over large scales greatly restrict the content of communications and are much less sensitive to social values and factual truth;
- Communicative mechanisms play critical roles in the generation of the social power needed for success in the first point above. This leads to systematic distortions in communications;
- Institutions are expressed, reproduced and marginally changed by micro-level behaviours and can be analysed through the observation of the norms guiding such behaviours

Science is an institution that relies heavily on what Habermas calls rational communications. This communicative mechanism allows institutions to be sensitive to factual truth, about nature for example, but is poorly equipped for coordinating behaviour across large scales (Wilson 2003). Rational communications meet two conditions: there is no manipulation involved in the communication; and everything communicated is open to any question, from any participant, about its validity (White 1988). This model should not be thought of as an attempt to describe empirical conditions. It is a norm in the sense described below. People use it as a yardstick to evaluate the kind of communicative situation they find themselves in. While no one expects the conditions to be fully met, they have to be met to some degree if a convincing shared reality is to be produced (Habermas and Nielsen 1990). Science, as Merton (1968) pointed out, makes very heavy use of norms that seek to maintain rational communications because institutional sensitivity to the truth requires, more than any other communicative facility, an openness to the raising and evaluation of any claim without *a priori* constraints.

The terms 'institution' and 'norm' need a bit of fleshing out here. Following Scott (1995), we see institutions as patterned social interactions with regulative, cognitive and normative dimensions. This is not the more common 'rules of the game' approach. The emphasis is on shared meanings that define behaviours and cognitions as fitting or not fitting particular normative patterns. Hence, where other scholars may choose to analyse institutions by examining the written rules that emerge from formal bargaining, we feel that it is necessary to delve further into the link between institutions and actual behaviour.

In recent decades the term 'norm', for good reason, has gone out of fashion in sociology, because earlier uses implied values so widely-shared that they were seen as structural components of society. This idea had few empirical referents in a conflict-ridden world. Norms, as we use the term, are not structural but phenomenological. They do not define appropriate behaviour *a priori*, rather they are created through processes of deciding on, rationalising and accounting for behaviour (Heritage 1984). It is these processes that link shared meanings to behaviour. Norms are not empirical phenomena, they are cognitive phenomena, they are counterfactual ideals through which observed behaviours (including acts of speaking) are understood. This observing, rationalising and judging, however, reproduces institutions and has tremendous influence on subsequent behaviour.

This idea of norms as an analytic link between micro-level behaviour and institutions defines the method we take in this paper. Our central research question is with the kinds of influences exerted on scientific processes by other participants in the implementation of the CFP. In this chapter, we focus on the influences on these deliberations that are traceable to the needs of the fishing industry, sometimes directly, but usually channelled through the European Commission. We conceptualise and term these links as forms of 'accountability'. The main reason for using this basically positive term for what also might be called 'political pressure' is because we believe that participation, even when hampered by large scales, is helpful and necessary. Stakeholders such as fishers, managers and scientists have a right in democratic societies to hold each other accountable. The tools they currently have for doing so, however, are blunt and crude and have the possibility of hindering, as well as helping, processes of creating pictures of nature that are both accurate and shared. The point of this paper is to describe the effects of these forms of participation on the work of scientists in hopes of finding ways to improve these tools.

19.2.1 THE SPECIFIC RESEARCH CONTEXT

Research activities were carried out under the auspices of the International Council for the Exploration of the Sea (ICES) Working Group on Fisheries Systems. We observed in detail two scientific deliberations within the ICES system: the September 2003 meeting of Working Group on the Assessment of Demersal Stocks in the North Sea and Skagerrak; and the October 2003 meeting of the Advisory Committee on Fishery Management (ACFM). We also observed two meetings of the Scientific, Technical and Economic Committee for Fisheries (STECF). We carried out 23 formal and numerous informal interviews with fisheries scientists. Publicly available documents such as ICES and STECF reports and the Memorandum of Understanding between ICES and the Directorate General for Fisheries (commonly called DG Fisheries or DG XIV) were also analysed. Notes from observations, informal interviews, and original documents were analysed using NUD*IST textual data analysis software.

Fisheries advice for Europe is given through the ICES system. ICES was founded[1] in the late nineteenth century to investigate both natural and man-made causes for fluctuations in fisheries stocks (Rozwadowski 2002). ICES is an inter-governmental organisation that coordinates and promotes marine research in the North Atlantic, including adjacent seas such as the Baltic and North Seas. With more than 1600 marine scientists from nineteen countries around the North Atlantic, scientists working through ICES gather information about the marine ecosystem. This information is used to fill gaps in existing knowledge; it is also developed into unbiased, non-political advice. ICES advice is used by the nineteen member countries to help manage the North Atlantic Ocean and adjacent seas.

ICES has three advisory committees that provide advice on marine ecosystem issues. The committee of interest to this chapter is the one which provides advice on fish and shellfish stocks, the ACFM. The ACFM is the official scientific body providing advice to the Commission of the European Union in the form of DG Fisheries. Outside of ICES, DG Fisheries has its own advisory committee, STECF, which often consists of many of the same members as ACFM working groups. Unless otherwise specified, when this chapter refers to scientific advice it means the advice from ACFM and/or STECF scientists to DG Fisheries.

19.3 Some examples of the influence of stakeholder participation on fisheries science deliberations

Our examination of the relationship between fisheries scientists within both ICES and DG Fisheries' own STECF has uncovered a number of ways in which changes are being demanded of scientists. These changes are rooted in a desire for forms of scientific advice that facilitate managers' relationship with industry and other stakeholders. This tightened accountability can be directly traced to pressures stemming from the participation of the fishing industry, and to a lesser extent, conservation NGOs. While a substantial number of changes are being demanded, for the sake of space and thoroughness we discuss just three of them. The first two are channelled through DG Fisheries, while the third is experienced by the scientists as coming more directly from the fishing industry. These changes are:

1. A demand that advice shift from the fish stock to the fishery as its basic unit of reference;
2. A demand that advice not be open to different interpretation by the various stakeholders;
3. A demand that the results of existing technical fisheries management measures be examined when preparing advice.

We examine each of these kinds of changes in turn, asking what impact they are having on how advice is formulated and communicated.

[1] Though ICES officially designates its start as 1902, scientists were working in the decades prior to this to get the organisation up and running.

19.3.1 SHIFTING FROM THE FISH STOCK TO THE FISHERY AS THE UNIT OF REFERENCE FOR ADVICE

The institutional reality of fisheries management is driving ICES towards giving advice based on fisheries, many of which affect several fish stocks, rather than on the fish stocks themselves. Historically, ACFM scientists provided advice on fish and shellfish in the form of a single stock of a single species. Such advice is founded on well-established theory and practice in fisheries management based on the principle of density-dependent population regulation (Rosenberg *et al* 1993).

DG Fisheries, in response to the needs of the industry, are making the case for this shift from the stock to the fishery. As the STECF has said:

> Q1. ICES advice is explicitly single species though on occasions a comment is included that management should take the restrictions on another stock into account. Whilst it is helpful to have the issue stated, the failure to provide suitable advice is a problem. In the particular instance of recovery plans there [are] often extensive interactions between the catches of different species in the fishery. There is a need for fishery-based options, which provide matched restrictions to **mixed fisheries** at a range of rates of exploitation.[2] (STECF 2001:9)

From the industry and Commission perspective, single species advice ignores the basic reality that it is fisheries, that is, complexes of fishing ports, fishing boats, and fishing gears, that managers actually manage. The term 'mixed fisheries' in the quote above is the critical one because the problems that the fishing industry is holding the managers accountable for, arise where a 'fishery' is fishing for more than one 'fish stock'. For example, boats that fish for nephrops also catch cod as bycatch. If, as current ICES advice would have it, there was a complete ban on the capture of cod, these nephrops boats would be unable to fish for the nephrops that they would otherwise be allowed to catch. The 2002 report of the Working Group on the North Sea and Skagerrak explains the issue as follows:

> Q2. Current advice provided by ICES is mainly given in the form of fishing mortality limits and associated catch options, which are derived separately for individual fish stocks. This form of advice has two major disadvantages. **First, it takes little account of biological interactions**. Second, the stocks being analysed are often caught together in mixed-species fisheries, so the catches of species harvested by a given fleet are not independent of each other. This process is traditionally referred to as technical interactions. If, as currently, TAC [Total Allowable Catch] are set independently for each stock, fishing for one species may lead to discards and/or misreporting of another species, for which the TAC has already been reached… **The Commission has** on several occasions **acknowledged the need** to deal with technical interactions in ICES advice. This year, a request has been made to ICES to compile age-structured catch and effort data by fleet as appropriate, and to initiate multi-fleet multispecies short-term forecasts based on these data. (ICES 2003c)

---

[2] Bold emphasis has been added to quotations by the authors.

The bolded wording asserts that dealing with mixed fisheries is both a scientific issue and an ICES initiative. By beginning with a reference to biological interactions, they are constructing the managerial imperative to address mixed fisheries as a scientific requirement, and this has indeed been a concern of fisheries scientists since the 1970s, one that has been frustrated by the substantial data demands of multispecies modelling (Peterson 1993). Describing the addressing of technical interactions in mixed fisheries as an ICES initiative, which the Commission rather passively 'acknowledges', in the context of a response to the Commission's managerial requirements, has the effect of repairing the 'science boundary'. This is a well known concept in the sociology of science (Gieryn 1983), which points to the ongoing contest of deciding what is and is not 'science' and, often, who is and is not a 'scientist'. It emerged as a key theme in this research, particularly in the form of distinguishing between science and management.

In 2003, ICES initiated a Study Group for the Development of Fishery Based Forecasts to try to define fisheries and establish a framework for pulling together the necessary data. They chose to base their definition of fisheries on a combination of target species, gear, geographical area and season (ICES 2003b), hence combining the biological question of species with technical and geographical variables.

The move toward mixed fisheries influences the ways that scientific advice is produced in a number of ways:

- It requires that biological advice be fitted to a social unit rather than a biological one;
- It intensifies a norm of consistency in descriptions of scientific outcomes;
- It blurs even further the distinction between 'science' and 'management'.

We discuss each of these influences in turn.

19.3.1.1 *Fitting Biological Advice to a Social Unit*

A fish stock is a natural phenomenon while a fishery is a social one. The move from basing scientific advice on a natural concept to a social concept is a profound one. Social scientists have long understood that the difference between references to things in the natural world and the social world is a critical aspect of human communication (Festinger *et al* 1950). A reference to something in the natural world involves material substance that makes verification of the reference possible, at least in principle. The social world, as we define it here, is a communicative system made up of shared meanings that can only be interpreted and never directly verified. Whether or not one agrees intellectually with this definition of the boundary between the two worlds, as an empirical reality this distinction between assertions with and without a material reference is built into the most basic coordination mechanisms of many social institutions (Habermas 1987), including ones that are important for fisheries management (Wilson 2003).

Many scholars would question our assertion that we can meaningfully distinguish between a 'social unit' like 'fishery' and a 'natural unit' like 'fish stock' (Freudenburg *et al* 1995; Latour 1987). Their concerns are well-grounded and require a short aside. Their argument is based on a) the fact that institutions can only respond to ideas (social constructions) about nature rather than nature itself; and b) in our 'post-modern condition' it is practically impossible to verify the degree to which these ideas reflect nature. We argue here that a fishery is a complex that includes economic, social, biological and technical ideas that are linked together through negotiated shared meanings, and hence is social. Then we argue that a fish stock is a construct based on physical proximity and genetics and that shared meanings do not play an ontological role in the constitution of the stock. Hence the stock is natural. They would then rightly respond 'but the fish stock as it actually matters is really a linkage of a set of definitions of species, people in landing areas sorting fish, research vessel hauls, computer data bases and so on, it is just as socially constructed as the fishery.' And they would be right. From some theoretical perspectives this is a very useful insight. In the communicative systems theory we are using here, however, the ontological distinction is important because there is a difference in what is meant by 'being correct' in respect to communicative assertions about material things, and in respect to communicative assertions about shared meanings. In common language, we call this the difference between 'facts' and 'opinions' (Festinger *et al* 1950). The size of a fish stock is a (likely unknowable) fact; the boundaries of a fishery will always be an opinion, even if codified in law. This difference matters crucially to institutions because certain institutional communicative mechanisms, particularly those effective on large scales, depend on this distinction in the way they coordinate action. This leads to a constant pressure from large-scale social systems to reify social relations – to reconstruct social phenomena as natural phenomena amenable to technical control (Habermas 1987). An example of this is the legal codification of the boundaries of a fishery in an attempt to make it into a 'fact' in respect to regulation. The reification of social relations is often strongly resisted because it can violate nuances and meanings that are important to people. This accounts, for instance, for much of the current resistance on local scales to 'globalisation'. These systemic changes in shared meanings are an important part of how communicative systems theory understands institutions.

While there are certainly many scientific complexities around defining a fish stock – including genetic variation, migration, spawning behaviour – these are the kinds of questions that fisheries scientists are trained to resolve. Fisheries, on the other hand, are social units with porous boundaries that individual fishers can cross. In fact, fishers can unconsciously or deliberately blur the boundary between various fisheries. Fisheries compete with one another among ports and nations, and have lobbyists and politicians that speak for them.

This shift changes the way that the fish themselves are understood, as classifications are driven by social rather than biological concerns, which become more important in the discussion. During the ACFM meeting, scientists reported that fishers had communicated concerns to them about scientists examining catch composition and making judgements about which species were targeted and which ones were bycatch, an economic distinction having little to do with fish biology. Nevertheless, if scientific advice is to attach to fisheries, then the fisheries must be defined as precisely as possible. The required precision, of course, stems from the fact that managers must apply often costly and contested regulations to fisheries. If it is not completely clear who or what is in that fishery, then regulations cannot be implemented.

ICES is in the process of trying to develop models for mixed fisheries that would allow managers to predict the complex outcomes for many fleets fishing for several fish stocks. We observed some of this process at the 2003 meeting of the ICES Working Group on the North Sea and Skaggerak (WGNSSK) where the scientists were concerned with building such a model focused on the North Sea cod, including other fish stocks caught by fleets that caught cod both as a target species and as bycatch. They defined the fishing fleets based on gear and mesh size and included the fleets for which they had the necessary data for a total (during this meeting anyway) of 83 fleets. The definitions of fleets used were a controversial point among the scientists. Getting useful and comparable data for all these fleets was perhaps the greatest challenge and they were glad to have the data they had, even though it was only for one year. Collating the data for the eventual use of this or similar models for management advice was going to require the attention of several ICES working groups. More aggregate data – total landings by fleets and countries – was available than data that reported the age composition of the catch. Age composition is an important aspect of stock assessment models so whether or not to include the simpler, age-aggregated data was a point of discussion. It was clear to the scientists that these data problems precluded any use of the model in decision-making, and they were concerned that this would be misinterpreted. The model's results were very sensitive to decisions about how the fishing fleet were defined and combined, which suggested that the model should be set up to aggregate fleets together as little as possible. This, however, meant that the model would be even more demanding of good data. Another problem was how to handle the question of relative stability – the principle in the CFP that the relative shares of fisheries enjoyed by countries does not change through management decisions.

In the Autumn 2003, ACFM scientists were confronted with having to figure out how to generate fisheries-based advice based on this new unit (fisheries, rather than fish stocks) in the face of the extremely serious situation with cod in the North Sea. The seriousness showed itself both in terms of the low numbers of fish – the biological reality of the stock – and economic implications for the people making their living in the related fisheries. Their data about the condition of the stock led to an unquestioned consensus that fishing on cod needed to be reduced to zero. As one scientist put it at ACFM: "are we giving stock or fisheries advice, are we bound because we give fisheries advice to ignore that this stock is near commercial extinction" (from observers' notes). But what did this mean for fisheries on other healthy stocks that could not avoid catching a few cod? They could not simply say 'a few cod would be alright' a few cod from many other fisheries would be many cod. They were loath, however, to put fishers fishing mainly on healthy stocks out of work.

The scientific decision they were being asked to make was unavoidably also a political one as soon as the focus was shifted to fisheries. An exchange at ACFM:

> Q3. **Session leader:** The problem is the linkage of stock and fisheries advice, and that is a problem. We should not say 'closure of all fisheries' but 'a zero catch of cod' then we raise the question of closing the fisheries. But we **have to keep the fisheries and the stock separate things**…[further discussion]. **Scientist One**: We don't want to take away the strong message [about the cod, but] we are giving unclear advice that says you can have fishing and not, we cannot escape criticism. **Session Leader** this is moving in the right direction, we must anticipate that criticism with some text. (From observer notes at the ACFM meeting in October 2003)

The advice they felt most comfortable giving was advice about cod fish, not about fisheries. In this case, the scientists were unable, in the end, to shift the scientific advice from the natural unit to the social one. The official outcome read as follows:

> Q4. It is not currently possible to provide analytical forecasts for input into mixed fishery evaluation models. The main obstacle is that ICES does not have access to discard data for most fisheries. Development of such capability furthermore requires better catch monitoring, fishery analyses, and management decisions. The lack of such mixed fishery forecasts necessitates the development of complementary processes that do not require analytical short-term forecasts. ICES has in this report taken a first step towards the formulation of advice in a mixed fisheries context…ICES acknowledges that defining relevant allocation scenarios places difficult demands on managers and that mixed fishery advice in particular will require interactive communication between scientists and managers. EC DG Fish has indicated to ICES some scenarios that would be of interest for managers. However, mainly because discard data for most fleets are not available, ICES is unable to provide the required scenarios at this time. (ICES 2003a:5-6)

The main problem cited was the technical question of inadequate data. But constraints stemming from the organisation of management and better interactions between scientists and managers, interactions which presumably help clarify the science-management boundary in relation to particular scenarios, are clearly important to ACFM.

19.3.1.2 *Fairness in Fisheries Management: Consistency and Scientific Advice*

Required to give advice for multiple fisheries, scientists are very concerned about consistency within advice for stocks and fairness among sectors. The source of this concern is the participation of the industry. It is not a new thing. Scientists have always been concerned about consistency in terms of making consistent use of the best information and methodologies. They have also been concerned about treating different fisheries consistently long before the fisheries-based advice became an issue. Nevertheless, the turn to fisheries-based advice intensifies this desire for consistency.

As the following quote indicates, DG Fisheries is held accountable by the industry for fairness among member states and fishing sectors:

> Q5. When ICES advises a closure for cod, haddock, and whiting and not for plaice, sole, and nephrops, there is a perception in the whitefish sector that the flatfish sector is not taking up its share of the conservation burden. **We need equitable and credible mixed-fishery advice**. The advice given for one may be in conflict with advice for other stocks, limiting the credibility of the advice. (An ICES official quoting a concern expressed by the Commission)

DG Fisheries wants ICES' scientific advice to be equitable and credible. Credibility is a clear enough idea from a scientific perspective, science is about credibly explaining how you know what you say you know. But equity? Equity is about distributive justice, it enters fisheries management through the desire of various user groups to be treated fairly (Loomis and Ditton 1993). How can scientific advice be equitable? The closest a scientist can come to 'equitable advice' is to be as consistent as possible in the ways they analyse and describe the various fish stocks (and now fisheries). Yet consistency of outcomes and descriptions of those outcomes is not a scientific value. In fact, good science tends to uncover differences. ICES' response to the particular DG Fisheries concern quoted above illustrates this:

> Q6. The situations were not similar as the fisheries in these areas were not identical, they take place on different grounds, cod is more in the north of the North Sea while plaice and sole are generally more southerly. ICES attempts to point to critical links between fisheries and provide good current advice, but the situation may change from year to year [*nevertheless*]...ICES has started to move toward fleet-based advice. (The same ICES official describing the ICES response)

To raise an argument such as 'we have to do it this way for sole because we did it this way for sprat' is to draw on other norms than scientific ones. Furthermore, as it is accountability from the industry that is driving this need for consistency, it is in the public face of the advice that the consistency is most imperative, leaving open the possibility that publicly offered explanations of conditions of fish stocks will be simplified to the point where differences are no longer apparent (see the comment of Scientist One below in quote Q8).

This norm of consistency has a strong influence on scientists' deliberations. During the ACFM discussion of cod in October 2003, the scientists' desire to be consistent about advice for cod influenced their interactions many times. In our first example, they were discussing how to deal with the ways that underreporting of catch (from both discards at sea and unrecorded landings) influenced stock assessment outcomes. Some amount of cod was removed from the stock by the fishery over-and-above the removals that the scientists had information about. This difference was serious enough so that it was one of two prominent reasons (the other being that the extremely small size of the cod stock itself introduced uncertainties beyond any scientist's experience) that the WGNSSK, the working group that does the cod assessment, had declined to make forecasts about the future of the stock based on their assessment. This decision led to a number of discussions at ACFM including the following:

> Q7. **Scientist One**: I had real concern about landings in 2001 about 2002 I don't know, I feel that 2003 will be weak again. In 2001 there was a change in F of 50 per cent, there were reports by social scientists that misreporting was going on. [In a recent meeting with fishers] we were raked over the coals by the industry by suggesting it was a problem. If we are going to reject this we will reject every assessment as the basis of a forecast, this is no less inconsistent, but for this stock in particular it will not change the advice so customers may say 'you must have a forecast' but for advice we don't need the forecast [cod was so low that no prediction was necessary about the impacts of fishing in the coming year as it was clear to them that no amount of fishing could be considered] . **Scientist Two:** if we are going to do a forecast I would rather it not be us. We can't correct for a bias in landings for an analytical forecast, let those who want massaged figures

> to do the massaging...**Scientist Three**: we need a decision**, the biggest argument for getting a forecast is consistency**, where does the burden of consistency fall, in sub groups [i.e. on judgements about individual stocks] in plenary [where the general advice is formulated]? **Scientist Four:** the art is for the sub-group to begin consistency by being internally consistent, but we need to be consistent about how we deal with language about underreporting. (From observer notes at the ACFM meeting in October 2003)

In spite of the fact that they already agreed that scientific advice for the cod stock had to be zero catch, and that the cod stock was in such a condition that singling it out for special treatment, by not offering a forecast, was justified scientifically, they were still very concerned that they describe the cod in a manner consistent with other stocks. This concern was driven, in this case directly, by the response of the industry to the issue of underreporting. The conclusion was to be careful that underreporting was dealt with the same way for each species within the language of the official advice.

Another exchange took place a while later. It illuminates the strength of the desire to be consistent:

> Q8. **Scientist One**: Don't write anything, leave it, it is too complicated, just say they [biological reference points for cod] have been updated. **Scientist Two**: We agreed that we could not do forecasts, so if we change reference points based on the same assumption, here we say we can revise a reference point in the medium term when we said we could not for haddock. **Scientist Three**: Is this repeating the medium term exercise? **Scientist Two**: We should be consistent. **Scientist Three**: Yes, but what is the Fpa based on. **Scientist Four:** The algorithm was run again at the same age range. **Scientist Three**: So it is technically the same. **Scientist Five**: A couple of well crafted sentences about changing age ranges and rescaling the reference points to make it clear what we have done in the introductory pages, otherwise I agree with Scientist One. **Scientist Two**: I am just saying for cod we concluded one thing and for haddock another because of the selectivity pattern. **Scientist Four**: The concern is starting stock sizes and that does not matter in the long term, in haddock it is the exploitation pattern and that matters in the medium term. **Scientist Two**: I don't want to complicate things, but if you then go to sole and only look at Floss the revised reference point for sole was only 0.56. **Scientist Three**: It was updated in different ways and was supposed to reproduce what was done, but it doesn't?? [Scientist Two is outnumbered and gives up with body language clearly suggesting dissatisfaction with the outcome.] (From observer notes at the ACFM meeting in October 2003)

During this exchange, no scientist questioned the idea that the assumptions underlying the identification of reference points for cod and haddock should be the same, even though there were important differences in the condition and available information about the stocks, as evidenced by Scientists Four and Five's comments. Scientist Two, however, was emotionally committed to the idea of consistency and pushing for it to an extent that the other scientists all thought would make the advice unnecessarily complex.

19.3.1.3 *Mixed Fisheries-based Advice and the Line between Science and Management*

The boundary between science and management is seen by scientists as a critical one.

Indeed, it is the basis of how scientists understand their role in fisheries management: scientists describe what is true about nature and then managers decide what to do about it (Wilson and Degnbol 2002). In practice this is a very hard line to maintain. All stakeholders at least nominally support this clear line between science and management. As a DG Fisheries official told us in an interview: "advice should tell managers **what** to do, not **how** to do it." Moving toward fisheries-based advice, and toward advice dealing with mixed fisheries in particular further obscures this already porous boundary.

The following exchange took place at the WGNSSK among a group of scientists working on the development of a fisheries-based model to aid managers in understanding interactions between different fisheries. The exchange illustrates two interesting things. The first is how the scientists, particularly on the level of a sub-group working on the nuts-and-bolts of figuring out how to meet the needs of managers have to feel their way into the details of a leading-edge question like mixed fisheries without having a very clear idea of what their work is going to be used for. Along with this is the real concern they feel that their work is going to be misinterpreted or misused by managers and other stakeholders who will be reading it. Particularly the suggestion by Scientist Six at the end of the exchange tells us something of the level of this concern:

> Q9. **Scientist One**: When we have completed this data base what shall we use it for**?** **Scientist Two**: Are we using this to produce alternative advice? **Scientist One**: Yes. **Scientist Two**: It will be used as an example. **Scientist Three**: ACFM wants to see this kind of thing. **Scientist Two**: It is illustrative, management will not be based on it this year. **Scientist Four**: We should use the 2004 data so people don't pick it up and use it as something real...**Scientist Five**: What I thought I would do is to use the data from last year's STECF meeting and do an exploratory analysis with data sets that are not proper enough for good results, we will use the analysis to explain what the model is doing and how it can be used...**Scientist Three**: Ideally it would be better to use the same data set. **Scientist Four**: But if it is just an example it won't matter. **Scientist Six**: Maybe you should use bad data so no one is tempted to use it for something inappropriate. (From observer notes at the WGNSSK meeting in September 2003)

In the plenary later on, this same model was evaluated as much in terms of its management implications as its technical characteristics:

> Q10. **Scientist One**: This is dangerous, let me give you an extreme example, a fleet is catching 100kg of cod and no other species. Another is catching 1000 kg of cod and 10,000 plaice. It is the first that will have to stop fishing! **Scientist Two**: No, that is why you have option P1 and P2[3], so that managers can make decisions like this. **Scientist Three**: We need to put in all the calculations, we can't put forward only one analysis. **Scientist Four**: You just suggested we put forward a scenario, while I thought this was just a sensitivity analysis. If you suggest options, one may be taken up, but this sensitivity analysis shows that this model is very sensitive to how it is set up. **Scientist Five**: But that is a political

[3] Ps refer to the fact that the model gives managers the option of reducing each fleet's catch equally or in proportion to the species composition of a fleet's catch, or in proportion to the portion of the catches of all fleets combined.

> decision…If we don't think we can explain this we should not put it forward. **Scientist Four**: After this discussion about Scientist One's point it sounds like we can't really explain this model **Scientist One**: Instead of naming the fleets give them a code or something so they can't use the data except for sensitivity. **Scientist Five** [visibly frustrated]: We step forward and as soon as it becomes a little political we say let's cover it up so you can't see it. (From observer notes at the WGNSSK meeting in September 2003)

Each of the scientists is coming at defining the line between science and management in respect to mixed fisheries in a different way. Scientist One wanted to make sure that nobody was going to be able to use the model to make decisions while there were possibilities of 'dangerous' inequities in results. He wanted to put the data in a code that obscured the identity of the fleets so that it could never be misused. This idea led to Scientist Five making strong objections. Scientist Two (explaining the approach taken by the sub-group, which included Scientist Five) wanted to give the managers options based on pre-programmed model parameters, so that 'managers can make decisions' within these predefined options. The options were likely intended to help the managers avoid, or at least deal with, the inevitable political wrangling between fleets as they competed to avoid having their portion of the mixed fishery cut back as little as possible.

The scientists at ACFM pick up this discussion of the mixed fisheries from the WGNSSK. They had put aside the model being developed at the WGNSSK both because it was not fully developed and tested and because there was insufficient data. They were still forced to deal with the underlying issue. What follows is an excerpt from their discussion:

> Q11. **Scientist One:** I think really we need to say there is not a science-based way of establishing what minimum [bycatch of cod] means and how it should be distributed among the fisheries. The managers have to deal with the ratio between the fisheries. [Extended discussion followed of the seriousness of the cod problem and the need for a zero catch.] **Scientist Two**: We could have an opening statement saying the catch should be 0 and all fisheries closed, then continue with this text [saying that bycatch should be minimised]. **Scientist Three**: I agree to a large extent, but it should be made conditional on the implementation of the cod recovery plan [a plan under consideration at that time by the Council of Ministers] as that would take account of the mixed fisheries. **Scientist One**: The evaluations of the recovery plan last year shows that that would take 8 years. **Scientist Three**: That may be acceptable to managers. **Scientist Four**: Yes, but to the stock. **Scientist Five:** This is, of course, a management decision, but you need to decide if you are giving stock advice or fisheries advice, this is the mixed fisheries issue. You stated in your evaluation of the recovery plan that you said it would work, so why say 0 here? **Scientist Three**: We are saying that we should give advice contingent on recovery plan. We need input from managers in priorities if we give fisheries-based advice. (From observer notes at the ACFM meeting in October 2003)

The scientists, without the possibility of a 'science-based way' or even a mathematical description of how bycatch could be distributed, continued to struggle with what their advice should be and what role it should play in the midst of a broad and confusing set of possible management scenarios for cod. They felt the need for a dialogue with the managers, with input about priorities, but this contradicted the formal role they are supposed to play to provide 'objective advice'. The main outcome of the debate is the following text from the ACFM Report:

> Q12. ...for the mixed demersal fisheries catch options must be based on the expected catch in specific combinations of effort in the various fisheries. The distributions of effort across fisheries should be responsive to objectives set by managers, but also must result in catches that comply with the scientific advice presented above...An evaluation of how any combination of effort among fleets would affect depleted stocks would require that the catch data on which such estimates were based included discard information for all relevant fleets. Such data have been collected for many fisheries, but have not been made available to ICES. Therefore, ICES is not in a position to present scenarios of the effects of various combinations of fleet effort. However, if reliable data on all landings and discards by fleet were available, it would be possible to present forecasts based on major groupings of fleet/fisheries, and evaluate the impacts on cod and other rebuilding species of various distributions of effort among fleets. If management were to allow any demersal fisheries in 2004...some catch of cod would be inevitable, and therefore the fisheries would be inconsistent with the ICES advice. It is obvious that the larger the catch of cod the larger the risk that the stock will decline even further, and the greater the discrepancy from the ICES advice...However, the data...do not make it possible to calculate the true catches (and hence the impact on the stocks) by fleet or fishery. Therefore, there is no defensible basis for suggesting what fishing opportunities would still ensure no catch of cod and few discards of plaice and sole. (ICES 2003a:222)

The text reflects the discussions. The inability of ACFM to resolve their dilemma is placed squarely on the data problems. The lack of clarity about the use, misuse and meaning of the work is no longer directly evident, though its shadow can be seen in the careful use of language. This choice of emphasis re-establishes the boundary between science and management and portrays this line once again as a clear one. Once a model has been developed that allows a mathematical description of the distribution of by-catch, and adequate data collated to run the model on the actual fleets involved, the model itself will stand on, define, and will likely in some fashion, such as the P options described above, hide the porousness of that boundary.

19.3.2 DEMANDING THAT ADVICE NOT BE OPEN TO DIFFERENT INTERPRETATIONS

The second type of influence we would like to discuss also arises because scientists are being held accountable for providing advice in a form that facilitates managers' relationship with industry and other stakeholders. DG Fisheries expects ICES to provide (fisheries) advice on sustainability and yield, and to provide managers with a range of options and their consequences with respect to the advice. While a range of options is desired, DG Fisheries does not want this advice to be open to different interpretation by various stakeholders. 'Stakeholders' in this instance refers to the fishing industry and conservation NGOs, as well as national interests.

DG Fisheries has been critical of the advice provided by ICES. This criticism was described in an interview with an ICES official, where it was commented that the DG Fisheries stated:

> Q13. Advice for a number of stocks leaves room for interpretation. Advice must be clear and understandable. (Interview with ICES official)

The same ICES official went on to explain that DG Fisheries felt, for example, that:

> Q14. ICES advice on cod included a short-term catch option table that was interpreted by industry to mean a moderate reduction in TAC would result in significant increase for biomass – such forecasts did not seem to fit the need for a cod moratorium (Interview with ICES official)

This interviewee agreed that ICES advice should not be open to interpretation. He further stated that they (ICES) remained interested in working together with DG Fisheries on this issue:

> Q15. Advice should be clear and [we] will continue to work on this along with the Commission observers at the ACFM meetings. (Interview with ICES official)

However, the ICES official did also defend the advice provided to the Commission:

> Q16. It was clearly stated that even if the indicated improvement could be realised, it was insufficient to rebuild the stock in a short time. (Interview with ICES official)

Much of DG Fisheries' desire for clear advice that is not open to interpretation revolves around data tables. According to some scientists, DG Fisheries seems to only read the tables and not all of the information presented to them. As one scientist noted in STECF, "We're in the difficult position again…people (managers) just look at the numbers and not the health warnings." In this case, 'health warnings' refers to the caveats in the explanatory text, which is an important part of the advice because it describes the limits of the knowledge. Such information, it is clear from the scientists' discussions, is not meant to be supplemental to tables, but to explain the full picture of the fishery. Scientists state that managers must read this text to get the full advice since the caveats given in the written text help describe some of the uncertainty and other issues important to take into consideration when making management decisions. There had been some discussion surrounding the idea of a separate ACFM report, or section of the report, written for laymen, such as the fishing industry. It was thought that this would be one means of mitigating part of **the** problem of differing interpretations by the various stakeholders. In the end, however, it was decided that having advice described in too many different ways could aggravate the problem rather than diminish it and this idea was shelved, according to one of our respondents who is an ICES official.

DG Fisheries demands that advice not be open to differing interpretations by various stakeholders. ICES agrees the advice should be clear and understandable. Yet, scientists have pointed out that the advice should be read in full by managers. Choosing not to do so may increase the number of differing interpretations with each interpretation being made by stakeholders to support their own position. The following exchange reflects the level of the scientists' concerns:

> Q17. **Scientist One**: We should stop pretending we know how many fish are out there**. Scientist Two**: That is where we are going. The trend is there, the scale is wrong**. Scientist One:** The system will use it at the Council of Ministers. **Scientist Two**: That is why I want all these caveats. (From observer notes at the WGNSSK meeting in September 2003)

Hence, ICES' most important technique for addressing uncertainty is writing textual caveats around the tables of numbers they are asked to provide. Managers seek to be objective and fair in making their decisions and to have such decisions be transparent and resting on the 'best available science.' As pointed out by Porter (1995), the best data for being objective and fair in holding people accountable is quantified data. This need for quantification to achieve non-scientific as well as scientific goals raises important conundrums for ICES scientists. The scientists are beginning to directly address these issues, as is evident in the report from the Working Group on Fisheries Systems (2004), for example, which questions whether such advice provided by ICES really is transparent, accountable, and of high quality:

> Q18. Just picking a number to express a piece of qualitative information is often not adequate and estimates are based on expert judgement rather than strictly objective criteria. Examples are choices of models and sub-models, generalisations and at times personal weighting of time series for the tuning (which is a quantification of qualitative knowledge or impression). Such choices are necessary but given the interpretive flexibility in the data, a single quantity decreases the transparency and the accountability in science. Two scientists do not necessarily produce identical assessments with the same assessment tool because the best choices in running the model are not always obvious. (ICES 2004:20)

Thus, though quantitative information is perceived as being of higher quality, the scientists providing the information realise this is not necessarily the case:

> Q19. In the text of the ACFM report some caveats may be addressed that still are not taken into account in the calculations. When the advice is presented in a precise way, it may thus look like the problem is not a significant problem, as the precision of the knowledge is not affected. (ICES 2004:20)

In order to insure that the advice remains 'scientific', in the sense that any assertion that something is truth can be backed up with an explanation of how it is known to be true, the scientists must provide these caveats. They further insist that DG Fisheries must read the caveats to know that the caveats are as much a part of the full advice as the tables. This reliance on qualitative explanation, however, makes it that much more difficult to reach DG Fisheries' ideal of information that is not open to various interpretations.

19.3.3 CONSIDERATION OF TECHNICAL MEASURES

The third demand for change in the way scientific advice is provided that we will examine is that scientists evaluate the effects of technical measures used in fisheries management. Technical measures include not only gear restrictions such as mesh size on fishing nets, but also measures such as area closures. The fishing industry feels strongly that scientists need to consider these measures as they formulate their advice. As related in one interview:

> Q20. Fishers and others believe that environmental changes, pollution, and **management measures, closed areas** and other things are important. These need to be described in the advice in a way that indicates they have been considered. (Commission view quoted in an interview with ICES official)

This form of accountability from the fishing industry is both directly felt and not entirely welcome, as the following exchange indicates:

> Q21. **Session Leader**: I wonder if all these technical details may not be overkill on this assessment, we can calculate the potential benefits of technical changes, but never demonstrate them. We can now do this and say 50 per cent and it will mean X, but in the past we have never observed any changes because of these technical measures…**Scientist One**: If we do a forecast and don't take into consideration these technical measures it is going to cause trouble back home, at least for us. **Scientist Two**: and when we have to take this to the North Sea Commission [Fisheries Partnership] **Session Leader**: How straightforward is it? **Scientist Two**: It is easy to put in a forecast if we know what it is, they have the multipliers for the selection patterns and they could look at a range of values. They could do a table showing how much gain you would get for the uptake, but our scenarios fall with this. **Scientist Three**: This is fine for whiting or haddock, but the expert group has said not to do this with cod. **Session Leader**: So we can refer to this expert group we were all in. **Scientist One:** It should not be ignored…I would much prefer it be looked at. Apart from temperature it will be the first thing the fishers will pick up on. (From observer notes at the WGNSSK meeting in September 2003)

Why is there such interest by the industry in technical measures? The industry feels technical measures need to be considered because these are questions they feel very directly and about which they can draw on a good deal of their own knowledge. Putting the advice in terms of technical measures translates it into a form they can directly and immediately understand and which may give results directly observable through changes in catch. It will also tell them if there will be an economic effect in terms of outlay for new gear if there is a technical measure change. Technical measures are things that fishers can do on the local scale. They are measures that the fishers understand the reasons for and can see whether and how they are having an impact. Technical measures are also politically easier than other kinds of measures (Wilson 2000). For example, requiring everyone to use a net of a certain size is something that can be seen to be implemented fairly much more easily than dividing a fishing quota among a large number of boats.

What seems clearer on a local scale, however, can seem much murkier to those who are observing from higher scales. The impacts of technical measures in the aggregate are far from clear:

> Q22. Technical measures are not supported by scientists but are popular with other stakeholders. The technical measures are so complicated and cannot really be understood, and the change in fishers' behaviour does not seem to be great and there seems to be little change in actual selectivity in the fish catch. (DG Fisheries representative)

and

> Q23. Predictions of the impact of measures are predicated on everything else being equal, and then fishers change their behaviour in response to the measures and they mention other reasons – they are long term, they have been known not to be effective, people cheat – that predicting measures is hard. Against this backdrop you will realise that the scientists are **reluctant to base advice on non-demonstrated effects of new management measures**. (ICES official)

Despite this lack of support for technical measures by scientists, the issue continues to be raised when scientists get together to discuss fisheries advice, such as in working groups (including WGNSSK) and the ACFM meetings.

DG Fisheries is beginning to de-emphasise technical measures despite industry concerns. Many assessment scientists and managers, in viewing the results of such measures across the broad perspective of EU fisheries, feel that technical measures do not really work as a management tool. As the DG Fisheries official commented above, "the change in fishers' behaviour does not seem to be great and there seems to be little change in actual selectivity in the fish catch." This could include gear-type technical changes, but also, area closures. One such example is the emergency measures instituted before the cod recovery plan was accepted. Some believe that industry supports technical measures simply because they have greater control and can 'tweak' the systems. For example, alterations can be made to fishing gear, which close gaps and decrease the mesh size. Though DG Fisheries is slowly de-emphasising technical measures, it is still an important issue for stakeholders and as such, one issue in which scientists are made accountable and pressure is applied on them to continue considering such measures.

19.4 Conclusion

Do these few examples of impacts on scientific deliberations from stakeholder participation tell us anything about how to 'carefully institutionalise' public accountability in science for policy decisions across large scales? This paper has considered just a few aspects of the impacts of certain types of participation, and then only on the formal scientific processes involved in describing the knowledge base for fisheries management. Some tentative lessons may be drawn, but a good deal of further research and reflection is required to understand these linkages.

The most telling point that has emerged in the interactions examined here is the degree to which demands for accountability from stakeholders are pushing the fisheries scientists to continually redefine their work to consider more and more extra-biological questions. This is pressure that works in direct contradiction to nearly everyone's expressed desire to keep the science objective and 'scientific'. The pressures, however, are very real. Their most direct expression is the need to move toward fishery-based advice, especially in mixed-fishery situations. This need reflects the reality that fisheries management itself is a social activity, empowered by politicians and implemented through agencies and organisations that work both with and within these indistinct social-technical-biological complexes we call fisheries. The scientists try to respond with complex quantitative models with immense data requirements, requirements that they are not currently able to meet, along with calls for more intensive communications between themselves and the managers to help define exactly what the managers need for particular situations.

The scientists experience this accountability both directly and consciously, and much of their time is spent discussing how to respond in a responsible, i.e. 'scientific' way. They have to do this without having a good picture of how their work is going to be used, and they experience a real fear, based on experience, that their work will be misused. This leads to acute concern with the presentation of their work. They worry about consistency beyond what is scientifically required. They spend a great deal of time writing caveats in hopes that their results will not be misused or overdrawn. They worry about the appropriate level of complexity, considering what should be included or what should be left out in the interest of simplicity. Some of this leads to strong and emotional disagreements.

The scientists constantly seek to repair the tattered boundary between what is and is not science. This science boundary, as Jasanoff (2002) and many others have pointed out, is never as clear as people would like it to be. Even in the most esoteric laboratory, objectivity is never perfect and social considerations influence results (Collins and Pinch 1998). Within fisheries science the essential link between science and management and the constant demand for scientific answers to management questions makes a clear distinction between science and not-science impossible. Scientists go to great lengths to maintain this boundary intact. The power they derive from professional prestige and solidarity that allows them to carry out their professional roles and privileges depends on that boundary being strong. This can lead to distortions in communications when non-scientific decisions are hidden within results or information is suppressed for fear of misuse. Their major tool is the caveat writing where they seek to distinguish as precisely as possible between what they are willing to call 'science' and what they are not. Managers' power relies on bureaucratic rule making, which directly depends on clear decision rules that trigger legal actions. They want scientific advice to provide this clarity, often in the form of a number on a table that does or does not exceed some predetermined threshold, while simultaneously and somewhat paradoxically they want the advice to give them flexible options to attain their policy goals. Tables surrounded by extensive qualitative caveats do not provide such clarity, creating the temptation to distort the information by pointing to the table while ignoring the caveats. Indeed, if simultaneous clarity and flexibility is the goal, complex models giving point estimates surrounded by pages of caveats seem to be almost the opposite of what is required.

In principle, nearly everyone wants the science boundary to be clear so that management negotiations can be based on realistic information about what is happening with the fish.

It is this constant need for objective information and clear decision rules about both the natural and social aspects of the fishery that makes it so difficult in practice to separate the scientific analysis from the rest of the management system. Managers want to be able to point to some kind of objective support for all their decisions, even those which are more about allocating the impacts of management than directly about how many fish are to be killed. This leads more and more to demands for science to address social realities. Adding social science activities to the broad-scale scientific deliberations would likely not be of much help. The decisions to be made are about complex biological-technical-economic-political interfaces where different aspects of the problem are appropriately approached with different methods and kinds of measurement. Hence, any cross-disciplinary 'model' purporting to provide a clear answer would likely be so abstracted from on-the-ground realities as to be useless. What would better address the problem would be allowing the science to have a more concentrated focus on biological, and ecological phenomena by creating more participatory and open institutions to address management questions in a less bureaucratic way that is less demanding of precise findings to underlie every decision.

The way participation is presently structured through large-scale lobbying and political pressure, particularly as it is channelled into bureaucratic requirements through DG Fisheries, does little to relieve the demand on science to find answers to every question. A less hierarchical approach to management would allow more locally tailored decision-making less dependent on exact findings. This could be a cooperative effort to repair the science boundary by finding "serviceable truths" (Guston 2001) that allow management to move forward without seeking great precision. Approaches involving simple indicators of ecosystem health are one good example. Third party certifiers that negotiate sustainability goals with managers of individual fisheries are another. Serviceable truths would be a more helpful way to approach the repair of the science boundary than present practice, which sometimes calls forth purely defensive reactions among scientists (and fishers) trying to maintain professional prestige.

A shortening of the chain of accountability that currently must run all too often through Brussels would allow richer communications. More direct ties between fishers and scientists would avoid the formalisation that defines too precisely what scientists should do – formalisation that leads too often to both overwrought and data hungry models on the one hand and an excessive concern with presentation and the appearance of consistency on the other. A network of fishers, conservationists, managers and scientists working on multiple problems at multiple scale levels may offer more flexible fisheries governance as well as better science. Institutionalised in **this** way, public accountability and transparency in fisheries science deliberations could make possible both a richer knowledge base and greater public support for management decisions.

References

Bernstein, BB and Iudicello, S (2000) *National Evaluation of Cooperative Data Gathering Efforts in Fisheries: A Report to the National Marine Fisheries Service*, National Fisheries Conservation Center

Collins, H and Pinch, T (1998) *The Golem: What You Should Know about Science*, Cambridge, Cambridge University Press

Festinger, L, Shachter, S and Back, S (1950) *Social Pressures in Informal Groups*, Stanford: Stanford University Press

Freudenburg, WR, Frickel, S and Gramling, R (1995) 'Beyond the nature/society divide: learning to think about a mountain' *Sociological Forum* 10(3):361-392

Gieryn, TF (1983) 'Boundary-work and the demarcation of science from non-science: strains and interests in professional ideologies of scientists' *American Sociological Review* 48:781-795

Guston, DH (2001) 'Toward a 'best practice' of constructing 'servicable truths'" in M Hisschemoller, R Hopper, WN Dunn & JR Ravetz (eds) *Knowledge, Power and Participation in Environmental Policy and Risk Assessment*, New Brunswick NJ, Transaction Press

Habermas, J (1984) *The Theory of Communicative Action: Volume I Reason and the Rationalization of Society*, Boston, Beacon Press

Habermas, J (1987) *The Theory of Communicative Action: Volume II Lifeworld and System: A Critique of Functionalist Reasoning*, Boston, Beacon Press

Habermas, J and Nielsen, TH (1990) 'Jurgen Habermas: Morality, society, and ethics - an interview with Torben Hviid Nielsen' *Acta Sociologia* 33:93-114

Heritage, J (1984) *Garfinkel and Ethnomethodology*, Cambridge, Polity Press

ICES [International Council for the Exploration of the Sea] (2003a) *ACFM Annual Report*, Copenhagen, Denmark

ICES [International Council for the Exploration of the Sea] (2003b) *Report of the Study Group on the Development of Fishery-based Forecasts*, ICES CM 2003/ACFM:08 Ref. D

ICES [International Council for the Exploration of the Sea] (2003c) *Report of the Working Group on the Assessment of Demersal Stocks in the North Sea and Skagerrak*, ICES CM 2003/ACFM:02

ICES [International Council for the Exploration of the Sea] (2004) Working Group for Fishery Systems (WGFS) *Annual Report*, Copenhagen, Denmark

Jasanoff, S (2002) 'Citizens at risk: cultures of modernity in the US and EU' *Science as Culture* 2(3):363-380

Latour, B (1987) *Science in Action*, Cambridge, Harvard University Press

Loomis, DK and Ditton, RB (1993) 'Distributive justice in fisheries management' *Fisheries* 18(2):14-18

McEvoy, A (1986) *The Fisherman's Problem*, Cambridge, Cambridge University Press

Merton, RK (1968) 'Science and democratic social structure' in RK Merton (ed.) *Social Theory and Social Structure*, New York, The Free Press

Ostrom, E (1990) *Governing the Commons*, New York, Cambridge University Press

Peterson, MJ (1993) 'International fisheries management' in PM Haas, RO Keohane and MA Levy (Eds) *Institutions for the Earth: Sources of Effective International Environmental Protection*, Cambridge, MIT Press

Porter, TM (1995) *Trust in Numbers: the Pursuit of Objectivity in Science and Public Life*, Princeton, Princeton University Press

Rosenberg, AA, Fogarty, MJ, Sissenwine, MP, Beddington, JR and Shepherd, JG (1993) 'Achieving sustainable use of renewable resources' *Science* 262:828-829

Rozwadowski, HM (2002) *The Sea Knows No Boundaries: A Century of Marine Science under ICES*, Copenhagan Denmark, The International Council for the Exploration of the Sea

Scott, R (1995) *Institutions and Organizations*, Thousand Oaks CA, Sage Publications

STECF [Scientific, Technical and Economic Committee for Fisheries] (2003) *Plenary Report #17*, European Commission, Brussels

White, S (1988) *The Recent Work of Jurgen Habermas: Reason, Justice and Modernity*, Cambridge, Cambridge University Press

Wilson, DC (1999) 'Fisheries science collaborations: the critical role of the community', a keynote presentation at the *Conference on Holistic Management and the Role of Fisheries and Mariculture in the Coastal Community*, November, Tjärnö Marine Biological Laboratory, Sweden

Wilson, DC and McCay, BJ (1998) 'A governance model for environmental policy', presented at *Crossing Boundaries*, The Seventh Conference of the International Association for the Study of Common Property, Vancouver BC, 10-14 June

Wilson, DC and Jentoft, S (1999) 'Structure, agency and embeddedness: sociological approaches to fisheries management institutions' in D Symes (ed) *Alternative Management Systems for Fisheries*, Oxford, Blackwell Science

Wilson, DC (2000) 'Fisheries Management as a Social Problem' in C Sheppard (ed) *Seas at the Millennium Vol III*, London, Elsevier

Wilson DC, McCay, Bonnie J, Rowan, V and Grandin, B (2002) 'Institutional differences among marine fisheries scientists' views of their working conditions, discipline, and fisheries management' *Fisheries* 27(8):14-24

Wilson, DC and Degnbol, P (2002) 'The effects of legal mandates on fisheries science deliberations: the case of Atlantic Bluefish in the United States' *Fisheries Research* 58:1-14

Wilson, DC (2003) 'Conflict and scale: a defence of community approaches in fisheries management' in DC Wilson, JR Nielsen and P Degnbol (eds) *The Fisheries Co-management Experience: Accomplishments, Challenges and Prospects*, Dordrecht The Netherlands, Kluwer Academic Publishers

CHAPTER 20
PARTICIPATORY FISHERIES GOVERNANCE – THREE CENTRAL THEMES

TIM S GRAY
School of Geography, Politics and Sociology, Newcastle University, Newcastle upon Tyne, NE1 7RU, UK

Abstract

Three central themes about participatory fisheries governance that lie at the heart of the chapters of this book are discussed in this final chapter: the benefits and deficiencies of stakeholder participation; the relationship between stakeholder participation and the ecosystem-based approach; and the role of fishers' knowledge in fisheries governance

20.1 Introduction

In this concluding chapter, by way of summarising the findings of the previous chapters, I would like to discuss three major themes that have emerged. The first theme is the justification for stakeholder participation: do the benefits outweigh the costs? The second theme is the relationship between participation and the ecosystem-based approach: is it an essential link? The third theme is the role of fishers' knowledge alongside fisheries science: does it improve decision-making? My overall conclusion is that although contemporary currents in fisheries governance suggest an affirmative answer to each of these questions, we must reserve judgement on how far such affirmations rest on rhetoric rather than conviction.

20.2 The benefits and deficiencies of stakeholder participation in fisheries governance

The deficiencies of stakeholder participation in fisheries governance have already been touched on directly by Coffey and Knapman, and indirectly by Hatchard. They are not, however, discussed at length in the fisheries governance literature, largely because participation is seen to be such an obviously 'good thing'. As a result, most of the debate has been about what kind of participation to choose, not whether there should be participation. By contrast, in the development governance literature, the value of participation, which was equally taken for granted during the 1980s, has been increasingly questioned since the mid-1990s, and there is now a lively debate about its deficiencies (Hickey and Mohan 2004a:3). This backlash against the participatory mode of governance in the development sector has come from both the right and the left wings of the political spectrum. From the right, critics have argued that participation has gone too far; from the left, critics have argued that participation has not gone far enough. In this section, I want to apply to fisheries governance this twin critique of the participatory mode in the development governance literature. I will first rehearse the benefits of participation; then I will discuss the backlash from both the right and the left. I will

T. S. Gray (ed.), Participation in Fisheries Governance, 343–356.

conclude that if advocates of the participatory mode in fisheries governance are to withstand such criticisms, they must address them much more directly than at present.

20.2.1 THE BENEFITS OF PARTICIPATION

In fisheries governance, as in development governance from the mid-1980s, "participation has…become an act of faith…something we believe in and rarely question" (Cleaver 2001:36). Indeed, the term has assumed the status of a 'hurrah-word' – a "'warmly persuasive' word which seems 'never to be used unfavourably'" (Hildyard *et al* 2001:58). Advocates of participation appear to "sit on some moral high ground and as such are immune to criticism" (Hailey 2001:97). Henkel and Stirrat (2001:168) describe participation as the "new orthodoxy", noting that it "is now difficult to find a development project that does not in one way or another claim to adopt a 'participatory' approach involving 'bottom-up' planning, acknowledging the importance of 'indigenous' knowledge and claiming to 'empower' local people." The same is now happening with fisheries management regimes.

Why did participation become such a buzz-word in the development sector, and now in the fisheries sector? Cleaver (2001:36) refers to the "heroic claims" that have been made for participatory approaches, some of which are rehearsed by Coffey and by Hatchard in this volume. Participation is held to be "intrinsically a 'good thing'" (Cleaver 2001:36). According to Cooke (2001:104), the argument that participation is an "end in itself…means giving people control over development processes from which they had traditionally been excluded." This would enable "people who are often marginalised by their…isolation from the…formation of policies…to be included in decisions that apply to their lives" (Kothari 2001:139). Moreover, participation is said to empower people, confirming their status as rational and moral beings, with a capacity for self-determination (Meeuwig *et al* 2003:208).

Also, it is claimed that participation leads to more efficient and effective policies, because people who are subjected to regulations, know which policies work, and which do not work; that participation results in policies that are fairer and more equitable between different groups, because all of the groups can voice their claims; and that participation ensures a higher rate of compliance with the rules, because people are more ready to accept rules that they have participated in formulating (Hall-Arber, this volume).

As we shall see, many of these claims are contested, but, if they are taken at face value, it is not surprising that one commentator refers to "the 'benign virus'…of participation" (Williams 2004:95). Let us now turn to the criticisms of participation.

20.2.2 THE DEFICIENCIES OF PARTICIPATION

During the last ten years, there has been a backlash against participation in the development sector, with criticism coming from both the right and the left. Many of these criticisms are applicable to participation in the fisheries sector, and, in this section, I consider the most important of them.

20.2.2.1 *Backlash from the Right*

Critics on the right argue that participation is unnecessary, and that participatory advocates "are often naïve as they overstate the value of 'local knowledge' and local potential to self-determination…the experts and the 'state' actually do know better" (Henkel and Stirrat 2001:171). Critics also claim that managers "need freedom to act quickly" (Taylor 2001:138) and to exercise their powers flexibly, but participation makes this difficult. Moreover, right wing critics assert that people are "more interested in short-term substantive livelihood improvement than participation" (Taylor 2001:138). Another criticism is that participation costs a lot of time, money and energy to organise (Williams 2004:100), but there is little evidence in the development sector that the results of participation justify that expenditure of resource: "Many claims about participation…remain unproven" (Cleaver 2001:53). There seems to be an assumption that the activity of participation is itself evidence of the success of a project – as though participation is self-validating (Mosse 2001:30).

However, the experience of participation in the fisheries sector would seem to withstand at least some of these criticisms, in that at least some participatory arrangements – notably co-management – appear to work efficiently and cost-effectively. Moreover, as we shall see in the final section, there is much evidence of the value of fishers' participation in research collaboration with scientists. On the other hand, it is true that many fishers are less interested in participation than in economic gain.

Perhaps the most serious criticism from the right, however, is that participation subverts the system of electoral representative democracy which is already in place in fisheries management (at any rate in developed countries). Although the advocates of participation claim that it supplements, rather than bypasses, the electoral system by enabling voters to exercise control between elections (Brown 2004:241-2), critics of participation argue that it undermines the legitimacy of the official representative system. This criticism is particularly telling in relation to the environmental stewardship type of the participatory mode of fisheries governance. For example, industrial fishers regard it as illegitimate for environmental non-governmental organisations (ENGOs) such as Greenpeace to arrogate to themselves the moral authority to confront Danish trawlers who are, under the Common Fisheries Policy (CFP) (a policy voluntarily agreed by democratic Member States), perfectly legally catching vast numbers of sandeels off the north east coast of Scotland. In other words, participatory governance mechanisms run the risk of seeming to justify illegal direct action.

20.2.2.2 *Backlash from the Left*

Critics on the left argue from the opposite perspective – that the moves towards participation have not proceeded far enough. The nub of this criticism is that, as Henkel and Stirrat (2001:171) put it, "participatory policies often do not really lead to participation and empowerment." Participation may lead simply to reproducing the dominant power structures which exist, both in the local communities and in the wider world. With regard to the local power structure, critics point out that advocates of participation have a roseate picture of local communities, seeing them as beacons of equality and harmony, but the fact is that they are often riven by inequality and discord (Kothari (2001:141). Because participative processes often "rely on a small sample of self-selecting participants", and "power often lies in the hands of the most articulate or politically adept", the outcome is to "reinforce the status and power of existing cliques within the community" (Hailey 2001:94). The case of Sea Fisheries Committees (SFCs) in England and Wales (Knapman, this volume) confirms this tendency. The current controversy over a Regulating Order in the Highlands of Scotland, is further evidence of deep divisions within and between fisheries-dependent communities.

It is with regard to the power structure in the wider world, however, that left wing critics make their most severe charges. They claim that participatory processes are manipulated by developers to ensure support for their projects, "rather than seeking 'real' participation from the affected community" (Henkel and Stirrat 2001:172). This means that 'empowerment' is an illusion: stakeholders "are being empowered to be elements in the great project of 'the modern'…'empowerment' is tantamount to what Foucault calls subjection" (Henkel and Stirrat 2001:182). Participatory decisions merely "reinforce the interests of the already powerful" (Cooke and Kothari 2001:8). Cooke (2001:121) refers to "participation as an instrument for control." According to Taylor (2001:125), participatory engagement is a deliberate technique used by elites to divert attention from their dominance: it gives the "'sense' and warm emotional pull of participation without its substance." It "masks continued centralisation in the name of decentralisation" (Cooke and Kothari 2001:7). Participants are politically anaesthetised by being co-opted into the management agenda, and participation is taken by management to legitimise their subjection (Williams 2004:93). This legitimising element is identified by the left wing critics as the whole purpose of the participatory approach. On this view, participation is a "largely cosmetic" exercise (Taylor 2001:136), designed to acquire legitimacy for decisions that have already been reached – "merely tokenistic" (Hickey and Mohan 2004b:161). Participants even begin to internalise this oppression, accepting "external needs **as** local needs, dominant interests *as* community concerns" (Mosse 2001:22).

To what extent do these harsh left wing criticisms of the manipulative nature of the participatory mode of development governance apply to the participatory mode of fisheries governance? Within Europe, the experience of the CFP reform's focus on the participatory mode gives some grounds for anxiety that it is largely rhetorical (Gray and Hatchard 2003). Although the Commission made much of its commitment to stakeholder consultation, it has ensured that, under the new European Union (EU) constitution, overall responsibility for fisheries in EU waters will remain with Brussels, and that the new regional councils (RACs) will be only advisory. This means that the Commission can milk the RACs of all their participatory worth, while retaining the real power. Some fishermen's organisations (such as the UK's Fishermen's Association Limited (FAL)) refuse to take part in the RACs for this reason, arguing that the RACs merely serve to exemplify dependence on the EU's centralised and politicised agenda for fisheries. Roddy McColl (Secretary of FAL) derided consultation exercises: "Stakeholder consultation is now the flavour of the day. It is not hard to be cynical and think 'yet another meeting to ask the industry which rope we would like to have placed around our necks to help ensure our own demise'" (*Fishing News* 27/8/04:3). Similarly, van Ginkel (this volume) argues that participatory co-management in the Netherlands, despite the hype, "remains, in essence, a command-and-control type of regulatory regime".

Outside Europe, there is similar evidence of cosmetic participatory practices. For instance, Glaesel and Simonitsch (2003:57) claim that in the USA and Kenya, although "there is a discourse of participatory marine management, the practice remains hierarchical." In Australia, according to Baelde (2003:82), "while government policies and legislation on resource management never fail to mention the importance of stakeholders' participation...(T)here are generally limited resources and expertise, and sometimes limited willingness, within government agencies to design and engage in effective consultation with the commercial fishing industry." The fact is, the charges of manipulation cannot be brushed aside by fisheries regulators anywhere, because power games may be played out wherever important decisions about fisheries governance are made (Murray *et al*, this volume).

However, what would these left wing critics put in the place of what they see as flawed participatory practices of fisheries governance? In the development field, they write of turning participation as 'tyranny' into participation as 'transformation', meaning endowing participants with civil and political rights. In the field of fisheries governance, this 'transformation' could be interpreted as genuinely integrating stakeholders and/or communities into the decision-making system. As we shall see in section 3, one important way of doing this would be to embed fisher's knowledge and expertise into the management process.

20.3 The relationship between stakeholder participation and the ecosystem-based approach

The second theme of this chapter is the relationship between participation and the ecosystem-based approach (EBA). For many people, stakeholder participation and the EBA are both desirable objectives, and, therefore, they must go together. However, just because participation and the EBA are deemed to be good things in themselves, doesn't mean to say that they are compatible with each other. As Isaiah Berlin pointed out, we may regard both liberty and equality as good things, but we don't deny that there may be tension between them. So why should we deny that there may be tension between participation and the EBA? In this section, I examine the arguments for and against the claim that participation in fisheries governance enhances the prospect of the EBA prevailing.

In many recent documents on fisheries governance, both stakeholder participation and the EBA are advocated. For example, in a UK government report, the Department for Environment, Food and Rural Affairs (DEFRA) (2004:94) states that: "The ecosystem approach should consider all forms of relevant information, including scientific and indigenous and local knowledge, innovations and practices...[and] should involve all relevant sectors of society." Similarly, in a publication entitled *Safeguarding Our Seas*, setting out its marine conservation strategy, DEFRA declares that its policy of fisheries management is rooted in "a commitment to adopting an ecosystem based approach, founded on stakeholder involvement", and that an EBA approach "means...full stakeholder involvement" (DEFRA 2002:19;7). Likewise, for Sissenwine and Mace (2001:1), the participatory mode of governance forms part of the very definition of the ecosystem approach, in that "decision-making that is participatory and transparent" is one of the six "key elements" of the ecosystem approach. Also, Meeuwig *et al* (2003:208) assert that "Stakeholder involvement in the planning and implementation of conservation initiatives is considered fundamental to the achievement of resource management objectives". However, in none of these publications is there any attempt to *demonstrate* how participation and the EBA are consistent with each other (Stead; Frid, both this volume).

Such an explanation is, however, provided by the US Ecosystem Principles Advisory Panel (EPAP), in claiming that the EBA depends on participation: "Ecosystem approaches to management rely on the participation, understanding and support of multiple constituencies" (EPAP 1999:20). EPAP (1999:37) insists that the public has a role in deciding what constitutes a healthy ecosystem: "society as a whole, will be increasingly challenged to help define ecosystem health and the limits of acceptable change in marine ecosystems, while still allowing sustainable fishing practices." Similarly, a recent US report on ocean policy makes the point that while scientific experts can tell us what the consequences would be of prioritising this or that use of marine resources, they cannot take the ethical decision of which priority to choose. That decision is a choice that must be made by "community judgement": "the critical process of setting goals to guide management will require active participation by many different stakeholders with divergent views" (USCOP 2004:36).

So this is one explanation of the link between EBA and participation – that only the public as a whole has the moral right to decide what the priorities should be in the EBA approach: for example, whether the balance should be tipped towards preservationism or towards sustainable development. This is an important normative point, but it begs the empirical question of whether the public will endorse the EBA in the first place. It is true that the public are keen on conserving the marine environment: opinion polls indicate a high level of support for moves to reduce marine pollution, especially in coastal areas, and for measures to protect marine mammals (such as seals, whales and dolphins), coral reefs and sea birds. However, are the public interested in preserving less charismatic species and habitats, particularly where there could be negative impacts on fishing opportunities and employment?

Of course, one way to close this gap between public participation and the EBA is to deliberately forge links by a process of public education (Frid, this volume). This is the solution put forward by the UK government: "Government should establish a mechanism to raise awareness of the marine environment with a wide range of stakeholders to encourage their participation in the delivery of marine nature conservation objectives'' (DEFRA 2004:54). A similar solution is proposed by the US Commission on Ocean Policy (USCOP 2004:39): "Instilling a stewardship ethic in the American public is an important element of a national ocean policy." The UK Royal Commission on Environmental Pollution (RCEP) (2004:para 10.14) implies that greater public participation will not lead to greater support for the EBA "unless the value of the seas is better understood." However, such a strategy seems more like indoctrination than education, and is open to the objection that the resulting participation would be merely rubber-stamping top-down initiatives. A reply to this charge could be that the government is not so much manipulating the public as awakening the latent environmentalism that exists in everyone, waiting to be stimulated into actuality. Alternatively, the gap between public participation and the EBA might be bridged by a Habermasian process of "collaborative learning" (Vodden *et al*, this volume).

A stronger link may be found between the EBA and particular segments of the public – such as the statutory conservation agencies (NCAs) (Eno and Gray, this volume) and the environmental NGOs (Dunn, this volume) – whose participation in the processes of fisheries governance has clearly enhanced the prospect of fisheries regulators employing the EBA. But the NCAs are sometimes regarded as an extension of the arm of the executive, while the ENGOs' claims to be the voice of the public, may ring hollow at times. ("Greenpeace are simply criminals" (*Fishing News* 18/03/05).)

Perhaps community partnerships have a more persuasive claim to environmental stewardship in that they involve the whole of a local population, which may be assumed to have a high regard for the environmental well-being of its coastal area (Poepoe *et al* 2003). However, a local community may not always fit very well with the EBA, because it covers too small an area of the marine environment. The smart money on the EBA at the moment is concentrated on large marine ecosystems (LMEs), in order to take account of the widest range of environmental dynamics (Pauly and Maclean 2003:4-7; USCOP 2004:34). Symes (this volume) considers "regional seas" such as the semi-enclosed Baltic, North and Irish Seas, as the appropriate scale of "self-contained ecosystems". However, by definition, local communities cannot cover such large areas. Perhaps the answer is to switch the smart money to local ecosystems, as Jim Wilson (this volume) suggests.

What about marine resource users? Participation in fisheries governance by recreational sea anglers is likely to improve the environmental quality of decisions, because anglers argue for less intensive methods of fishing, which inflict less harm on the marine environment. But participation by commercial fishers is much more problematical. On the one hand, there is considerable anti-environmental feeling among commercial fishers, because many of them (including, according to Symes (this volume), "the majority opinion on the RACs") view environmental restrictions as a threat to their livelihoods. On the other hand, they may see environmentalism as the lesser of two evils: as Melvin and Parrish (2003:225) point out, the key condition for getting fishers to take conservation issues seriously, is a crisis in the fisheries, which imperils their economic survival.

More positively, some fishers see the EBA as an opportunity rather than a threat. For instance, Wilson (2003a:171) points out that EBA fits in with fishers' wish to shift attention away from over-fishing to environmental factors: ''The perceived need to manage fisheries as part of a broader ecosystem is intuitively appealing, especially to fishers. It resonates with their common argument to focus more on non-fishing related causes of decline in fish stocks." Thus the more far-sighted fishers have seen the writing on the wall, and have realised that environmentalism is here to stay, and that their best strategy is to learn to work within a green agenda. Williams and Bax (2003:243) state that another incentive is that fishers' participation helps the industry to respond to the torrent of environmental legislation that is coming its way, and also to improve its battered public image (Dunn, this volume).

Other fishers have a disinterested (public-interested) desire to protect the marine environment, and voluntarily employ environmentally-friendly fishing gear (such as larger mesh sizes) to reduce discards (Catchpole, Frid and Gray forthcoming). Some fishers take part in schemes to monitor quality indicators of ecosystem health, such as recording sightings of marine mammals or seabirds. Moreover, some commentators have argued that the future for the environmental movement is to recruit fishers to the cause. Baelde (2003:82), for instance, criticises conservation agencies for failing to "recognise and promote the role that fishers could play in protection of the marine environment." Bird *et al* (2003:179, 182) describe a successful research project in Mexico to conserve sea turtles that involved a community partnership in which fishers were prominent players: "Placing value in the opinions, experiences, and knowledge of the fishers, and involving them directly in the project from the first step may form strong conservation alliances…In this way, fishers are viewed, and view themselves, as an integral part of the conservation team contributing valuable knowledge and ideas."

The conclusion reached at the end of this section, therefore, is that while we cannot simply assume a connection between participation and the EBA, there is a plausible normative argument for linking them (that the public has the right to determine whether the EBA shall be adopted, and what its priorities should be), and there is some empirical evidence that the public (and especially some influential segments of it) are committed to the environmental management of fisheries, and that many fishers are becoming reconciled to this eventuality.

20.4 The role of fishers' knowledge in fisheries governance

Let us now turn to the last of our three themes – the role of fishers' knowledge (FK) in fisheries governance – which is discussed directly in four chapters (Hawkins, Rice, Murray *et al*, and Wilson and Delaney), and indirectly in three others (Frid, Wilson, and Vodden *et al)*. There is also a vast literature on the subject. This third theme is closely linked to both the first and second themes. The link to the first theme lies in the fact that where fishers' knowledge is contributed to fisheries governance, it is a form of participation. However, the two themes are nevertheless separate: a participatory governance regime may or may not include fishers' knowledge. Indeed, according to Bird *et al* (2003:180), few cases of community-based fisheries regimes have integrated fishers' knowledge into their management systems. Similarly, Phelan (2003:100) points out that in Australia, while there is an increasing number of co-management regimes, "the value of collaborative research partnerships in fisheries research has often been ignored" (an assertion that, however, seems contradicted by Williams and Bax (2003:238) who claim that "Australia involves fishers at all stages of the fishery assessment and management process"). There are many reasons for this reluctance to embrace FK, but as more fisheries face crises, there is a greater willingness to consider it (Baelde 2003:83). The link between the third theme and the second theme is that fishers' knowledge and experience may improve our understanding of the way in which the marine ecosystem works (Ames 2003; Poulsen 2003).

In what follows, first, I explain the distinction between FK and fisheries science (FS), and second, I discuss the three main ways in which their relationship has been understood: political hierarchy; epistemological synthesis; and social integration. In this discussion, we can see a distinction between two sorts of questions: the methodological and the organisational. The methodological question is how can the two sets of knowledge be integrated? This is the question that occupies the chapters by Hawkins, and (to some extent) Frid. The organisational question is how can the two groups of people – fishers and scientists – be integrated? This is the question that occupies the chapters by Rice, Wilson and Delaney, and (to some extent) Murray *et al,* and Vodden *et al.*

20.4.1 THE DISTINCTION BETWEEN FISHERS' KNOWLEDGE (FK) AND FISHERIES SCIENCE (FS)

FK is derived from the practical experience of fishing activity over many years – even many generations (Lessard *et al* 2003:42) – some of which may be recorded in log books and other documents, but much of it is stored in fishers' memories. This knowledge ranges widely over fish species, marine mammals, habitats, seabirds, benthos, climate, and topography, as well as techniques of fishing, effectiveness of regulations, and marketing of catches. It is generally small-scale, qualitative and empirical knowledge, sometimes intuitive, invariably subjective, always particularistic if not *ad hominem*, and often anecdotal. Stanley and Rice (2003:44) claim that: "individuals who are intimately associated with the resource have a wealth of knowledge that can enhance research and improve management." But scientists criticise FK for what they claim is its unreliability: because it is unquantifiable and unsystematic; because it cannot be replicated for testing; because it is biased by vested interests and therefore self-serving (Pido *et al* 2003:247; Ames 2003:187); and because fishers' presence would politicise stock assessments (Rice, this volume).

FS is derived from sophisticated computer modelling of survey, catch, and landing data, which yields calculations of stock assessments, and is used by managers to base their total allowable catch (TAC) decisions upon. This is generally large-scale (Degnbol 2003:34), quantitative and rational knowledge, largely statistical, ostensibly objective, apparently universalistic, and highly systematic. Wilson and Delaney (this volume) point out that regulators value FS for precisely these qualities. But fishers criticise FS for what they claim is its irrelevance: it has a high margin of error (up to 50 per cent (Daw and Gray 2005)); it focuses too narrowly on single stocks; it uses inadequate trawling techniques in its research surveys; its data is generally two years out of date; it has no peer review system; its activities are often wrapped in secrecy; it cannot take into account the economic and social consequences of its advice; and it lacks expertise on management measures (Hawkins, this volume).

20.4.2 THREE WAYS OF RELATING FK TO FS

There are three main ways in which FK has been related to FS: political hierarchy; epistemological synthesis; and social integration (Gray 2002). First, the **political hierarchy** mode entails that there is a stand-off between these two irreconcilably different forms of knowledge, and that one or other must prevail. According to some writers, fishers have completely different mindsets from scientists, operating on distinct wave lengths, and, therefore, there is no possibility of reconciliation between their two types of knowledge. In 'western' countries, FS invariably triumphs over FK, because of the high prestige that modern science enjoys in the minds of politicians, and because scientists present their case in clear, unambiguous, systematic form, thus facilitating closure of decision-making. However, because of the failures of contemporary science-based fisheries management, there are increasing calls for a reversal of this rank order, and the acceptance of FK as the dominant guide for fisheries managers (Johannes 2003:18).

The **epistemological synthesis** mode entails that it is possible to avoid both of these extremes, and reconcile the two sets of knowledge. Palsson (1998) sees them as complementary and interactive, not as mutually exclusive forms of knowledge. Wilson (1999) describes this mode as one in which FK is regarded as just as valid a form of knowledge as FS. On this view, FK provides information that can be integrated with FS, plugging gaps and indicating long-term trends (Rice, this volume; Williams and Bax 2003). Such syntheses of FK with FS provide "a fuller understanding of the natural environment and more complete information for management decisions" (Gosse *et al* 2003:25). For instance, Phelan (2003:104) points out that, "when access to data is otherwise not available…oral history proves an invaluable tool in establishing a retrospective analysis of resource use". Kalikoski and Vasconcellos (2003:454) refer to the resulting composite form of knowledge as "civic science".

There are many examples of FK being incorporated into FS. For instance, Lydon and Langley (2003) explain how it is done in New Zealand; Lessard *et al* (2003:42) report that, in their study of the Vancouver goose barnacle fishery, "a large amount of anecdotal information provided tendencies and directions that would have taken years to assimilate in a scientific study"; Meeuwig *et al* (2003:208) show how local fishers' knowledge has been "harnessed in the management of South Pacific fisheries"; and since 2000, an ICES North Sea Demersal Working Group has been collecting information about the fish stocks by an annual questionnaire administered to fishers "complementing the information provided by commercial landings records and scientific research vessel surveys." This last exercise is aimed at ensuring that "fishermen's knowledge is considered during the development of scientific management advice" (*Fishing News* 15/10/04:3).

However, there is a danger of the integrity of FK being devalued by being used to improve FS (Murray *et al*,. this volume). The language sometimes employed to describe this 'synthesis' demonstrates its one-sidedness (Rice, this volume). For instance, Gosse *et al* (2003:26) refer to the "rigorous procedures that allow scientists to test some of the assumptions found in harvesters' knowledge and the validity of their interpretations." Garcia-Allut *et al* (2003:229) state that "After filtering, systemising and formalising fishers' ecological knowledge, it can contribute to broaden our understanding." USCOP explains how FK must be processed through a scientific lens before it can be utilised: "anecdotal or traditional information was not unconditionally accepted…informal information can only be used in decision making after it has been tested and verified according to a methodical, scientific process" (USCOP 2004:227). USCOP appears to see no problem with such a reductionist approach to the use of fishers' knowledge, but it is, in fact, extremely controversial, because at stake is the authenticity of fishers' knowledge and experience.

In relation to the development sector, Cooke and Kothari (2001:12) point out that "participatory research 'cleans up' local knowledge through mapping and codification, and marginalises that which might challenge the status quo or is messy or unmanageable." Similarly, Kothari (2001:147) claims that: "The use of participatory techniques often requires the taking out of anything complicated…a process of controlling to produce the norm, the usual and the expected." Indeed, the very definition of what "counts as 'local knowledge' is very often the effect of specific kinds of techniques of power, of regulation and of normalisation" (Kothari 2001:152).

These charges are equally applicable to the fisheries sector. Palsson (1998) has argued strongly that processing of fishers' knowledge to make it fit scientific methodologies is to devalue it, forcing it into an alien strait jacket. He claims that fishers' knowledge is fluid and intuitive, rather than passive and quantitative, and that it cannot be disaggregated into neat little boxes. By attempting to quantify it, he says, we perpetuate its alleged inferiority to so-called hard science. Stanley and Rice (2003:45) make a similar point: "the 'data collection' model…assumes that for local knowledge to contribute it must be systematised, stored, manipulated, and made intelligible to others in a manner similar to treatment of data from conventional monitoring sources…Although there is a place for this model, it represents an appending of fishers to conventional scientific research as junior partners. It maintains for researchers, the 'we *vs.* they' …dichotomy" (cf Wilson; Murray *et al*, both this volume)

The **social integration** mode is the most recent way of relating FK and FS, and avoids the reductionist criticism that FK is treated as a mere add-on extra to FS. Stanley and Rice (2003:44) argue that it is a mistake to confine fishers to the role of "simply data collectors or knowledge sources, thereby ignoring their skills in hypothesis formulation, research design, and interpretation." Social integration entails two processes: dialogue and teamwork. Dialogue means mutual learning through a genuine meeting of minds as fishers and scientists share their understandings on particular issues (Vodden *et al*, this volume. Teamwork means full collaboration between fishers and scientists in joint research activity: including the proposal of hypotheses to be tested; the design of projects; the management of investigative work; the analysis of data; the discussion of outcomes; and the process of agreeing on the implications for fisheries management measures. Wilson (2003b:275) describes this as moving "beyond fishers acting as research assistants to truly collegial relationships".

An example of such teamwork is given by Gosse *et al* (2003:34), who state that in their research project on coastal cod in Newfoundland and Labrador ('Coasts Under Stress'), "We're continually interacting with fishers…The project is not to just take knowledge and run away with it." In such teamwork, the fishers are "active collaborators" rather than just "information providers" (Baelde 2003:79-80). Phelan (2003:103,101) similarly testifies to the value of collaborative research in a project in Northern Australia: "community members were involved in the design and implementation of the project, as well as the interpretation of results…The research described in this study gained tremendously from its participative approach and clearly demonstrates the benefits that may result from collaborative partnerships…the strong management outcomes [a two year ban on fishing] which resulted from this research would not have been achieved without this interaction."

In the UK, Arnold Locker, the National Federation of Fishermen's Organisations (NFFO) President, expressed the wish that the Fisheries Science Partnership (which is a UK government-funded partnership between the fishing industry and the fisheries scientists to work together to build consensus on the condition of the stocks) would graduate from the synthesis stage to the collaborative stage: the synthesis stage is "only a beginning. We hope that fishermen will in future be fully integrated with the whole process of design, implementation and interpretation of assessment data" (*Fishing News* 6/8/04:8). NFFO's chief executive, Barrie Deas, appears to believe that the partnership has already begun the move from synthesis to collaboration: "The industry has a say now in what research is undertaken – CEFAS, DEFRA and the NFFO can agree on projects and priorities" (*Fishing News* 18/2/05:20).

This is not to deny that there are difficulties with the social integration mode. For one thing, it entails arduous processes of collaborative learning (Vodden *et al*, this volume). For another, according to Baelde (2003:84), the greatest obstacle to this collaborative approach is the socio-cultural divide between fishers and scientists, which results in completely different mindsets. However, Wilson (2003a:164) argues that this cultural explanation, focusing on immovable cognitive barriers, which make fishers and scientists "mutually incoherent", is less convincing than is the institutional explanation, which focuses on organisational barriers between fishers and scientists that can be removed. Wilson (2003a:163) points out that there is more than one culture of FK, and more than one culture of FS; and that the cultural explanation threatens to reify the two categories, and "to underestimate the degree to which the rules governing management and stakeholder interactions create these apparent gaps in how the world is seen" (Wilson 2003a:164). If we change the way in which meetings and agendas are organised, and the protocols of research collaboration, then much of the so-called mutual incoherence could disappear.

In conclusion, if fisheries governance is to take full advantage of the contribution that fishers can make, then both scientists and fishers have to change their attitudes towards each other. Scientists have to recognise that fishers can provide not only valuable information that will improve our understanding of the marine eco-system, but also critical insights into research direction and design. But fishers', too, must recognise that they have to change. They can no longer escape responsibility for helping scientists to conserve the marine environment; they must recognise that they have a duty to society to play their full part as environmental stewards, and that this means regarding scientists as partners in a joint endeavour, not as enemies preventing them from fishing as they please. Participation brings with it responsibilities as well as rights, and the principal responsibility for all participants is to safeguard the health of the marine ecosystem.

20.5 Conclusion

Summing up the three themes of this chapter, and the book as a whole, we may conclude that participation, despite its flaws, is an inescapable part of fisheries governance; that the link between participation and the ecosystem-based approach, notwithstanding its fragility, is a crucial assumption of fisheries governance; and that fishers' knowledge and collaboration, however problematic, are increasingly employed in fisheries governance. Such conclusions could not have been reached five years ago, which is a measure of the extent to which the parameters of fisheries governance have recently shifted in the direction of participation. However, how far this acceptance of participation is genuine, rather than rhetorical, remains an open question.

References

Ames, T (2003) 'Putting fishermen's knowledge to wotk: the promise and the pitfalls' *Fisheries Center Research Reports* 11(1):184-188, Vancouver BC, University of British Columbia Press

Baelde, P (2003) 'Using fishers' knowledge goes beyond filling gaps in scientific knowledge – analysis of Australian experiences' *Fisheries Center Research Reports* 11(1):78-86, Vancouver BC, University

of British Columbia Press
Bird, KE, Nichols, WJ and Tambiah, R (2003) 'The value of local knowledge in sea turtle conservation: a case from Baja California, Mexico' *Fisheries Center Research Reports* 11(1):178-183, Vancouver BC, University of British Columbia Press
Brown, D (2004) 'Participation in poverty reduction strategies: democracy strengthened or democracy undermined?' in S Hickey and G Mohan (eds) *Participation – from Tyranny to Transformation?,* London, Zed Books
Catchpole, T, Frid, CLJ and Gray, T (forthcoming) 'Discarding in North Sea fisheries: causes, consequences and solutions' *Marine Policy*
Cleaver, F (2001) 'Limits of participatory approaches' in B Cooke and U Kothari (eds) *Participation – the New Tyranny?*, London, Zed Books
Cooke, B (2001) 'The social psychological limits of participation?' in B Cooke and U Kothari (eds) *Participation – the New Tyranny?*, London, Zed Books
Cooke, B and Kothari, U (2001) 'The case for participation as tyranny' in B Cooke and U Kothari (eds) *Participation – the New Tyranny?*, London, Zed Books
Daw, T and Gray T (2005) 'Fisheries science and sustainability in international policy: a study in failure in the EU's Common Fisheries Policy' *Marine Policy* 29:189-197
DEFRA [Department for Environment, Food and Rural Affairs] (2002) *Safeguarding Our Seas*,,London, DEFRA
DEFRA [Department for Environment, Food and Rural Affairs] (2004) *Review of Marine Nature Conservation*, London, DEFRA
Degnbol, P (2003) 'Science and the user perspective: the gap co-management must address' in Wilson, DC, Nielsen JK and Degnbol, P (eds) *The Fisheries Co-management Experience*, Dordrecht, Kluwer
EPAP [Ecosystem Principles Advisory Panel] (1999) *Ecosystem-Based Fishery Management*, Washington DC, Department of Commerce
Fishing News (2004-5) various articles, London, Informa
Garcia-Allut, A, Freire, J, Barreiro, A and Losada, DE (2003) 'Methodology for integration of fishers' ecological knowledge in fisheries biology and management using knowledge representation [Artificial Intelligence]' *Fisheries Center Research Reports* 11(1):227-237, Vancouver BC, University of British Columbia Press
Glaesel, H and Simonitsch, M (2003) 'The discourse of participatory democracy in marine fisheries management' *Fisheries Center Research Reports* 11(1):54-67, Vancouver BC, University of British Columbia Press
Gosse, K, Wroblewski, J and Neis, B (2003) 'Closing the loop: commercial fish harvesters' local ecological knowledge and science in a study of coastal cod in Newfoundland and Labrador in Canada' *Fisheries Center Research Reports* 11(1):25-35, Vancouver BC, University of British Columbia Press
Gray, T (2002) 'Fisheries science and fishers' knowledge' *ENSUS 2002,* International Conference on Marine Science and Technology for Environmental Sustainability, Newcastle University, 18 December
Gray, T and Hatchard, J (2003) 'The 2002 reform of the Common Fisheries Policy's system of governance – rhetoric or reality?' *Marine Policy* 27:545-554
Hailey, J (2001) 'Beyond the formulaic: process and practice in South Asian NGOs' in B Cooke and U Kothari (eds) *Participation – the New Tyranny?*, London, Zed Books
Henkel, H and Stirrat, R (2001) 'Participation as spiritual duty: empowerment as secular subjection' in B Cooke and U Kothari (eds) *Participation – the New Tyranny?*, London, Zed Books
Hickey, S and Mohan, G (2004a) 'Towards participation as transformation: critical themes and challenges' in S Hickey and G Mohan (eds) *Participation – from Tyranny to Transformation?,* London, Zed Books
Hickey, S and Mohan G (2004b) 'Relocating participation within a radical politics of development: insights from political action and practice' in S Hickey and G Mohan (eds) *Participation – from Tyranny to Transformation?,* London, Zed Books
Hildyard, N, Hedge, P, Wolvekamp, P and Reddy, S (2001) 'Pluralism, participation and power: joint forest management in India' in B Cooke and U Kothari (eds) *Participation – the New Tyranny?*, London, Zed Books
Johannes, RE (2003) 'Fishers' knowledge and management: differing fundamentals in artisanal and industrial fisheries' *Fisheries Center Research Reports* 11(1):15-19, Vancouver BC, University of British Columbia Press
Kalikoski, DC and Vasconcellos, M (2003) 'Fishers' knowledge role in the co-management of artisanal fisheries in the estuary of Patos Lagoon, southern Brazil' *Fisheries Center Research Reports* 11(1):445-455, Vancouver BC, University of British Columbia Press
Kothari, U (2001) 'Power, knowledge and social control in participatory development' in B Cooke and U Kothari (eds) *Participation – the New Tyranny?*, London, Zed Books

Lessard, J, Osborne, J, Lauzier, R, Jamieson, G and Harbo, R (2003) 'Applying local and scientific knowledge to the establishment of a sustainable fishery: the current west coast Vancouver Island goose barnacle fishery experience' *Fisheries Center Research Reports* 11(1):36-43, Vancouver BC, University of British Columbia Press

Lydon, GJ and Langley, A (2003) 'How local fishers' knowledge improves the management of fisheries in New Zealand – a seafood industry perspective' *Fisheries Center Research Reports* 11(1):357-365, Vancouver BC, University of British Columbia Press

Meeuwig, JJ, Samoilys, MA, Erediano, J and Hall, H (2003) 'Fishers' perceptions on the seahorse fishery in central Philippines: interactive approaches and an evaluation of results' *Fisheries Center Research Reports* 11(1):208-223, Vancouver BC, University of British Columbia Press

Melvin, EF and Parrish, JK (2003) 'Focusing and testing fisher know-how to solve conservation problems' *Fisheries Center Research Reports* 11(1):224-226, Vancouver BC, University of British Columbia Press

Mosse, D (2001) '"People's knowledge", participation and patronage: operations and representations in rural development' in B Cooke and U Kothari (eds) *Participation – the New Tyranny?*, London, Zed Books

Palsson, G (1998) 'Learning by fishing: practical engagement and environmental concerns' in F Berkes, C Folke and J Colding (eds) *Linking Social and Ecological Systems*, Cambridge, Cambridge University Press

Pauly, D and Maclean, J (2003) *In a Perfect Ocean: the State of Fisheries and Ecosystems in the North Atlantic Ocean*, Washington, Island Press

Phelan, MJ (2003) 'Sciaenid aggregations in Northern Australia: an example of successful outcomes through collaborative research' *Fisheries Center Research Reports* 11(1):100-109, Vancouver BC, University of British Columbia Press

Pido, MD, Valentine, PS and Fenton, M (2003) 'Sustainability vectors as guides in fisheries management: with examples from net fisheries in the Philippines and Australia' *Fisheries Center Research Reports* 11(1):246-258, Vancouver BC, University of British Columbia Press

Poepoe, KK, Bartram, PK and Friedalander, AM (2003) 'The use of traditional Hawaiian knowledge in the management of marine resources' *Fisheries Center Research Reports* 11(1):328-339, Vancouver BC, University of British Columbia Press

Poulsen, AA (2003) 'Integration of fishers' knowledge into research on a large tropical river basin, the Mekong river in southeast Asia' *Fisheries Center Research Reports* 11(1):198-207, Vancouver BC, University of British Columbia Press

RCEP [Royal Commission on Environmental Pollution] (2004) *Turning the Tide: Addressing the Impact of Fisheries on the Marine Environment*, London, RCEP

Sissenwine, MP and Mace, PM (2001) 'Governance for responsible fisheries: an ecosystem approach' *Reykjavik Conference on Responsible Fisheries in the Marine Ecosystem*, Reykjavik Iceland, 1-4 October

Stanley, RD and Rice, J (2003) 'Participatory research in the British Columbia groundfish fishery' *Fisheries Center Research Reports* 11(1):44-56, Vancouver BC, University of British Columbia Press

Taylor, H (2001) 'Insights into participation from critical management and labour perspectives' in B Cooke and U Kothari (eds) *Participation – the New Tyranny?*, London, Zed Books

USCOP [US Commission on Ocean Policy] (2004) *Preliminary Report of the US Commission on Ocean Policy – Governor's Draft*, Washington, DC

Williams, A and Bax, N (2003) 'Integrating fishers' knowledge with survey data to understand the structure, ecology and use of a seascape off southeastern Australia' *Fisheries Center Research Reports* 11(1):235-245, Vancouver BC, University of British Columbia Press

Williams, G (2004) 'Towards a repoliticisation of participatory development: potential capabilities and spaces of empowerment' in S Hickey and G Mohan (eds) *Participation – from Tyranny to Transformation?,* London, Zed Books

Wilson, DC (1999) *Fisheries Science Collaboration: the Critical Role of the Community, Research Report No. 45*, Hirtshals, Denmark, IFM

Wilson, D (2003a) 'Examining the two cultures theory of fisheries knowledge: the case of the northwest Atlantic bluefish' *Fisheries Center Research Reports* 11(1):163-177, Vancouver BC, University of British Columbia Press

Wilson, DC 2003b) 'Fisheries co-management and the knowledge base for management decisions' in Wilson, DC, Nielsen, JK and Degnbol, P (eds) *The Fisheries Co-management Experience*, Dordrecht, Kluwer

INDEX

Reviews: Methods and Technologies in Fish Biology and Fisheries

1. J.R. Sibert and J.L. Nielsen (eds.): *Electronic Tagging and Tracking in Marine Fisheries*. 2001 ISBN 1-4020-0125-8
2. D. Symes and J. Phillipson (eds.): *Inshore Fisheries Management*. 2001 ISBN 1-4020-0128-2
3. C.O. Cunningham (ed.): *Molecular Diagnosis of Salmonid Diseases*. 2002 ISBN 1-4020-0506-7
4. T.S. Gray (ed.): *Participation in Fisheries Governance*. 2005 ISBN 1-4020-3777-5